International Political Economy Series

Series Editor
Timothy M. Shaw
University of Massachusetts Boston
Boston, MA, USA

Emeritus Professor
University of London
London, UK

The global political economy is in flux as a series of cumulative crises impacts its organization and governance. The IPE series has tracked its development in both analysis and structure over the last three decades. It has always had a concentration on the global South. Now the South increasingly challenges the North as the centre of development, also reflected in a growing number of submissions and publications on indebted Eurozone economies in Southern Europe. An indispensable resource for scholars and researchers, the series examines a variety of capitalisms and connections by focusing on emerging economies, companies and sectors, debates and policies. It informs diverse policy communities as the established trans-Atlantic North declines and 'the rest', especially the BRICS, rise. NOW INDEXED ON SCOPUS!

More information about this series at
http://www.palgrave.com/gp/series/13996

Paul Midford · Espen Moe
Editors

New Challenges and Solutions for Renewable Energy

Japan, East Asia and Northern Europe

palgrave
macmillan

Editors
Paul Midford
Department of Sociology and Political Science
Norwegian University of Science and Technology (NTNU)
Trondheim, Norway

Espen Moe
Department of Sociology and Political Science
Norwegian University of Science and Technology (NTNU)
Trondheim, Norway

ISSN 2662-2483 ISSN 2662-2491 (electronic)
International Political Economy Series
ISBN 978-3-030-54513-0 ISBN 978-3-030-54514-7 (eBook)
https://doi.org/10.1007/978-3-030-54514-7

This Palgrave Macmillan imprint is published by the registered company Springer Nature Switzerland AG
The registered company address is: Gewerbestrasse 11, 6330 Cham, Switzerland

Paul Midford dedicates this book to his mother, Carol Erickson Midford
Espen Moe dedicates this book to his son, Ansel Idris Moe

Preface

This is the Third Norwegian University for Science and Technology (NTNU) Japan Program Policy Study. The first two Japan Program Policy Studies were published with Palgrave Macmillan in 2014, and focused on *The Political Economy of Renewable Energy and Energy Security: Challenges and National Responses in Japan*, and *Eldercare Policies in Japan and Norway: Aging Societies East and West*. Like those studies, it is hoped that this study, and the others that follow, will contribute to understanding the major policy issues facing Japan and their relevance for other advanced industrial democracies, and indeed for the global community as a whole. Japan faces several policy challenges in common with other advanced industrial democracies, especially those in Europe. The focus here is on using common values as the basis for cooperation to overcome common challenges. Renewable energy is one such policy area, and the subject of this Third NTNU Japan Program Policy Study.

This study is very much a follow-on and update of our first volume on renewable energy and energy security from 2014, but one that focuses on the new "second-stage" challenges and opportunities that renewable energy faces in order to become a mainstream power source that replaces fossil and nuclear fueled alternatives. These second-stage challenges and opportunities focus on grid capacity, both extensively and in terms of smart flexibility (i.e., smart grids), the need for storage assets to fully utilize and integrate variable solar PV and wind power on the grid, and electricity market reform and liberalization.

The NTNU Japan Program originates in the 1980s and early 1990s, when a number of NTNU scientists and engineers conducted research at Japanese universities as visiting scholars. Based on their very favorable experiences and interest from Norwegian industry, NTNU established its Japan Program in 1998. Since the Program's establishment, it has offered courses on Japanese language, society, and politics, and on East Asian politics. Another hallmark of the Program is its annual Japan Seminar, which has become a leading venue for presenting and promoting the latest research on Japan and East Asia in North Europe and beyond. It also is a cross-disciplinary Seminar, and especially promotes cross-disciplinary cooperation between engineering and natural sciences on the one hand and the social sciences on the other.

The present volume emerged from three NTNU Japan Seminars focusing on renewable energy. The first, held in April 2016, was held in conjunction with the annual conference of the Nordic Association for the Study of Contemporary Japanese Society (NAJS). The second was held at the Norwegian Embassy in Tokyo in March 2017. This was the first NTNU Japan Seminar held in Japan, and featured an impassioned keynote speech by Kōno Tarō, a representative of the Lower House of the Japanese Diet, who has subsequently served as Japan's Foreign Minister, and then as Defense Minister. The third NTNU Japan Seminar where the penultimate drafts of the chapters in this book were presented was held in Trondheim in October 2018. This seminar featured a keynote address by Steffen Møller-Holst, Chairman of the Norwegian Hydrogen Forum, on "Norway's Green Hydrogen Strategy."

The central idea underpinning this work, and one of the main missions of the NTNU Japan Program itself, is to bring together insights from engineering and the natural sciences on the nature of technological change together with social science insights on how technology affects society, and how society affects the development of technology, its diffusion and use. Renewable energy illustrates both the opportunities and the necessity for promoting this collaboration. For social scientists it seems to go without saying that we depend on engineers and natural scientists to get a clear and accurate picture of the current state and changing nature of technology, a picture that is an absolute prerequisite for us to understand how technology and technological change affects society, economics, and politics. Our success in understanding all these fields is thus increasingly tied to understanding technological change. On the other hand, the funding, success, and diffusion of innovative technology are not always simply a function of the degree of innovation and the characteristics of the technology in question. Often, the success of new technology follows a social logic more than a technological logic. Among other factors, human perceptions of risk and benefit, economic and political interests can either promote or inhibit the success of any technology, regardless of technological merit.

The area of renewable energy offers rich examples of this. The interplay of perceptions of risk and benefit, and of economic and political interests, has exercised a powerful impact on which types of energy technology are adopted and diffused, and which are not. The interplay of these social factors plus the objective strengths and weaknesses of various technologies is a theme vividly illustrated in the chapters of this book.

We can see this interplay in debates about the comparative merits of various forms of renewable energy, such as wind and solar, and in the debate about the merits of nuclear power as an alternative. It is striking how often these debates often boil down to assertions about comparative technical feasibility: Is it easier to secure nuclear power plants against any kind of earthquake or tsunami, or to deal with the inevitable power flux/intermittent nature of solar and wind energy, and to build smart grids and grid-scale storage of electricity? Indeed, in an era of rapid technological change it is not hard to imagine that given a long-enough time horizon, all these goals could perhaps be largely achieved, which means that these debates are perhaps not ultimately about technological feasibility, even though that is how these debates are often portrayed.

Rather, the technical issues in these debates often mask the real issues at stake, which are more about values, interests, identity, and perceptions of risk that in significant respects are independent of science and technology. Ultimately, these debates will have to address values, interests, identity, and perceptions more directly in order to reach successful conclusions. What forms of energy will best represent our values, material interests, identity, and risk tolerance? In short, half of the answer to this question will be technological, and the other half will be social. We believe this volume makes an important contribution to addressing both sides of this question, the technical and social aspects of renewable energy.

We would like to thank the Faculty of Social Science and Technology Management (SVT), and the Faculty of Social and Educational Sciences (SU), the successor to SVT, and the Toshiba International Foundation (TIFO) for generously funding the three NTNU Japan Seminars from which this book emerged. We would like to thank the Royal Norwegian

Embassy in Tokyo for allowing us to use their premises for the 2017 NTNU Japan Seminar, for help with organizing that Seminar, and for so many other forms of support. We would like to thank Japan's Ambassador to Norway Tauchi Masahiro, and Tord Tukun, Charge d'affaires at the Royal Norwegian Embassy in Tokyo in 2017 for their remarks at two of our Seminars.

We owe thanks to a great many people who contributed directly or indirectly to this volume. Especially we would like to thank participants from the three NTNU Japan Program Seminars from 2016 to 2018 from which this volume emerged, including all the chapter authors. We would like to thank Kōno Tarō, Steffen Møller-Holst, as well as those by two chapter authors in this volume, Koichi Hasegawa and Hiroshi Ohta, for their intellectually inspiring keynote addresses at our three Seminars. We would like to thank several experts whose presentations made significant knowledge contributions to this volume, including Geir Martin Haarberg of NTNU and his presentations on the Kyoto International Forum on Environment and Energy (KIFEE) and green hydrogen, and Takeshi Bessho of Toyota on the smart use of energy and materials. We would also like to thank Hiroshi Okamoto, President of TEPCO Research Institute, Mika Obayashi, Director, Renewable Energy Institute, Ali Izadi-Najafabadi, Bloomberg New Energy Finance, Japan, and Tetsuya Azuma, Deputy Director, Electricity Market Surveillance Commission for the insights they provided at our March 2017 Seminar on the state of renewable energy policy and its prospects in Japan. We would like to thank Petter Nekså of NTNU and SINTEF, and Eric Zusman of the Institute for Global Environmental Studies (IGES, Hayama Japan) for serving as able discussants at two NTNU Japan Seminars.

We would like to thank Svein Grandum and Hiroshi Matsumoto of Innovation Norway and the Royal Norwegian Embassy in Tokyo for their expert advice and support for the 2017 NTNU Japan Seminar and for this project generally. At Palgrave Macmillan and Springer, we would like to thank Anca Pusca, Thangarasan Boopalan, Preetha Kuttiappan, Katelyn Zingg, and Timothy M. Shaw, International Political Economy Series editor at Palgrave Macmillan, in which this volume is included. Paul Midford would like to thank the Konrad Adenauer Foundation and Paul Linnarz (then head of KAS's Tokyo Office), for inviting Midford to present on renewable energy in Scandinavia and lessons for Japan in a series of seminars at the Tokyu Capitol Hotel in Tokyo, at Ritusmeikan University in Kyoto, and at Kwansei Gakuin University in

September–October 2014. The experience and the feedback Midford received inspired the focus of this volume.

Paul Midford would like to dedicate this book to his mother, Carol Erickson Midford. Espen Moe would like to dedicate this volume to his son Ansel Idris Moe.

Trondheim, Norway

Paul Midford
Espen Moe

Contents

Notes on Contributors

Pami Aalto is Jean Monnet Professor, Faculty of Management and Politics at the University of Tampere, Finland, and Consortium leader at Transition to a Resource Efficient and Climate Neutral Electricity System (EL-TRAN).

Luis Boscán is Head of Section at the Danish Utility Regulator (DUR), where he develops a data-driven approach to the surveillance of Danish electricity and gas markets. In parallel, he is Guest Researcher at the University of Southern Denmark (SDU) and Associate Editor at the journal Energy Reports, published by Elsevier.

Huyen Van Bui is an Associate Professor and currently the Director of Institute of Economics, Ho Chi Minh National Politics Academy. His research covers market economy, environmental economics focusing on natural capital and industrial economy. Dr. Huyen has published 50 peer-review journal articles and is the co-author of 20 books and book chapters. In 2016, Dr. Huyen was selected as elite member of Vietnam's Central Communist Party Economics Advisory Board.

Binh Van Doan is currently the Director of the VAST Institute of Energy Science, Vietnam Academy of Science and Technology. His research interests include energy security, energy and power system. He is national leading expert in national and regional Power Development Plan and author of more than 60 national and international peer-review articles.

Anton Finenko is Sales Operations Manager at Siemens Power and Gas, and a former Research Associate at ESI. His area of expertise includes power and renewables sector. Anton was part of the team advising policy-makers of Singapore on economics of the domestic solar PV industry. His work has been published in several academic journals.

Gang Chen is Assistant Director and Senior Research Fellow of the East Asian Institute (EAI), National University of Singapore. His monographs include *The Politics of Disaster Management in China* (New York: Palgrave Macmillan, 2016), *China's Climate Policy* (London: Routledge, 2012), and *Politics of China's Environmental Protection: Problems and Progress* (Singapore: World Scientific, 2009).

Susanne Therese Hansen is a postdoctoral fellow at the Norwegian University of Science and Technology (NTNU), Norway, Department of Sociology and Political Science. She obtained her doctorate in Political Science from NTNU in 2016. She has been a university teacher at NTNU's European Studies program, and a visiting researcher at the Stockholm International Peace Research Institute. Her current research interests include renewable energy policy between Norway and the EU, prospects and barriers for renewable energy transition, and the formation and effectiveness of norms and international law in various policy areas, including climate and the international arms trade. She has published in journals such as *European Journal of International Relations* and *European Security*.

Koichi Hasegawa is Professor in the Graduate School of Arts & Letters at Tohoku University. He serves as the President of the ISA's Research Committee on Environment and Society, and Vice-president of Japan Sociological Society. He received his Ph.D. from the University of Tokyo. He has published many articles on environmental sociology and social movements.

Gautam Jindal is a Research Fellow at the Energy Studies Institute, National University of Singapore. He holds a Masters in Carbon Management from the University of Edinburgh. Gautam's areas of research include renewables in electricity markets, carbon market mechanisms, and phase-out of high GWP refrigerants.

Brooks A. Kaiser is Professor and Head of the Management and Economics of Resources and the Environment (MERE) research group at the Department of Sociology, Environmental and Business Economics. She is a resource economist and economic historian.

Sarah Kilpeläinen is a Ph.D. candidate in the Faculty of Management and Business (International Relations) at Tampere University, Finland. She works on energy policy and sustainable energy transitions.

So Young Kim is Head of the Graduate School of Science and Technology Policy at KAIST with research interests in R&D funding and evaluation, science & engineering workforce, science-based ODA, and governance of emerging technologies. As a public intellectual, she sits on several high-level committees including the World Economic Forum's Global Future Council.

Juha Kiviluoma is a Senior Scientist, VTT Technical Research Centre of Finland.

Eirik Hovland Kjær is Higher Executive Officer in the Office of the Auditor General of Norway, with an MA in Political Science from the Norwegian University of Science and Technology (NTNU). He specializes in public policy and administration, with a focus on climate change, environmental politics, and renewable energy transitions. His MA thesis addressed the question of why Norway built two power cables, NordLink and North Sea Link, and the implications of these cables for the Norwegian renewable energy sector.

Florentine Koppenborg earned her Ph.D. in Political Science from the Free University Berlin. In 2017, she became a postdoctoral fellow at the School of Governance (Hochschule für Politik, HfP) at the Technical University Munich. Her research interests are within the area of energy and climate policy, particularly energy transitions ("Energiewende") and interactions with climate policy. She authored several peer-reviewed articles and book chapters on Japan's nuclear energy and climate policy. Currently, she is preparing a book manuscript on "Nuclear crisis and policy change: Safety regulation as a game changer." She is also embarking on a new research project comparing energy transitions, and the concomitant phase-in and phase-out of technologies, in Japan and Germany.

Quyen Le Luu focuses her research on the development of energy systems with consideration to environment protection and social inclusion. She is now working for the up-scaling of renewable energy and promoting sustainable production and consumption. Quyen was awarded an Australian Award Scholarship for her M.Sc. degree in 2014, and a YSEALI fellowship for professional development in environmental sustainability in 2015.

Paul Midford is Professor, and Director of the Japan Program, at the Norwegian University for Science and Technology (NTNU) in Trondheim. Midford received his Ph.D. in Political Science from Columbia University in 2001. He is author of *Rethinking Japanese Public Opinion and Security: From Pacifism to Realism?* (Stanford University Press, 2011), and co-editor with Espen Moe of *The Political Economy of Renewable Energy and Energy Security: Common Challenges and National Responses in Japan, China and Northern Europe* (Palgrave, 2014). Midford has published in *International Organization*, *International Studies Quarterly*, *Security Studies*, *The Pacific Review*, *Asian Survey*, and *Japan Forum*.

Espen Moe is Professor of Political Science at the Norwegian University of Science and Technology (NTNU), Norway. He obtained his doctorate in Political Science from UCLA. He has been a JSPS Fellow at the Kwansei Gakuin University in Japan, visiting professor at Alpen-Adria Universität in Vienna and at Beijing Normal University. His research centers on structural economic change, with a focus on the prospects for a renewable energy transition. He is the author of *Governance, Growth and Global Leadership* (Ashgate, 2007) and of *Renewable Energy Transformation or Fossil Fuel Backlash* (Palgrave Macmillan, 2015), as well as editor of *The Political Economy of Renewable Energy and Energy Security* with Paul Midford (Palgrave Macmillan, 2014). He has published in journals such as *Energy*, *Energy Policy* and *Energy Research & Social Science*.

Nam Hoai Nguyen is Vice Director at VAST Institute of Energy Science. He obtained his Ph.D in Economics in 2018, specializing in energy and electricity markets. He has also the background in Sustainable Development, Environmental Management, and Electrical Engineering in various graduate schools in Vietnam and Australia. Since 2009, Nam's major research topic is energy systems reliability and planning employing various integrated energy models. Energy economics is also of Nam's interest for academic research including theories and practical pathways for electricity market. Nam has published more than 30 articles in peer-reviewed journals in energy systems and economics of renewable energy.

Hiroshi Ohta is Professor at the School of International Liberal Studies (SILS), Waseda University. He received his Ph.D. from the Department of Political Science of the Graduate School of Arts and Sciences of Columbia University. Ohta's recent works include "EU and Japanese climate and energy security" (with Katja Biedenkopf), in Emil Kirchner and Han Dorussen eds., *EU-Japan Security Cooperation: Trends and Prospects* (London and New York: Routledge, 2019); *Kankyoto enerugie-wo meguru shuyōkoku-no hikakuseiji: Jizokukanōshakai heno sentaku* (*Comparative Politics about the Environmental and Energy Policies of Major States: Make a Choice for A Sustainable Society*) (Tokyo: Toshindō, 2016: 536pp.); and "Saving the Kyoto Protocol: What Can We Learn from the Experience of Japan-EU Cooperation?" (Yves Tiberghien), in P. Bacon, H. Mayer, and H. Nakamura, eds., *The European Union and Japan: A New Chapter in Civilian Power Cooperation?* (Surrey, UK: Ashgate, 2015: 169–184).

Lars Ravn-Jonsen is Associate Professor at the Department of Sociology, Environmental and Business Economics, University of Southern Denmark (SDU). He is a resource economist affiliated with the Management and Economics of Resources and the Environment (MERE) research group.

Inkyoung Sun is a researcher at the Science and Technology Policy Institute (STEPI). In her doctoral dissertation *Energy Innovator without Energy*, she investigated multiple factors influencing energy technology policy and co-evolutions of energy industry, technology and policy. She holds a M.A. in Political Science from University of Pennsylvania.

Yuki Takebuta is an executive at Minna Denryoku electric power company.

Jacqueline Tao is a Research Associate in the APAC Gas and LNG team at Wood Mackenzie. She has worked on several government-funded projects in the area of energy and climate change, including gas, renewable energy, and energy finance. She has also co-authored various papers published in peer-reviewed journals such as Energy Policy.

Robert M. Uriu is Associate Professor of Political Science at the University of California, Irvine. His earlier research has involved different aspects of Japanese industrial policy and trade relations. His current research project is on the industrial policy promotion of hydrogen fuel cells in Northeast Asia.

Abbreviations

ANRE	Agency for Natural Resources and Energy (Japan)
BEV	Battery Electric Vehicle
CAES	Compressed Air Energy Storage
CCS	Carbon Capture and Storage
CHP	Combined Heat and Power Plants
DoD	Department of Defense (US)
DPJ	Democratic Party of Japan
EASE	European Association for Storage of Energy
EPCO	Electric Power Company
EVN	Electricity of Vietnam company
FCCJ	Fuel Cell Commercialization Conference of Japan
FCEV	Fuel Cell Electric Vehicle
FIT	Feed-in Tariff
FYP	Five Year Plan
GHG	Green House Gas
GIZ	German Agency for Development Cooperation
HRS	Hydrogen Refueling Station
IAEA	International Atomic Energy Agency
ICE	Internal Combustion Engine
IEA	International Energy Agency
ISEP	Institute for Sustainable Energy Policies
JAEA	Japan Atomic Energy Agency
KEPCO	Kansai Electric Power Company
kWh	Kilowatt hour
LCOE	Levelized Cost of Energy
LDP	Liberal Democratic Party (Japan)

LNG	Liquefied Natural Gas
LO	Norwegian Confederation of Trade Unions
METI	Ministry of Economics, Trade, and Industry (Japan)
MOIT	Ministry of Industry and Trade (Vietnam)
MOPE	Ministry of Petroleum and Energy (Norway)
MOX	Mixed Uranium-plutonium Oxide Fuel
NDRC	National Development and Reform Commission (China)
NEDO	New Energy and Industrial Technology Development Organization (Japan)
NEDS	National Energy Development Strategy (Vietnam)
NEMS	National Electricity Market of Singapore
NGOs	Non-governmental organizations
NHO	Confederation of Norwegian Enterprise
NIMBY	Not in My Back Yard
NMEP	National Master Energy Plan (Vietnam)
NRA	Nuclear Regulatory Agency (Japan)
OCCTO	Organization for Cross-Regional Coordination of Transmission Operators (Japan)
OECD	Organization of Economic Cooperation and Development
OPEC	Organization of Petroleum Exporting Countries
PDPr	National Master Power Development Plan (Vietnam)
PHEV	Plug-in Hybrid Electric Vehicle
PPA	Power Purchase Agreement
PPC	Provincial People's Committee (Vietnam)
PWR	Pressurized Water Reactor
RES	Renewable Energy Strategy (Vietnam)
RPS	Renewable Portfolio Standard
SDPC	State Development and Planning Commission (China)
TEPCO	Tokyo Electric Power Company
TOE	Tons of Oil Equivalent
UHV	Ultra-high voltage
UNEP	United Nations Environment Program
UNFCCC	United Nations Framework Convention on Climate Change (Paris Conference, COP21),
WE-NET	World Energy Network
WTP	Willingness To Pay

List of Figures

List of Tables

CHAPTER 1

Introduction

Paul Midford

Overview

A global energy shift, if not an energy revolution, is currently underway. Global investment in environmentally sustainable renewable energy generation assets has already far overtaken combined new investment in fossil and nuclear fuel powered electricity generation infrastructure. In 2017 global investment in renewable energy reached $333 billion, versus only $144 in fossil fuel and nuclear generation assets. This gap is expected to continue growing, with $7.3 trillion in renewable energy investment globally by 2040, by which point renewable energy is expected to make up nearly 50% of installed capacity. The collapse of fossil-fuel prices, even into negative pricing at some points, during the COVID19 pandemic is both a reflection of the slipping position of fossil fuels vis-à-vis renewables and a sign of further decline to come as investment in unprofitable fossil fuels is likely to plunge.[1]

P. Midford (✉)
Norwegian University of Science and Technology (NTNU), Trondheim, Norway
e-mail: paul.midford@ntnu.no

P. Midford and E. Moe (eds.), *New Challenges and Solutions for Renewable Energy*, International Political Economy Series,
https://doi.org/10.1007/978-3-030-54514-7_1

This shift toward renewables is in part driven by falling costs. For example, the unsubsidized levelized cost of utility-scale solar PV projects fell by 89% from 2009 to 2019, and wind power fell by 70% during the same period (Lazard 2019, slide 9; Jackson 2018). Already by 2014 total installed wind-power capacity had surpassed installed nuclear capacity, with more than 500 GW in wind capacity in 2017 (Wind Power Monthly 2017), and 651 GW by 2019 (GWEC 2020), versus 392 GW for nuclear power (Power Technology 2018). The cost of related technologies is also falling dramatically as well. Most notably, the levelized cost of storage batteries fell by nearly 50% between 2018 and 2020 to USD150/MWh (Eckhouse 2020). In 2020, despite the pro-coal policies of the Trump administration, renewable energy is eclipsing coal as source of electricity in 2020, a development powered by a drop in the USA of more than 40% since 2010 in the cost of building new wind farms, and a drop in solar costs of more than 80% (Plumer 2020).

Although renewable energy technology has advanced remarkably in the past decade and this new technology has rapidly spread, adoption of renewable energy has varied markedly by country. For example, renewable energy already constitutes a large percentage of electricity generation in Denmark, yet remains exceptionally modest in South Korea, despite the latter's lack of domestic fossil-fuel resources. This book tackles this puzzle of why the diffusion of essentially the same renewable energy technology globally is producing very different outcomes across nations. It does so by focusing on the emerging bottlenecks and opportunities for the rapid diffusion of renewable energy that have arisen as the technology has developed and become increasingly economically competitive, despite the recent environment of declining fossil-fuel prices. Specifically, this book focuses on what can be called the stage-two challenges and opportunities facing renewables.

Stage one was about commercializing newly developed renewable technologies, especially solar PV and large-scale wind power, driving down their costs and making them a viable power source. If one policy encapsulates this period, it is the Feed-in-Tariff (FIT), that guaranteed above market rates for renewable electricity over multiple years, but with the rate for new installations gradually declining over many years. FIT schemes and related policies have been very successful in many countries, so much so that new challenges have emerged to the further diffusion of renewable energy, starting with the ability of grids to absorb all the variable electricity being produced.

This book is about stage two: the challenge of overcoming limits to the electricity grid to allow for variable or intermittent electricity produced by renewables to move from being a niche source of power to being a mainstay that increasingly replaces fossil and nuclear fueled sources. Stage two challenges and opportunities include the need for extensive grid development, especially increasing tie-lines among regional and national grids, smart grid development and deployment, increasing electricity storage infrastructure, accelerating the emerging hydrogen economy, and promoting electricity market liberalization for facilitating the entry of smaller renewable energy producers into electricity markets while decoupling control (and even ownership) of the grid from previously incumbent (often monopoly) generators of electricity, who mostly rely on fossil and nuclear fueled sources of electricity generation.

Explaining National Variation

This volume focuses on how the adoption of renewable energy technology is inhibited or promoted by three social factors: vested interests, public opinion, and strong states. First, there is the presence of vested interests in the form of incumbent industry stakeholders who have an incentive to try to use their political and structural influence to stop up-and-coming newer and more efficient industries from displacing them through what Joseph Schumpeter calls "waves of creative destruction."[2] New industries rise when they provide more efficient ways of providing goods and services, and they fall when what was once technologically revolutionary becomes commonplace and obsolescent and new industries based on new technologies arise that offer more efficient ways to provide goods and services (including new and better goods and services). This produces waves of industrial rise and fall: innovation, stagnation, "creative destruction." Periods of stagnation correspond to what could be called a silting up the political economy as vested interests expand their influence over government, and use that influence to stop rising industries from destroying older industries through competition in the marketplace (Moe 2015; Olson 1982).

Vested interests are thus currently an inhibiting factor for renewable energy, as existing energy producers, especially the fossil fuel and nuclear industries, can, and have in many instances, exercised political influence and even structural influence through their control of the electricity grid and existing electricity storage infrastructure (e.g., pump-hydro) to limit the adoption of renewable energy. Structural influence in turn reflects what Unruh (2000) calls a techno-institutional complex that perpetuates the existing energy infrastructure, effectively blocking alternative energy infrastructures from emerging. This can result in "carbon lock-in," with fossil-fuel industries dominant, and in some countries. This can be complemented by "nuclear lock-in." This structural power is centrally related to the focus of this book on the new challenges to further expanding renewable energy due to a lack of electricity grid and storage capacity.

Second public opinion, especially, but not exclusively, in liberal democracies, has a significant influence over the widespread adoption of renewable energy. The "rational public" forms coherent, if not rational, and stable opinions on policy issues based on underlying attitudes. These opinions respond in coherent ways to new information and are not easily manipulated or influenced by elites (Page and Shapiro 1992).[3] Even authoritarian states like China must be responsive to public opinion to some degree or risk public backlash and loss of regime legitimacy (Gries 2006; Tang 2005; Reilly 2012; Chen Weiss 2014). Especially in democratic systems but even potentially in authoritarian states, public opinion can be a force promoting renewable energy adoption in preference to alternatives such as continued fossil fuel or nuclear-power reliance, or public opinion can be a force inhibiting the spread of renewable energy through Not-in-My-Back-Yard (NIMBY) movements (see Chapter 12 regarding NIMBY opposition to wind power in Norway), or through a reluctance to accept and adjust to new technologies, such as smart meters.

Finally, strong states can play a major role in helping, or in some cases hindering, renewable energy as it faces the challenges of transitioning from being a niche supply of energy to replacing fossil and nuclear fueled power. A strong state is defined as possessing even control over its territory and significant bureaucratic capacity, and as having autonomy from non-state actors (Giraudy 2012, pp. 600–604). It is the latter two characteristics and especially autonomy from non-state actors that are key for this study. All the states examined in this book exercise effective control over their territory and have significant levels of bureaucratic capacity. However, autonomy from non-state actors varies a great deal, especially autonomy from vested interests and public opinion varies significantly.

The liberal democracies examined in this book, namely Japan, South Korea, Denmark, Finland, and Norway, are where we would expect public opinion to matter the most. By comparison, in more authoritarian regimes such as Singapore, Vietnam, and China public opinion should matter less. Vested interests can matter in both liberal democracies and authoritarian systems, although authoritarian regimes, such as South Korea from the 1960s until the 1980s (Woo-Cumings 1999; Evans 1995), may be expected to be somewhat less vulnerable to pressures from non-state actors. Yet, liberal democracies can also be relatively autonomous and act as strong states. Chalmers Johnson's study of Japan's Ministry of International Trade and Industry (MITI, now the Ministry of Economy, Trade, and Industry, or METI) and its role as the "pilot" organization directing Japan's industrialization from the 1950s to the 1980s is a classic example of a study that finds a highly autonomous state with great bureaucratic capacity (Johnson 1982).[4] The "Capitalist Development State" model Johnson posits has been applied (beyond Japan it is usually referred to simply as the "developmental state") to South Korea and several other East Asian states (Woo-Cumings 1999). Evans finds strong states that are not predatory, but stewards in promoting national economic development, must be autonomous but nevertheless have close links to society, a condition he labels "embedded autonomy" (Evans 1995).

Beyond combating human-caused climate change strong states have an incentive to promote renewable energy as an inexhaustible domestic (autarkic) source of energy that reduces dependence on energy imports or prepares for domestic fossil-fuel resource depletion in the case of exporters like Norway. In relatively more authoritarian states the presence of a strong state not captured by vested interests can override vested-interest groups, public opinion, NIMBY, and other forms of public opposition to renewable energy and related technology (e.g., smart meters) and infrastructure (e.g., grid expansion), and thus are well positioned to implement policies on grid development, storage infrastructure, and even retail market liberalization, that lead to the rapid diffusion of renewable energy. Of course, if a strong state does not see renewable energy as in its interest, it can act to limit or shut down renewable energy expansion. Russia might be a case of such a state (Chernysheva 2014).

The Focus of This Volume

This volume presents comparative cases that offer lessons on how to deal with the second-stage new challenges and opportunities this book highlights, beginning with a deep look at the world's third largest economy, Japan, but then looking at a variety of other cases, including China, the world's second largest economy, and several other national cases in East Asia and Northern Europe. As the third largest economy that is also fossil-fuel poor, Japan is an important case for understanding the barriers and opportunities for making the transition to renewable energy, yet a case that has received only limited analysis in terms of the second-stage challenges it faces. Most of the existing literature (see below) focuses on Japan's decision in the wake of the Fukushima nuclear disaster to switch from expanding nuclear power to expanding renewable energy through introducing a FIT and other means. The more recent bottlenecks, including a political backlash, facing the rapid expansion of renewable energy that has threatened grid access (exacerbated by a very underdeveloped and even segmented grid),[5] limited access to storage, and, on the other hand, the opportunity of electricity market liberalization, have received little attention to date. The Japan section well illustrates all the new second-stage challenges and opportunities facing renewable energy that this book focuses on: the new and growing issue of renewable energy (especially solar PV) curtailment of grid access, attempts to wrestle market and grid control from incumbent electricity generators, retail market liberalization, efforts to expand grid storage available to renewables and to build a hydrogen economy as ways to bypass entrenched energy sector interests.

These issues have come to the fore due precisely to Japan's success in rapidly expanding renewable energy capacity, although mostly this has been limited to solar PV, through relatively simple to devise first-wave policies such as FITs. From a pre-FIT annual installation rate of 250 MW for solar PV before 2010, Japan realized 10.5 GW in newly installed capacity in 2015. After 2015 growth in installed capacity slowed again, but rebounded to 7.5 GW of new capacity in 2019, with some market observers predicting 8 GW of new capacity in 2020. METI's target of solar PV accounting for 7% of the power mix by 2030 was surpassed already in 2019 (PV-Tech 2020; REN21 2019).

A significant cause of the slowdown in new installations of solar PV solar after 2015 was growing resistance from Electric Power Companies (EPCOs) to connecting this new solar capacity, because they claimed their grids were reaching the limits of how much more solar capacity they could absorb. Already in fall 2014 several regional electric companies suspended new connections with the support of METI. In 2015 new grid connections of mega-solar facilities resumed, but under new and less favorable rules allowing EPCOs greater scope for uncompensated curtailment of renewable energy producers' access to the grid (Watanabe 2014; Publicover 2016). Japan's new grid regulator, the Organization for Cross-regional Coordination of Transmission Operators (OCCTO),[6] and financial unbundling of the EPCOs' transmission operations from their generation business were supposed to help ameliorate grid capacity limits and ensure parity access to the grid for all producers. Yet, grid capacity, grid parity, along with storage and electricity market liberalization continue to pose challenges, but also opportunities, for renewable energy producers in Japan, as the chapters in the Japan section of this book demonstrate. These new stage-two challenges and opportunities are well illustrated by Japan, but are in fact common across many countries with growing renewable energy generation as this book makes clear.

The new challenges facing renewable energy as it transitions from a niche power source to replacing fossil fuels is not one of inadequate technology. Rather, these challenges stem from a lack of adequate policy, planning and investment, and ultimately from political, economic, and even social challenges. Japan, for example, possesses the world's largest pump-hydro storage capacity. However, most of this capacity is owned by EPCOs and dedicated to backing up inflexible nuclear power plants, and renewable energy producers have only limited access to this capacity. Grid battery storage technology has become not only a viable technology, but also one that can be very profitable as battery costs have fallen and its speed and efficiency in offering balancing services as well as storage has been demonstrated, as grid battery storage investments such as Tesla's 100-megawatt grid battery facility in Australia, have proven (Deign 2018; Gerdes 2018; O'Kane 2019). The hydrogen economy, on the other hand, is not as far along as battery storage (as is the case with electric vehicles where hydrogen fuel-cell powered cars continue to lag well behind battery powered cars), but even in the case of hydrogen, fundamental technological breakthroughs are no longer needed. Rather, what is needed are continued steady incremental improvements that keep the costs of this technology falling (see Uriu's Chapter 7 on Japan's hydrogen strategy).

After Japan we examine several countries in Japan's neighborhood of East Asia. In Northeast Asia, we analyze South Korea, like Japan a fossil-fuel poor highly developed economy with significant strengths in renewable energy technology and industry. South Korea boasts a far better developed national grid than what Japan possesses. Yet, South Korea is an economy that has made far less progress on renewable energy diffusion or market liberalization than has Japan. We analyze why we have had such different outcomes in similarly situated states, and briefly touch on a possible Japan-South Korea partnership in renewable energy through building electricity interconnections, something South Korea has proposed.[7]

We include a chapter on China because of its intrinsic importance as the world's second largest economy that has been more successful than almost any other country, including Japan, in promoting renewable energy, at least in terms of total installed capacity. By 2018 China possessed over a third of global installed capacity in wind and solar power: over 360 GW out of 1046 GW globally in 2018, and 416 GW out of 1202 GW globally in 2019, with offshore wind a rapidly growing sector (Korsnes 2020, pp. 3, 120; IRENA 2020). Yet, perhaps for this reason, China is also an economy that has suffered from far larger renewable energy curtailment problems than Japan and many other countries. However, the country is making efforts to reduce curtailment through grid expansion, and most recently through developing smart grids and storage, including hydrogen infrastructure. China's experience with renewable energy as an authoritarian state that might also be a relatively strong state and relatively less sensitive to public opinion and special intersts, thus offers an important comparison with Japan's progress in the same areas.

This volume includes chapters on two Southeast Asian nations: Vietnam and Singapore. Vietnam, is the least developed economy in this study, yet one that is experiencing rapid economic development and expansion of renewable energy. It has boldly abandoned plans to develop nuclear power and offers an important case of a developing country that is already starting to tackle the second-stage roadblocks to greater renewable energy adoption, notably grid development and storage. Finally, Singapore, as one of the most advanced economies in East, as well as Southeast, Asia, if not globally, offers a distinctive case where solar PV is expanding and obtaining additional grid and storage access despite the absence of subsidies and opposition from incumbent generators who already face overcapacity.

As Japan is a highly developed democracy with a poorly developed grid, we also compare it with highly developed countries with well-developed grids that in a number of ways have made more progress on renewable energy, specifically by examining three Nordic countries: Denmark, Finland, and Norway. These Nordic countries are especially relevant for second-stage renewable energy development and Japan, because they have liberalized their energy markets and established a national grid regulator, and Japan has consciously followed these Nordic models. Like Japan, grid development and storage in response to expanding renewable electricity generation has already become a key issue for Denmark and Finland.

There are several other reasons why the Nordic countries are a good fit for a book about renewables and the way ahead through second-stage challenges. Together they represent very different energy situations while being politically relatively similar. With respect to facing the future, Nordic countries represent very different situations. Denmark has the highest share of new renewables of any country in its electricity mix. It is thus also one of the countries that have come the farthest with respect to the new stage-two challenges and opportunities this book focuses on: electricity intermittency as a challenge for traditional and national grid systems.

Norway offers a distinctive case as a country with strong incumbent fossil-fuel stakeholders and renewable energy in the form of traditional large-scale hydro, and the challenges these interests pose for the expansion of other renewables, especially wind power. Norway has one of the highest shares of old renewables, in the form of large-scale hydropower, of any country in its electricity mix. Its problems are thus of a completely different kind. What is interesting here, is how a relatively slow renewable energy transition (wind power) is creating major popular opposition, from a wide range of groups—nature conservationists, NIMBY groups, anti-globalists, and far-left industrial interests. Norway may be a bit of a special case, as its wealth of hydropower allows anti-wind groups to argue that the country is already renewable as well as fully supplied in terms of energy. At the same time, it is also an example of why the next stage of the renewable energy transition is difficult, and the many interests that the politics of this transition needs to accommodate.

Finland is another distinctive case and a somewhat understudied one. Its energy mix relies far more on biomass, but it also has a considerable amount of nuclear power, and from an initially slow response with respect to renewables, it has set itself a very ambitious target of reaching negative emissions by the 2040s, relying to a major extent on large forest-based carbon sinks. Thus, Finland represents an uncommon mixed strategy with biomass far more important than in most countries.

This is of course not to say that looking at other European countries would have been uninteresting, or that a European focus exclusively on the Nordic countries is without weaknesses, but we believe that there are good reasons to focus on the Nordic countries, irrespective of everything else. The Norwegian challenge is intimately linked to the European Continent, as one of several points of contention is the extent to which Norway will build subsea cables to sell surplus Norwegian renewable electricity to the Continent. This illustrates one of the broader challenges facing a renewable energy transition in Europe, namely the lack of cross-border interconnectors, although the EU has ambitious goals. Specifically, the European Commission set a target for each member state to build interconnector capacity equal to 10% of their domestic installed generating capacity by 2020, a goal most members largely met, with a further goal of 15% by 2030. Spain would have been an interesting country in this respect, as the Iberian Peninsula is a bit of an energy island, with only weak interconnectors to France, but with a doubling of capacity realized in 2015 (planete energies 2016). This again highlights the centrality of grid development, along with storage and cross-border and liberalized energy markets as the defining challenges and opportunities for renewables moving forward.

Contribution

This book makes a distinctive contribution to our knowledge regarding renewable energy as it is one of the first to focus on new challenges to renewables, what we call second-stage issues. If first-stage renewable energy development is symbolized by wind turbines, solar PV panels, and the FIT, stage two is symbolized by grids, grid batteries, and hydrogen. There are currently almost no book-length studies that focus on these issues.

One book that begins to address these issues is Jinhui Duan's *Grid Connection of China's Renewable Energy and Its Energy Structure* (Duan 2015), although this work only focuses grid expansion, not smart grids, storage, hydrogen, or market liberalization. Another distinctive feature of the present work is that it has a strong focus on Japan's development of renewable energy. The only other volume that does so is our (Moe and Midford 2014) edited book, although that volume did not focus on the stage-two new challenges we examine in this book for the obvious reason that these challenges had not yet grown apparent at that stage. Instead, these are challenges that have moved to the forefront of renewable energy development since 2014.

A related distinctive feature is that while we focus on Japan, we also include intra-regional and cross-regional comparisons, whereas most other works focus on single countries, such as China (e.g., Lewis 2013; Korsnes 2020). The most similar book in this respect is Michaël Aklin and Johannes Urpelainen, *Renewables: The Politics of a Global Energy Transition* (2018), but this book focuses almost exclusively on the developed west (the USA and Western Europe, including 12 pages on Denmark, 3 on Finland), with less than twenty pages on China and India. Co-editor Espen Moe's (2015) book also includes cross-national comparisons, including chapters on China, Japan, USA, Denmark, Germany, and Norway. However, this book does not examine other Asian economies and does not focus on the new challenges to renewable energy that this volume focuses on.

Plan of This Book

This book is divided into three parts. Part one examines Japan, part two East Asia, and part three Nordic Europe. Chapter 2, by Koichi Hasegawa, examines Japan's Energy policy since the Fukushima Daiichi nuclear accident of March 11, 2011. In response to this accident Germany and several other governments adopted energy transition policies promoting energy efficiency, renewables, and denuclearization, while Japan's policy changed relatively little. This chapter addresses this puzzle by analyzing documents and media reports, and interviews with key players. The relative weakness of opposition political parties vis-à-vis the ruling coalition led by the Liberal Democratic Party (LDP), and relatively weak civil society provide the context for explaining Japan's relatively unchanged energy policy. An inner circle of vested interests called the "nuclear village" (*genpatsu mura*) still stands against real reforms. The balance of this chapter examines how grassroots community renewable-power movements, especially active in Aizu and other areas of Fukushima Prefecture, reveal new ways that the Japanese public is mobilizing to support an energy transition.

Chapter 3 by Hiroshi Ohta asks why is Japan so reluctant to take a leadership role in global climate change negotiations, and is trailing in the development of renewable energy? It argues that the source of Japan's inaction in climate diplomacy arises from its energy policy, which had long discouraged the extensive development of renewables. Japan's energy security concerns that dominated its energy policy since the two oil crises of the 1970s are the root cause. Since then, the Japanese government has promoted nuclear energy as the primary alternative to oil, and the solution to energy insecurity, not renewable energy, while also pursuing energy conservation. Climate change mitigation policy is closely tied to energy policy, over which the Agency for Natural Resources and Energy (ANRE) of the Ministry of Economy, Trade, and Industry (METI) has jurisdiction. The lack of strong political leadership on energy and climate policy leaves organized vested economic interests and METI as the most influential actors in setting policy. Thus, despite its tremendous potential to become a leader in renewable energy, Japan has effectively relinquished its leadership in climate diplomacy and the development of renewable energy.

In Chapter 4, Florentine Koppenborg analyzes nuclear power regulation, which as briefly discussed by Ohta, Hasegawa, and Midford in other Japan chapters, has a real influence on energy politics and policy in Japan, not least of all on renewable energy policy. A trade-off between promoting renewables and nuclear power is often visible in Japanese politics and energy policy. Koppenborg focuses on the Nuclear Regulation Authority (NRA) that was created in reaction to Japan's Fukushima Daiichi nuclear plant accident of March 2011, and its attempts to regain public trust by strictly regulating the safety of nuclear power. Defying expectations of collusion, the NRA has asserted itself as an independent regulatory agency. It warded off pressure from the Abe administration to speed up the restart process, and its enforcing of new safety standards has meant that plant operators have faced the need for expensive investments, expensive enough to render some nuclear reactors economically unviable. As a result, the maximum feasible share of nuclear power in the country's energy mix by 2030 will be approximately 15%, as opposed to the 20–22% the Abe administration set as a target. However, independent safety regulation has not been enough to convince the majority of Japanese that nuclear reactors should be restarted. Lawsuits by citizens challenging restarts have in some cases added to the already high costs of nuclear safety by prolonging the restart process. Overall, the NRA's independent nuclear safety regulation turned out to be a game changer for Japan's nuclear policy—and Japan's climate policy goals, unless the gap in electricity generation can be quickly filled with renewable energy sources.

Chapter 5 by Paul Midford examines the post-3–11 politics of nuclear power restarts versus renewable energy promotion. The Abe administration's support for restarting some nuclear power plants, based on safety authorizations given by the NRA, and Abe's personal pro-nuclear position, were often assumed to mean that the Abe administration was hostile toward renewables and was returning to the pre-3–11 goal of nuclear expansion. This chapter asks whether this is an accurate assessment of the Abe administration's energy policy. It finds that the answer is no. Rather, the Abe administration merely slowed down the phase out of nuclear power, essentially applying the long-standing Japanese "convoy" concept from economic policy, while continuing its predecessors' policies of promoting renewable energy through electricity market liberalization, promotion of storage capacity for renewables, unbundling of grid ownership from generation, and even the promotion of a hydrogen economy, which facilitates the further adoption of renewable energy.

In Chapter 6, Yuki Takebuta provides an account of the opportunities Japan's deregulation of its electricity market in April 2016 have created for new start-up entrants who are endeavoring to offer consumers the choice of buying only or mostly renewable energy, and thereby avoid buying energy produced with fossil fuel or nuclear power. She does so by introducing the experiences and strategy of one such company: Minna Denryoku, which has engaged in several innovative strategies to show the "face of electricity" on the other side of the socket to consumers.

Chapter 7 by Robert Uriu examines the Japanese government's unexpected announcement in 2014 of plans to realize a "Hydrogen Society" by 2050, making Japan the first major country to announce a national hydrogen strategy. At the heart of this strategy is the widespread use of hydrogen fuel cells as an important energy storage medium that allows for the large-scale expansion of renewable energy, as well as helping to decarbonize transportation and residential sectors. While business-government efforts in Japan on hydrogen fuel cells began in the early 1990s, the 3–11 crisis and the subsequent promotion of renewables has propelled this approach to the forefront. Nevertheless, it remains unclear when Japan can overcome the remaining technological and market obstacles.

Part II of this book examines four countries in East Asia. In Chapter 8, So Young Kim and Inkyoung Sun tackle the puzzling case of South Korea, which boasts global-scale investment in research and development for renewable energy, yet has one of the lowest levels of renewable energy adoption. They explore the political, economic, and technological dimensions underlying this puzzle, and pay close attention to political cleavages shaping renewable energy policy in South Korea. They also consider possible electricity market liberalization, grid capacity, and storage issues that could affect Korea's adoption of renewable energy moving forward.

In Chapter 9, Gang Chen considers how the Chinese government has been adjusting its renewable energy promotion policies in the wake of fierce competition among various renewable energy and low-carbon energy sources. Specifically, the government has been adjusting subsidy amounts, on-grid tariffs, and other financial incentives to support various non-fossil fuel energy sources. The central authorities have adjusted renewable energy targets in response to new industrial and market conditions, as well as in response to concerns from different interest groups. Hydrogen storage technology may be a future solution to the severe intermittency challenges facing renewable energy sources in China, especially for wind and solar power.

Chapter 10 by Gautam Jindal, Jacqueline Tao, and Anton Finenko examines the growth of solar PV in Singapore even in the absence of subsidies. Singapore's distinct profile as a densely populated city-state with little potential for land-intensive renewable energy sources such as wind turbines, and limited and crowded territorial waters and EEZ for offshore wind, makes solar PV the key renewable for Singapore. Remarkably, solar has been expanding despite a lack of subsidies. Nonetheless, policymakers need to address three vital challenges—limited space, intermittency, and no subsidies likely in the future, challenges that can restrict PV solar from playing a large role in Singapore's energy future. Grid capacity and storage also loom as future challenges to expanding solar PV generation.

In Chapter 11, Nam Hoai Nguyen and co-authors analyze how Vietnam's renewable energy policy has changed as it has achieved success in starting to significantly scale up renewables. The Vietnamese government has initiated systematic policies for promoting renewable energy since the early 2000s, and in 2017 the country became the first major emerging economy in East Asia to give up developing nuclear power. However, to meet the growing demand for lower emission electricity, Vietnam needs market-based policies facilitating a favorable environment for expanding renewable energy. With stronger commitments from the government, more effective policies such as Renewable Portfolio Standard (RPS) frameworks and auctions are expected to deliver more renewable electricity to customers. With increasing penetration of solar and wind power causing congestion on the grid, there are considerable opportunities for building local and smart grids that could facilitate more reliable and secure electricity services. Abundant hydro-electric capacity offers large potential storage capacity that can also facilitate the expansion of variable solar and wind renewable energy.

Part III of this book considers the very diverse experiences of three Northern European countries in navigating their transition to renewable energy. In Chapter 12, Espen Moe, Susanne Therese Hansen, and Eirik Hovland Kjær consider Norway as Europe's potential "green battery." From a climate perspective, the green battery idea is tremendously attractive, as Norway has some of Europe's best renewable energy resources, and domestic consumption of electricity is derived almost entirely from renewables. Utilizing Norwegian hydropower and wind to contribute to a European energy transition seems an obvious choice. Yet, despite building subsea cables to Germany, Netherlands, and Great Britain for renewable energy exports, and greatly increasing domestic wind power generation, enabled by a past vested-interest compromise between the power sector and energy-intensive industries, there is little political support for the green battery idea. While climate arguments are part of most actors' reasoning, these arguments seem to be secondary. Rather, the main motivation is increasing power exchange, with Norway providing power balancing services, selling excess renewable energy at a profit, and building cables not for climate reasons, but for profit. This chapter outlines a long-term future with some more undersea cables and wind power. However, in the short-term, the green battery is not edging closer to fruition. With energy-intensive industry interests currently dominating over power sector interests, no new cables will be built for now, while wind power has encountered significant political obstacles.

Chapter 13, written by Luis Boscán, Brooks A. Kaiser, and Lars Ravn-Jonsen, argues that Denmark can be considered a model of adoption and market integration of renewables, especially wind power. This success story has resulted from a favorable geographical location, cross-border cooperation with neighboring countries, and a well-struck balance between centralized planning and market-oriented policies. Nonetheless, the transition between present-day success and a future fossil-free Denmark presents several unresolved challenges. This chapter identifies four challenges facing Denmark's transition to even more renewables. The first is greening the transportation sector, and the second entails reforming the district heating sector while increasing its renewable-energy usage. The third challenge involves removing the cost barriers for end users to adopt electric transportation and heating. The final challenge involves meeting increasing flexibility requirements for the grid alongside dwindling domestic backup capacity. This chapter identifies lessons learned and experiences that are potentially transferrable to other countries, evaluating the extent to which the Danish success story can apply to other countries and regions.

In Chapter 14, Sarah Kilpeläinen and Pami Aalto analyze Finland's transition from a production-centric to a Smart Grid centric electricity system that can extensively utilize renewable energy. By the late 2010s, the idea of a full-scale energy transition was mainstreamed in Finnish society, including the expectation of renewable energy becoming the main production component in the Finnish energy system. This chapter argues that the increasing use of renewable energy sources is associated with the trends of electrification, decentralization, and variability. These trends contribute to a shift from a production-centric to a consumption centric energy system and require a focus on how flows of electricity are managed, stored, and redistributed, and how this affects the interests of the widening field of stakeholders. This chapter emphasizes the influence of stakeholders' interests as they navigate and respond to the trends associated with a higher share of renewable energy in the system with a special focus on grid development and energy storage. The analysis finds that although the need for a transition to higher shares of renewable energy has been mainstreamed, the policy development necessary for this transition is still in a formative phase as stakeholders struggle to balance and interlink their diverse interests.

In Chapter 15, Espen Moe summarizes the findings of this volume and takes stock of how we have progressed since the publication of our previous volume (Moe and Midford 2014). He finds that while the countries that figure both in this book and in our previous volume have changed, and moved on to face challenges that still lay in the future in 2014, the analysis we presented then is still recognizable. Fukushima triggered considerable change to Japanese energy policy, but arguably no energy revolution. China still has curtailment problems and coal is still an elephant in the room. Norway still does not really think of wind power in terms of climate policy. Moreover, Denmark is still the number one wind-power country in the world in terms of share of electricity consumption. Thus, in these four countries, as well in the countries that were not included in the 2014 volume, there is a significant amount of muddling through on the part of politicians. In other words, while this volume strongly suggests that the barriers and challenges facing today's policy-makers have changed, the surrounding politics really has not. Politicians are still faced with vested-interest problems (even if some of them have changed since 2014) and they still mostly seem to be solving problems as they appear rather than thinking beyond the horizon. With the costs of renewable energy continuing to drop, second-stage challenges to renewables will continue to hit ever more countries and make the international dimension to renewables ever more obvious.

Notes

1. Even with negative prices, the marginal costs of fossil-fuel consumption are not zero as they still require storage and transport, not to mention the environmental costs.
2. Regarding creative destruction, see Schumpeter (1942). Regarding the role of vested interests in preventing change, see Olson (1982).
3. This summarizes what is known as the pluralist school of public opinion. Other representative works include Nincic (1988), Jentleson (1992). The older elitist school of public opinion holds that public opinion is moody, unstable, often incoherent, but also subject to elite molding if not manipulation, and ultimately therefore is not a constraint on political elites. For representative works from this older school, see Lippmann (1925), Almond (1950), Ginsberg (1986).
4. For an account showing that the Capitalist Development State in Japan was captured by vested interests in the early 1970s, see Katz (1998).
5. Japan's grid is divided into two incompatible parts: Eastern Japan, including Tokyo, uses power with a frequency of 50 Hz, while western Japan, including Osaka and Nagoya, uses 60 Hz, with 1.2 GW of conversion capacity between the two regions by 2015 and more planned.
6. 電力広域的運営推進機関 (denryoku kōkiteki unei suishin kikan) Regarding the OCCTO, see OCCTO (2020).
7. The president of Softbank, Son Masayoshi, has proposed a Northeast Asian super-grid that would link Japan, Korea, and China, with vast wind farms in Mongolia (Asia International Grid Connection Study Group 2017).

References

Almond, Gabriel. 1950. *The American People and Foreign Policy*. New York: Harcourt & Brace.

Asia International Grid Connection Study Group. 2017. *Interim Report* (April) at https://www.renewable-ei.org/en/activities/reports/20170419.html. Accessed 11 Apr 2020.

Chen Weiss, Jessica. 2014. *Powerful Patriots: Nationalist Protest in China's Foreign Relations*. New York: Oxford University Press.

Chernysheva, Svetlana. 2014. *The Develop of Renewable Energy in Russia: Challenges and Constraint*. Trondheim: MA thesis, Department of Sociology and Political Science, Norwegian University of Science and Technology (NTNU).

Deign, Jason. 2018. Did Tesla's Big Australian Battery Kill the Business Case for More? *Greentech Media*, May 18. Accessed at https://www.greentechmedia.com/articles/read/has-teslas-big-australian-battery-killed-the-business-case-for-more#gs.uml.AOfHlA8. 29 Feb 2020.

Duan, Jinhui. 2015. *Grid Connection of China's Renewable Energy and Its Energy Structure*. Riga, Latvia: Lambert.

Eckhouse, Brian. 2020. Solar and Wind Cheapest Sources of Power in Most of the World. *Bloomberg Green*, April 28. Accessed at: https://www.bloomberg.com/news/articles/2020-04-28/solar-and-wind-cheapest-sources-of-power-in-most-of-the-world. 11 May 2020.

Evans, Peter. 1995. *Embedded Autonomy: States and Industrial Transformation*. Princeton: Princeton University Press.

Gerdes, Justin. 2018. With Focus on the Model 3, What's Up with Tesla's Storage and Solar Businesses? Greentech Media, May 11. Accessed at: https://www.greentechmedia.com/articles/read/the-focus-is-on-the-model-3-whats-going-on-in-teslas-storage-and-solar-bus#gs.uml.LatLLUA. 1 Mar 2020.

Ginsberg, Benjamin. 1986. *The Captive Public: How Mass Opinion Promotes State Power*. New York: Basic Books.

Giraudy, Agustina. 2012. Conceptualizing State Strength: Moving Beyond Strong and Weak States. *Revista de Ciencia Politica* 32 (3): 599–611.

GWEC (Global Wind Energy Council). 2020. Global Wind Report 2019. Accessed at: https://gwec.net/global-wind-report-2019/#:~:text=Key%20findings%3A,per%20cent%20compared%20to%202018. 1 June 2020.

Gries, Peter Hays. 2006. *China's New Nationalism: Pride, Politics, and Diplomacy*. Berkeley: University of California Press.

IRENA. 2020. Country Rankings. Accessed at: https://www.irena.org/Statistics/View-Data-by-Topic/Capacity-and-Generation/Country-Rankings. 17 May 2020.

Jackson, Felicia. 2018. Renewables Investment Nudges Out Fossil Fuel and Nuclear. *Forbes*, May 15. Accessed at: https://www.forbes.com/sites/feliciaja

ckson/2018/05/15/renewables-investment-nudges-out-fossil-fuel-and-nuclear/#38ea5f503752. 29 Feb 2020.

Jentleson, Bruce W. 1992. The Pretty Prudent Public: Post Post-Vietnam American Opinion on the Use of Military Force. *International Studies Quarterly* 36 (1): 49–74.

Johnson, Chalmers. 1982. *MITI and the Japanese Miracle: The Growth of Industrial Policy, 1925–1975*. Stanford: Stanford University Press.

Katz, Richard. 1998. *Japan: The System That Soured*. Boulder, CO: Lynne Rienner.

Korsnes, Marius. 2020. *Wind and Solar Energy Transition in China*. London: Routledge.

Lazard. 2019. Lazard's Levelized Cost of Energy Analysis—Version 13.0. November 2019. Accessed at: https://www.lazard.com/media/451086/lazards-levelized-cost-of-energy-version-130-vf.pdf?fbclid=IwAR3vv8ADl2XB3zErld9JF0BuoOCgTLIdCYhBrDI9tqoNvRzjW0Eau7IRsTI. 13 Apr 2020.

Lewis, Joanna I. 2013. *Green Innovation in China: China's Wind Power Industry and the Global Transition to a Low-Carbon Economy*. New York: Columbia University Press.

Lippmann, Walter. 1925. *The Phantom Public*. New York: Harcourt, Brace.

Moe, Espen. 2015. *Renewable Energy Transformation or Fossil Fuel Backlash: Vested Interests in the Political Economy*. Houndsmill, Basingstoke: Palgrave Macmillan.

Moe, Espen, and Paul Midford (eds.). 2014. *The Political Economy of Renewable Energy and Energy Security*. Houndsmill, Basingstoke: Palgrave Macmillan.

Nincic, Miroslav. 1988. The United States, the Soviet Union, and the Politics of Opposites. *World Politics* 40 (4): 452–475.

O'kane, Sean. 2019. Tesla's Megapack Battery Is Big Enough to Help Grids Handle Peak Demand. *The Verge*, July 29. Accessed at: https://www.theverge.com/2019/7/29/20746170/tesla-megapack-battery-pge-storage-announced. 1 Mar 2020.

Olson, Macur. 1982. *The Rise and Decline of Nations*. London: Yale University Press.

Organization for Cross-Regional Coordination of Transmission Operators (OCCTO). 2020. About OCCTO. Accessed at: https://www.occto.or.jp/en/about_occto/index.html. 1 Mar 2020.

Page, Benjamin I., and Robert Y. Shapiro. 1992. *The Rational Public: Fifty Years of Trends in Americans' Policy Preferences*. Chicago: University of Chicago Press.

Planete Energies. 2016. The European Commission: Electrical Interconnectors (February 22, 2016). Accessed at: https://www.planete-energies.com/en/medias/close/european-commission-electrical-interconnectors#form_id=media_node_form. 14 May 2020.

Plumer, Brad. 2020. In a first, Renewable Energy Is Poised to Eclipse Coal in US. *New York Times*, May 13. Accessed at: https://www.nytimes.com/2020/05/13/climate/coronavirus-coal-electricity-renewables.html?action=click&module=Top%20Stories&pgtype=Homepage. 13 May 2020.

Power Technology. 2018. Global nuclear power capacity expected to reach 536GW by 2030. At https://www.power-technology.com/comment/global-nuclear-power-capacity-expected-reach-536gw-2030/. Accessed 29 Feb 2020.

Publicover, Brian. 2016. New Curtailment Rules Cast Pall over Japanese Renewable Industry. *Recharge News*, September 28. Accessed at: https://www.rechargenews.com/solar/new-curtailment-rules-cast-pall-over-japanese-renewables-industry/1-1-870172. 29 Feb 2020.

PV-Tech. 2020. Domestic Solar and Storage on the Rise as Japanese Market Bounces Back. January 22. Available at: https://www.pv-tech.org/editors-blog/domestic-solar-and-storage-on-the-rise-as-japanese-market-shifts-gear. Accessed 15 Mar 2020.

Reilly, James. 2012. *Strong Society, Smart State: The Rise of Public Opinion in China's Japan Policy*. New York: Columbia University Press.

REN21. 2019. *Renewables 2019 Global Status Report*. Paris: REN21 Secretariat.

Schumpeter, Joseph A. 1942. *Capitalism, Socialism and Democracy*. New York: Harper Torchbooks.

Tang, Wenfang. 2005. *Public Opinion and Political Change in China*. Stanford: Stanford University Press.

Unruh, Gregory C. 2000. Understanding Carbon Lock-In. *Energy Policy* 28: 817–830.

Watanabe, Chisaki. 2014. Clean Energy Boom Challenges Power Grid. *Japan Times*, October 2. Accessed at: https://www.japantimes.co.jp/news/2014/10/02/national/clean-energy-boom-challenges-power-grid/#.XryGTJlUtnA. 29 Feb 2020.

Wind Power Monthly. 2017. Connected Global Capacity Tops 500GW, Say WPI, October 18. Accessed at https://www.windpowermonthly.com/article/1447740/connected-global-capacity-tops-500gw-says-wpi. 29 Feb 2020.

Woo-Cumings, Meredith. 1999. Introduction: Chalmers Johnson and the Politics of Nationalism and Development. In *The Developmental State*, ed. Meredith Woo-Cumings. Ithaca: Cornell University Press.

PART I

New Challenges and Opportunities in Japan

CHAPTER 2

Japan's Energy Policy and Community Power Movement After the Fukushima Nuclear Accident

Koichi Hasegawa

"The Happy Land" Suddenly Turned to "the Land of Tragedy"

The earthquake of magnitude 9.0 and the tsunami disaster on March 11, 2011, is the largest disaster in Japan since World War II. More than 18,000 people died or are still missing. This is the third largest disaster in modern Japanese history after the Great Kanto Earthquake of 1923, which resulted in around 105,000 dead or missing, and the Great Meiji Sanriku Tsunami of 1896, when around 22,000 died or went missing (Samuels 2013, p. 47).

This disaster brought the Fukushima nuclear disaster, which is the second largest nuclear accident in the world after the Chernobyl disaster in 1986. Among the six nuclear power units of the Fukushima Daiichi Nuclear Power Station operated by Tokyo Electric Power Company (TEPCO), Units 1–3 were in operation, and Units 4–6 were out of

K. Hasegawa (✉)
Tohoku University, Miyagi, Japan

P. Midford and E. Moe (eds.), *New Challenges and Solutions for Renewable Energy*, International Political Economy Series,
https://doi.org/10.1007/978-3-030-54514-7_2

operation for regular maintenance at the time of the earthquake. Units 1–3 automatically shut down at the onset of the earthquake, but the external power supplies and almost all in-house AC power supplies were lost due to the earthquake and the tsunami. Reactors and spent-fuel pools lost their cooling capabilities as a result. Explosions of the reactor buildings occurred at Units 1, 3, and 4, which were caused by the hydrogen released from the damaged core and that filled the reactor buildings. The reactor core of Unit 2 also seems to have been seriously damaged. A large amount of radioactive materials has been released and spread from this accident.

The severity of the Chernobyl and the Fukushima accidents was rated seven on the International Nuclear Event Scale. The Fukushima nuclear accident has several historical "firsts." It is the first severe accident of a nuclear power station; the complex disaster being triggered by a large earthquake and tsunami. It is the first example of "Quake and Nuclear Disaster Complex" or "complex disaster" which a seismologist, Dr. Ishibashi Katsuhiko, has warned of since 1997. Second, four reactors were simultaneously endangered for the first time. In the midst of the accident, on March 15, only about 50 workers, the smallest number of workers needed to operate the plant, a group that has since come to be known as the "Fukushima 50" (the subject of a recent movie in Japan) had to operate under strong radiation and very unstable conditions, in darkness due to the loss of AC power and numerous after-shocks. Third, the stable cooling function started only after 4 months from the accident and the uncontrolled situation of the reactor meltdown continued for more than 9 months. The crisis of the Three Mile Island (TMI) nuclear accident in the USA in 1978 passed after the first 6 days. Even in case of the Chernobyl accident, a large amount of radiation was halted after the first 10 days. Fourth, it is the first severe accident of a nuclear power plant on the coast. The TMI and the Chernobyl stations were located inland. In Japan, all nuclear reactors are located on the coast so that cooling water can be sourced from the sea and heated effluent after cooling can be discharged into it. Contaminated water overflows into the ocean sometimes even still now when heavy rainfalls by typhoon or other kind of storms. Scientists are worried about serious contamination of seawater and damages to the ecosystem as the long-term effect.

The most tragic and painful cases are the 160,000 evacuees, the peak number that was registered in June 2012. These evacuees had been living near the melted-down nuclear plant. A zone of up to 20 km from the site was designated as an "Access-restricted Area" and no entry was been allowed unless authorized, a situation that continued until the end of March 2017. Some areas, like Iitate Village, outside the 20-km zone, have been also designated as "Deliberate Evacuation Areas." Even as of December 2019, more than eight years after accident, around 42,000 people were still evacuated, with 11,000 relocated within Fukushima Prefecture and another 310,000 outside the prefecture at their own volition (Reconstruction Agency 2019). Their situation has continued to be serious over these ensuing years. The central and local governments are giving a strong pressure to return to homes under the "the early return policy," because in most of the contaminated areas, the radiation level has already dropped under the level of 20 millisievert per year, the new standard for evacuation. The government has started to curtail the subsidies they provided for evacuees.

Evacuees criticize these policies and claim that the real intention of national and local governments in introducing this policy is to create the public image that the situation is already under control. This has irritated and caused mental pain for evacuees. They have had difficulty deciding what they should do, return soon, wait long years for the level of radiation to decrease, or give up returning. Within families, members often have different points of view depending on their age, job, and whether they are raising children. Older generation want to return if possible. Many of evacuees who are younger and raising children decided to give up returning because of their fear of being exposed to high radiation. They want to rebuild their life in a new place as soon as possible. The radiation level of 20 millisievert per year is 20 times higher than the former safety standard of one millisievert in place prior to the Fukushima accident. More than 40,000 people lost their homes, their beautiful farmland on which many generations of their ancestors had worked ardently, their fisheries, their customers, their workplaces, and their hope. Their communities were divided, and many had their family lives destroyed.

The Chinese character of "Fukushima" consists of "Fuku" literally meaning "happy" and "Shima" meaning "land" in this context. This area with the happy name, which produces a variety of delicious farming products, including famous peaches, and is also famous for its beautiful landscape suddenly turned to "the land of tragedy," like Hiroshima and Nagasaki where the US military dropped atomic bombs in 1945.

Japan's Curious Political Stability After the Accident

Remember what had happened after the 1986 Chernobyl accident. Mikhail Gorbachev, the last President of the Soviet Union, confessed "the accident at the Chernobyl nuclear power station was graphic evidence, not only of how obsolete our technology was, but also of the failure of the old system" (Gorbachev 1996, p. 189). From the accident in April 1986, it was only five and a half years until the collapse of the Soviet Union in December 1991 following the fall of the Berlin Wall in November 1989. The accident was a remarkable incident in world history that marked a major turning point that helped precipitate the end of the Cold War in Europe and the collapse of the Soviet Union "by literally knocking the country off its tracks" (Gorbachev 1996, p. 189). What changes have the Fukushima disaster caused for Japan's energy policies and those elsewhere throughout the globe?[1]

In July 2011, Germany adopted an energy shift policy that included abandoning all 19 nuclear reactors by the end of 2022. Switzerland also decided to abandon new nuclear reactor construction and close all 5 reactors after 50 years of operation, which means all reactors will be shut down by 2034. In East Asia, in January 2017, the new government of Taiwan decided to abandon all 6 reactors by 2025, and to this end a new law was passed (Chou 2018). In June 2017, the new Korean President, Moon Jae-in also declared to implement denuclearization over the next 40 years. Both South Korea and Taiwan are dependent on oil imports, like Japan. In South Korea, 30% of electricity is provided by nuclear power, which is similar to Japan's dependence before the Fukushima accident. In Taiwan, 19% of electricity is provided by nuclear power. In both cases, changes of government and the role of strong political leadership seem to have played a critical role in bringing about policy shifts toward denuclearization. Thus, denuclearization, and promoting renewables and energy efficiency are the new mainstream worldwide after the Fukushima accident. But, how about in Japan?

By contrast, politically, the current Liberal Democratic Party (LDP) government in Japan has curiously enjoyed stable public support and won all five national elections of the lower house and the upper house since it returned to power with the change of government in December 2012. The right-wing Abe administration is the one of front runners of the recent "We-firstism" trend (like Donald Trump's "American First") and is fortunate enough to be the first multi-year cabinet since the five-year Koizumi administration from April 2001 to September 2006. Following Koizumi each of the subsequent six prime ministers, including Abe during his first cabinet from September 2006 to September 2007, had to resign within approximately one year or less due to low public support. It seems that the Fukushima accident created greater political damage for the Democratic Party of Japan (DPJ), which was the ruling party from September 2009 to December 2012, rather than for the LDP, then an opposition party. The DPJ cabinet of Prime Minister Kan Naoto misled the public and created a lot of confusion by deliberating concealing information and delaying information disclosure (Hasegawa 2015, pp. 11–13). In fact, the LDP as the dominant ruling party over more than five decades bears most of the responsibility for promoting nuclear power and the policies that led to the Fukushima accident.

Nuclear Energy Debates After Fukushima

How did the Fukushima accident affect Japan's electricity supply? A drastic energy policy shift like that of Germany, which decided right after the Fukushima accident to abandon all nuclear reactors by the end of 2022, has not so far happened in Japan. All nuclear reactors in Japan were closed temporarily for 23 months between September 15, 2013, and August 11, 2015, due to the new legal and regulatory requirements as explained below. Nonetheless, there were no power outages or shortages during this period. Especially areas within Tokyo Electric Power Company (TEPCO), which is responsible for the Fukushima Daiichi nuclear accident, and Tohoku EPCO, where the 3.11 earthquake and the tsunami hit, there have had no nuclear reactors operating since the disaster. As of the end of December 2019, only nine reactors had been approved and restarted by the new regulatory agency, the Nuclear Regulation Authority (NRA). Due to the legal requirement for regular maintenance, a nuclear reactor in Japan must shut down temporarily after operating for at most 13 months. Up to 15 September 2013, all reactors had to close temporarily on a one-by-one basis for regular maintenance until Japan reached zero nuclear power generation. All units along the Pacific Ocean were forced to close because of direct damage sustained from the earthquake and the tsunami. The NRA was established on September

18, 2012, replacing the Nuclear Safety Commission and the Nuclear and Industrial Safety Agency, both of which had failed to prevent the Fukushima accident, and thereby revealed their malfunctioning. The NRA is expected to be independent of the national government.

After the accident, the focal point of the nuclear energy debate shifted (Hasegawa 2018b). The main justification for nuclear power, namely avoiding a short-term electricity shortage, lost persuasiveness. The amount of electricity supplied by nuclear power has been replaced by thermal reactors. The electricity rates increased by several percent, but the majority of people supporting decreasing reliance on nuclear power, as can be seen in Fig. 2.1.

Japan has three options for nuclear energy policy; (1) immediately shut down all nuclear reactors. Former Prime Minister Koizumi changed completely his mind after the Fukushima Daiichi accident, and now supports immediately and permanently shutting down all of Japan's nuclear reactors. (2) Reopening the nuclear reactors that can meet the new standards set by the NRA, strictly limiting each reactor to the legal limit of 40 years the operation and rejecting construction of new reactors, thereby placing Japan on a trajectory for denuclearization by the end of the 2030s. The former DPJ cabinet of Prime Minister Noda Yoshihiko set this policy, but the current Abe cabinet subsequently abandoned it. (3) Reopening nuclear reactors that can meet the NRA's new standards and aim to construct new reactors and replace old reactors. The Abe cabinet supports this policy and wants to keep the certain level of electricity produced by nuclear, around 20–22% of the electricity supply in 2030. Except for two reactors that were already under construction before the Fukushima accident, the Abe administration has not proposed the construction of any new reactors. Nonetheless, the real intention of the Abe administration is to construct nuclear reactors and is waiting for an opportune time to propose doing so.

In the Ministry of Economics, Trade, and Industry's (METI) long-term outlook published in July 2015, which outlines how Japan plans to achieve its 2030 Green House Gas (GHG) emission target for the UNFCC (United Nations Framework Convention for Climate Change, COP21) Paris Conference, METI stated that total electricity generation of 1065 billion kWh in the fiscal year 2030 will be composed of 22–24% of renewable energy, 22–20% of nuclear, 27% of LNG-fired thermal, 26%

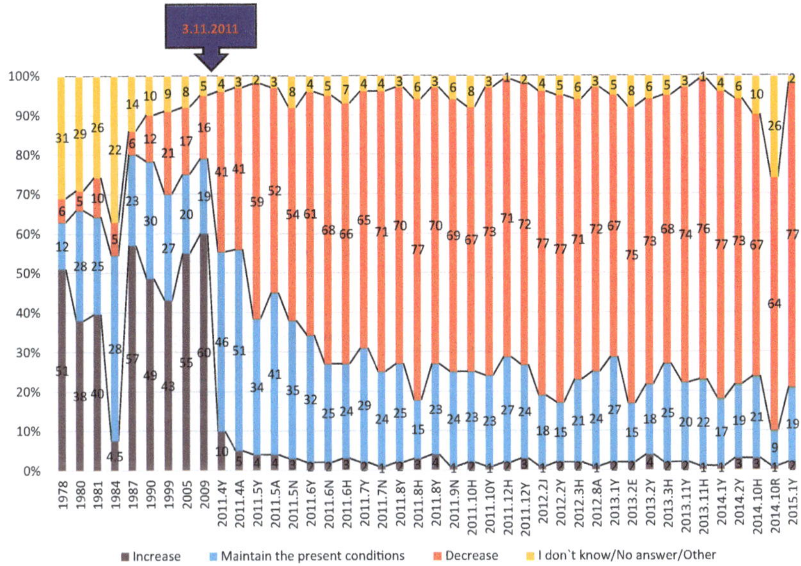

Fig. 2.1 Public opinion regarding increasing, decreasing, or maintaining the status quo of nuclear power (*Legend* Results between 1978 and 2009 are based on surveys conducted by Naikakufu daijin kanbō seifu kōhō shitsu, yoron chōsa shitsu under the title of "Genshiryoku ni kansuru yoron chōsa," and "Enerugi ni kansuru yoron chōsa." Surveys conducted by *Asahi Shinbun* are denoted by the letter "A" after the date and month, those conducted by National Institute of Environmental Studies are denoted with an "E," "H" denotes NHK, "N" denotes *Nihon Keizai Shinbun*, "J" denotes Japanese General Social Surveys [JGSS], "R" denotes the Japan Atomic Energy Relations Organization [JAERO], and "Y" denotes *Yomiuri Shinbun*. *Note* Author created figure, with assistance from Mathias Shabanaj Janklila, based on data from Iwai and Shishido [2015], JAERO [2014], Yomiuri Shinbun [2015])

of coal-fired and 3% of oil-fired (Fig. 2.2). This outlook assumes that economic growth will be continued at an average of 1.7% increase per year from Fiscal Year 2013 to 2030. The required power generation will be 1278 billion kWh in 2030, and it will be reduced to 1065 billion kWh by 17% of power conservation. See Fig. 2.2. What is wrong with METI's outlook?

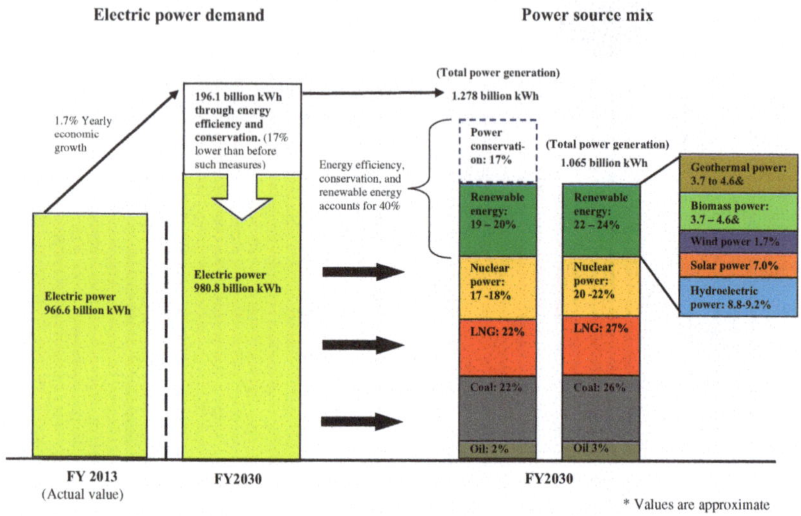

Fig. 2.2 Electric power demand and generation source mix in Japan (*Note* Author created figure, with assistance from Mathias Shabanaj Janklila, based on data from the Ministry of Economy, Trade, and Industry of Japan [http://www.meti.go.jp/english/press/2015/0716_01.html])

Figures 2.2 and 2.3 show that the peak of total power generation was 1030 billion kWh and was reached in FY 2007. It then declined 14.1% to 885 billion kWh by FY 2015. 1278 billion kWh in FY 2030 is a drastic overestimation, representing a 36% increase over the level of electricity supplied in FY 2013. It is important to keep in mind that Japan is already experiencing population shrinkage and has low economic growth averaging 0.7% per year between FY 2010 and 2015. The estimated share of 22–20% to be provided by nuclear, 213–234 billion kWh in FY 2030, can be replaced by power conservation with no carbon emissions. The decoupling of economic growth from electricity consumption, as is the case in Germany, has already started in Japan. Japan is decoupling of economic growth and GHG emissions without relying on nuclear power. The peak

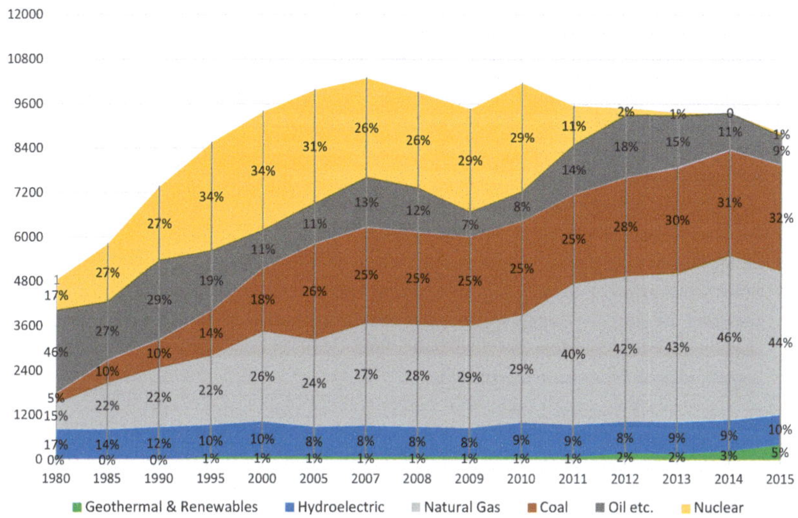

Fig. 2.3 Annual generation by volume and source in Japan: 1980–2015 (*Note* Author created figure, with assistance from Mathias Shabanaj Janklila, based on data from JAERO [ud])

of Japan's GHG emissions was reached in FY 2013, when Japan shut down all its reactors. Emissions then declined by 3.4%, 6.0%, and 6.2% in FY 2014, FY 2015, and FY 2016 (Fig. 2.4) despite no operating nuclear reactors in 2014, and extremely few in 2015 and 2016.

Since the Fukushima accident, many citizens worldwide are deciding that nuclear energy is unnecessary, not only in the short term, but also in the long term. Other major reasons for opposing nuclear energy in the Japanese context are (1) the high risk of accidents and disaster in a quake-prone archipelago, (2) financial risks, (3) the risks of handling spent nuclear fuel, and (4) the risk of nuclear proliferation.

Risk of Nuclear Accident in a Quake-Prone Archipelago

The Fukushima accident raised serious concerns that a major "quake-prone country" and nuclear power generation are incompatible. Japan is a very earthquake-prone country due to its location on the boundaries of

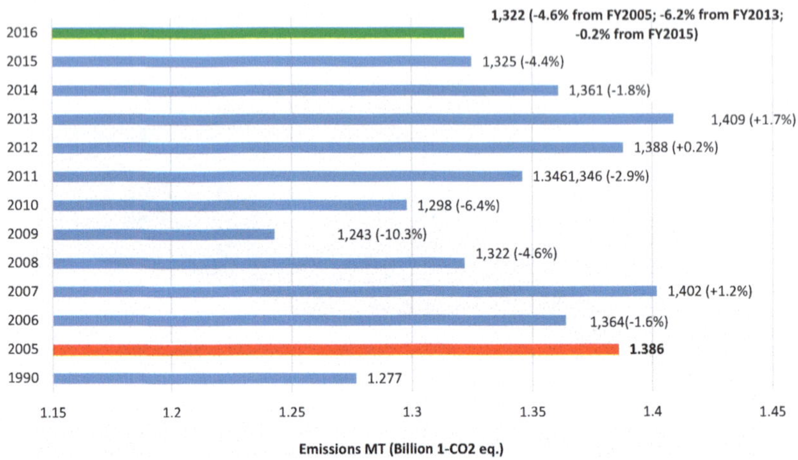

Fig. 2.4 Japan's greenhouse gas emissions: 1990–2016 (*Notes* (1) Percentages in parenthesis are changes from the base year of 2005. (2) Emissions are estimated based on annual figures from various measures. Regarding preliminary figures for FY 2016, some annual figures from FY 2015 were used in place of FY 2016 figures that have yet to be released. Also, some estimation methodologies are being reviewed to achieve greater accuracy in emissions estimations. Consequently, some of the figures released in April 2018 differ from the preliminary figures in this summary. Removals by forest and other carbon sinks are also estimated and announced along with the final figures. (3) Total GHG emissions for each fiscal year and percentage changes from previous years [such as changes from FY 2005] do not include removals by forest and other carbon sinks from activities under the Kyoto Protocol. (4) Author created figure, with assistance from Mathias Shabanaj Janklila, based on data from the Ministry of Environment [http://www.env.go.jp/press/files/en/750.pdf])

four tectonic plates. As Katsuhiko Ishibashi (2011) pointed out, fifty-four reactors (including the six at the Fukushima Daiichi plant), or 12.2% of the world's commercial nuclear reactors, are situated in this small island nation where about ten percent of the world's seismic activity occurs. Moreover, Japan has entered a seismically highly active phase. He observes that the Great East Japan Earthquake has raised the probability of major earthquakes in almost all parts of the Japanese Archipelago (Ishibashi 2011).

Financial Risk for Utility Companies and Plant Makers

Although TEPCO is still the largest private utility company in the world, since the Fukushima accident it has been facing the threat of bankruptcy, as it is obliged to pay a total of 8000 billion Yen (USD 72 billion), or annually 300 billion Yen (USD 2.7 billion) as the cost for decommissioning the Fukushima reactors, and a total of 3900 billion Yen (USD 35.1 billion), or 200 billion Yen (USD 1.8 billion) annually to compensate refugees and other victims. Since the accident the Japanese government has injected capital to prevent TEPCO from going bankrupt, and as a result has taken a temporary controlling share of the company. Citizens have come to see that if another nuclear accident happens the national government will have to assume financial and managerial responsibility for the company owning that plant. "Nuclear energy is safe, clean and cheap": The Fukushima accident revealed this claim to be nothing more than a myth and illusionary propaganda.

This accident also creates extra cost for implementing new and improved safety standards. Toshiba, one of Japan's leading giant electronics companies, also faced bankruptcy at of the end of April 2017 due to the manipulation of accounts to the tune of 230 billion Yen (2.1 billion Dollars), related losses of 500 billion Yen (4.5 billion Dollars) in FY 2015, and losses stemming from its purchase of a US nuclear reactor manufacturer, losses that totaled another 700 billion Yen (6.3 billion Dollars) in FY 2016. The trigger was buying the nuclear energy division of Westinghouse in 2006. Then Toshiba boasted that it had become the world's largest nuclear reactor manufacturer. However, this was misplaced pride and the beginning of Toshiba's downfall. Westinghouse used to be a leading electric appliance manufacturer in the USA that had been considered a major rival to GE and developed pressurized-water reactors (PWR) that were used by Kansai Electric Power Corporation (KEPCO) and some other Japanese Electric Power Companies (EPCOs). Westinghouse gradually declined and became defunct in 1999. It sold its nuclear energy division to BNFL in 1998. BNFL was driven to the brink of bankruptcy and its owner, the UK government, decided to let Westinghouse go. Toshiba won the bid for the company and paid five billion dollars, double the estimated value, in anticipation of "the nuclear renaissance." In 2009 Toshiba declared the goal of booking new orders for 39 reactors globally by 2015. In fact, Toshiba started construction of no more than eight reactors in China and the USA from 2006 to 2015. The Fukushima accident dealt the final blow to the nuclear renaissance, which was already under dark clouds before 2010.

Handling Spent Nuclear Fuel and Three Endemic Problems

The greatest dilemma facing every country relying on nuclear power is how to finally dispose of spent nuclear fuel. The extreme difficulties of spent-fuel disposal prompted a move away from nuclear energy in the USA and Europe. Japan has stuck to a closed nuclear fuel cycle policy under which all spent fuel is reprocessed and high-level radioactive waste is stored for cooling over a period of thirty to fifty years in a vitrified form and eventually disposed of by deep burial in a geological formation 300–700 meters below ground. Handling high-level radioactive waste is extremely cumbersome and dangerous due to extremely high levels of radioactivity, large amounts of heat generation, high toxicity, very long half-lives, and the presence of a wide variety of elements. It needs to be isolated from the human living environment for approximately 100,000 years, and hence requires extremely long-term storage facilities. Every nuclear power generating country agonizes over the location of final disposal sites. Only Finland has started construction, since 2012, of a final spent nuclear fuel repository at Onkalo. Sweden also decided the place for its repository site in June 2009 (Hasegawa 2015, p. 48).

Japan's difficulties disposing of spent nuclear fuel are compounded by three other endemic problems. First, it is questionable as to whether there are any suitable sites for final disposal in such an earthquake-prone country, as we were reminded on 3.11, 2011, with interlaced active fault lines, a limited land area, high population density, and many subterranean water veins. In July 2017, the government published the "Nationwide Map of Scientific Features for Geological Disposal," which indicates that about 30% of the Japanese Archipelago is suitable on scientific grounds for final disposal of high-level radioactive waste outside of a radius of 15 km from volcanoes, and which is not located close to any active faults (Agency for Natural Resources and Energy 2017). This map shows some sites are favorable from the viewpoint of waste transportation also. However, the next step forward, which is to start researching suitable sites, is not clear. No cities or towns have yet shown any interest in hosting this site (Hasegawa 2017).

The second problem is the issue of surplus plutonium. Measured by its plutonium stockpile of 45.7 ton as of the end of 2018 (Office of Atomic Energy Policy of Cabinet Office 2019), Japan has been estimated to have the highest potential capability for nuclear armament in the world by the International Atomic Energy Agency (IAEA) which has been very concerned by the proliferation risk posed by this stockpile. For this reason, in 1991 the Japanese government announced its commitment to not hold surplus plutonium. For Japan to adhere to its two principles, namely, the full reprocessing of all spent nuclear fuel and zero surplus plutonium, it needs to strictly maintain the demand and supply balance for plutonium extracted by reprocessing. The dilemma is that reprocessing needs to be stopped in order to prevent the accumulation of surplus plutonium, but this cessation will breach the principle of full reprocessing of all spent nuclear fuel. The Fukushima accident makes this dilemma much more severe (see Hasegawa 2015, pp. 51–53). The Japanese term "pluthermal" signifies the use of mixed uranium-plutonium oxide fuel (MOX) in thermal reactors (light water reactors). After the Fukushima accident, the pluthermal program is operating only at reactor number 3 of the Genkai nuclear power plant, and reactors 3 and 4 of the Takahama nuclear plant, which together burn a total of 1.5 tons of plutonium per year (Office of Atomic Energy Policy of Cabinet Office 2019). At this current rate, it is hard for Japan to fully reprocess all spent fuel and maintain no surplus plutonium.

The third problem is the difficulty of changing policy despite the high cost of reprocessing. Why is Japan so obstinately fixated on reprocessing? The ulterior motive may be to retain international rights relevant to the potential military use of nuclear resources in the future. After Germany abandoned its program to extract plutonium, Japan is the only non-nuclear-weapon state allowed to possess technologies for uranium enrichment, reprocessing and fast breeder reactors. The fundamental premise of the nuclear fuel cycle was to continue to expand nuclear power generation. However, the Fukushima accident prompted Japanese society to radically rethink to the point of abandoning the nuclear fuel cycle program. In December 2016, Japanese government finally decided to abandon the multi-decade troubled and highly costly fast breeder reactor, *Monju*. The nuclear fuel cycle program has thus lost one of its essential parts: the fast breeder reactor. This in turn raised further doubts about the purpose of Japan's reprocessing program.

Potential Nuclear Deterrence and the Risk of Nuclear Armament

The final reason Japan continues to promote nuclear power generation is that maintaining nuclear power technology, especially nuclear fuel cycle capability, which is considered to have value as a potential nuclear deterrent, or what is known as recessive nuclear deterrence. The Japan-US Nuclear Cooperation Agreement that was revised in 1988 gave Japan "blanket consent" regarding spent nuclear fuel reprocessing to any amount up to a certain limit. When this agreement expired in July 2018, both governments reached an agreement for subsequent automatic renewals. How significant is it for Japan to have the potential to arm itself with nuclear weapons? Japan's nuclear cooperation agreements with the USA, Canada, and Australia limit the use of nuclear fuel to "peaceful purposes." The official position adopted by Japan's past cabinets has been that "possession [of nuclear weapons] does not violate the Constitution if it is limited to... a bare minimum required for self-defense." Legal restrictions on the possession of nuclear weapons domestically include Article Two of the Atomic Energy Basic Law, which states that it is "... limited to peaceful purposes," and the so-called three non-nuclear principles adopted by the Diet in 1971 (Ministry of Foreign Affairs of Japan 2014). In terms of international law, Japan is bound by nuclear cooperation agreements with the USA and other countries as well as by the Nuclear Non-Proliferation Treaty. Thus, Japan is restricted from nuclear armament by domestic and international laws.

The USA would not readily agree to Japan's nuclear armament as it would provoke Japan's neighbors such as South Korea, North Korea, and China and might lead to nuclear proliferation in the region. South and North Korea, China, and Russia would disapprove, and Japan would draw criticism from the rest of the world as well. Japan is also extremely vulnerable to nuclear attack due to the high population density and concentration in Tokyo, Osaka, and a few other cities. In other words, it is not realistic to think that Japan could have a nuclear deterrent on its own. Trying to develop one would be an extremely destabilizing move that would only increase the risk of nuclear war. Nuclear deterrence is especially unrealistic for Japan.

Reasons to Keep Nuclear Power Plants for Power Companies

Why cannot Japan relinquish its nuclear power plants? The Fukushima accident has clearly shown that there is no such thing as absolute safety in terms of nuclear power generation, and that risks remain especially high for an earthquake-prone country like Japan. It has also demonstrated that Japan's electricity supply is not adversely affected by the absence of nuclear energy, as all nuclear power plants were offline for twenty-three months from September 2013 to August 2015.

Electric power companies are afraid that if they decide to change their business plan to zero nuclear energy, they will suffer extraordinary losses as they lose asset value in terms of their nuclear power generation plants, equipment, and nuclear fuel. If this extraordinary loss exceeds a company's net asset value, the company technically becomes insolvent and fails. As an extraordinary loss cannot be included in the costs used to calculate electricity rates, the companies cannot cover the loss by way of higher electricity charges. Thus, the electric power companies that are highly dependent on nuclear power are against denuclearization for fear of corporate failure (Citizens' Commission on Nuclear Energy ed. 2015, pp. 198–199).

In order to mitigate such effects, changes to the current accounting system will be needed. It should be allowed to depreciate nuclear power generation plants for a certain period after decommissioning. In fact, some changes were made to the accounting system in October 2013. We need to work out a transparent mechanism with a clear division of responsibilities between the national government and electric power companies in order to avoid the disruption of insolvency and business failure.

Rush of New Coal-Fired Plants Across the Nation

After the Fukushima accident, the electricity market was finally deregulated, which had been a major challenge for the government and power companies since the late 1990s. However, one result of the introduction of deregulation has been to stimulate a rush of new coal-fired power plants across Japan. After 2012, the construction of 50 new plants was announced (approximately 43 GW of capacity). This will double the total electricity output of coal power plants in Japan. If all these new plants start operations, Japan's estimated total GHG emissions in 2030 will reach

1110 billion tons, which is beyond its target level of 1042 billion tons set by the Paris Agreement. This is 5% higher than the target, and means it will be impossible to meet the target of a 26% decrease from the level of FY 2013 emissions. This is a nightmare for climate change protection in the world trying to keep global temperatures from exceeding two degrees centigrade of warming under the Paris Agreement. Coal-fired plants also emit other kinds of air pollution, notably SOx, NOx, PM 2.5, mercury, and water pollution through sewage emissions. As of the end of December 2019, 15 plants already started to operate, another 15 plants are under construction and another 13 projects were canceled or converted to wood biomass-fired plants. The start of construction on the rest of the projects has been delayed primarily due to financial reasons.[2]

After the Paris Agreement was enacted the worldwide situation regarding coal-fired plants has changed drastically. In European countries and the USA, a total of 64 GW of operating coal-fired plants were retired in 2015 and 2016 due to the impact of the Paris Agreement in 2015. In even China and India, a total of 68 GW of coal-fired plant capacity under construction is now frozen at over 100 project sites. As of January 2017, the amount of coal-fired power capacity in pre-construction planning fell to 570 gigawatts (GW) from 1090 GW in January 2016 in China and India. In October 2017, the United Nations Environment Program (UNEP) recommended an end to the construction of new coal power plants and an accelerated phasing out of existing plants as key steps toward achieving the goals of the Paris Agreement. Recently, the fossil-fuel divestment movement is very active in the USA and Europe to reduce carbon emissions, applying public pressure to stigmatize "fossil fuel companies" that are currently involved in fossil-fuel extraction to move investments away from fossil fuels and toward renewable energy. UN Secretary General Antonio Guterres criticized Japan's "coal addiction" at the UN Climate Change Conference, COP25 in Madrid in December 2019 (Mainichi 2019).

Nonetheless, the Japanese Ministry of Economy, Trade, and Industry (METI) has been supporting construction of coal-fired plants, both domestically and abroad. Companies are eager to build coal-fired plants due to its cheaper fuel price compared to natural gas in Japan, in contrast with the USA and Europe where natural gas is cheaper than coal. It seems that power companies and the national government are strategically forcing citizens into an unpalatable choice between reopening nuclear plants and constructing coal-fired plants. This is in clear contrast with Germany, which decided to close all nuclear plants by the end of 2022, and is also moving to phase out all 84 coal plants by 2038, which provides 35% of electricity, based on the policy recommendation of a government commission that were released on January 26, 2019 (Vaughan 2019). Germany is the leader of energy transition in terms of denuclearization and decarbonization.

The fossil-fuel divestment movement has criticized large Japanese banks, especially Mizuho Bank and Mitsubishi UFJ Bank, for being the largest and the second largest lenders in the global banking industry to coal plant developers (BankTrack 2018). The city of Sendai, only a few years after the devastation of the 2011 tsunami, approved construction of a new coal-fired power plant in the municipality. The company constructing the plant cleverly exploited a policy loophole to avoid making an environmental assessment that would have delayed the project for an additional two years. In March 2017, another company announced it was planning another coal plant in Sendai. Subsequently, this project was converted to a wood biomass-fired plant in reaction to the criticism received about building another coal-fired plant in Sendai.[3] Citizens are afraid and angry that this breaks Sendai's proud coal-free history of the past 10 years. There are approximately 150,000 residents, and 32 schools including, 17 primary schools, within a 5-km radius of the proposed plants, where children will be at risk from fine particulate and NOx pollution. Ignoring the global imperative to fight climate change and transition to renewables, Japanese corporations that have invested heavily in coal abroad need an outlet for their dirty fuel. They are taking advantage of the economically challenged post-tsunami region of Tohoku, specifically Sendai. The electricity produced will go to Tokyo, profits will go headquarters in large cities, while the air pollution will choke residents locally, and the climate risk will burden today's young generation (Hasegawa 2018b).

In September 2017 citizens in Sendai filed a suit against the company starting to operate a coal-fired power plant in Sendai. Following this suit, citizens in Kobe and Yokosuka filed suits against companies and the national government to block the construction of coal-fired plants in those cities (Hasegawa 2018a).

The Tough Challenge of Promoting Wind Turbines in Japan

Worldwide, since right after the Fukushima accident, wind and solar power have dramatically increased. About 26.5% of electricity came from renewables by the end of 2017 (REN21 2018). 2014 constituted a milestone as wind generating capacity surpassed nuclear capacity for the first time, as depicted in Fig. 2.5. *Windpower Monthly* observed that global grid-connected wind capacity passed the 500 GW milestone in June 2017 (Windpower Monthly 2017).

Nonetheless, Japanese EPCOs have been reluctant to take up wind power generation before, and even after, the Fukushima accident. In fact, they have been actively promoting nuclear power generation, as described above, by claiming that wind power generation is unreliable. The main reason for these companies' policy is that they prefer monopolizing electricity generation based on supply-side management to meet a given electricity demand. They prefer large-scale power generating facilities like nuclear power and coal-fired plants.

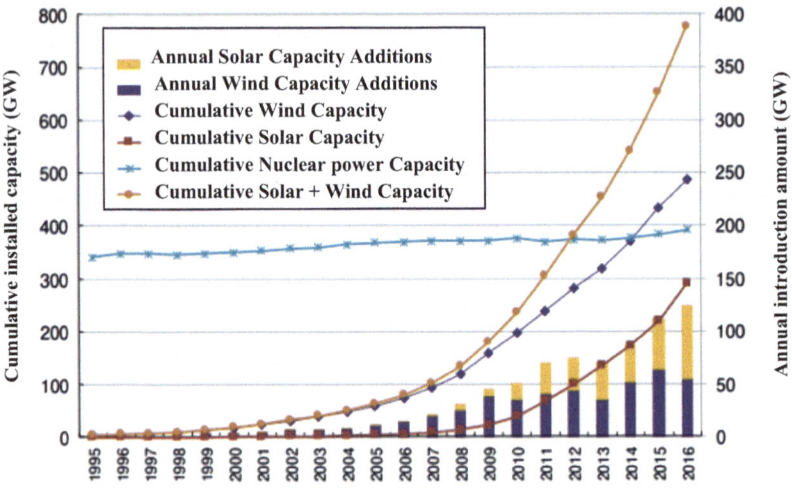

Fig. 2.5 Global installed cumulative generating capacity of nuclear energy and renewable energy (*Note* Author created figure, with assistance from Mathias Shabanaj Janklila, based on data from ISEP [2018])

Japan's nuclear supporters, including METI and the EPCOs, are afraid of, and have attempted to limit, renewable energy capacity. A typical example of this is the 2003 New Energy Act. The FIT (feed-in-tariff) system was finally enacted in July 2012, after the Fukushima Daiichi accident, under the DPJ Kan cabinet, following a multi-year dispute between environmental NGOs, METI, and the EPCOs. This led to a rapid increase in renewables, especially PV solar. The share of renewable energy sources in total electricity generation was 17.4% in 2018. 7.8% came from large-scale hydro, and another 6.5% came from solar (see Fig. 2.6). A symbolic milestone was reached when 100% of electricity supply came from renewable energy in the Shikoku EPCO service region (on the island of Shikoku) from 10:00 a.m. to noon on May 20, 2018. Regarding solar PV, Japan has 49 GW of cumulative installed capacity, which is the third largest after only, China, which had 131 GW of capacity, and the USA, which had 52 GW of capacity as of 2017. See Fig. 2.7. Japan's installed solar PV capacity has increased six times since 2012, when the FIT was introduced. However, the generation capacity of Japan's wind turbines remains low, with a total of only 3.65 GW in 2017, even five years after introducing the FIT. Japan's wind capacity is 19th largest in the world, although it was eighth in 1995 (ISEP 2019).

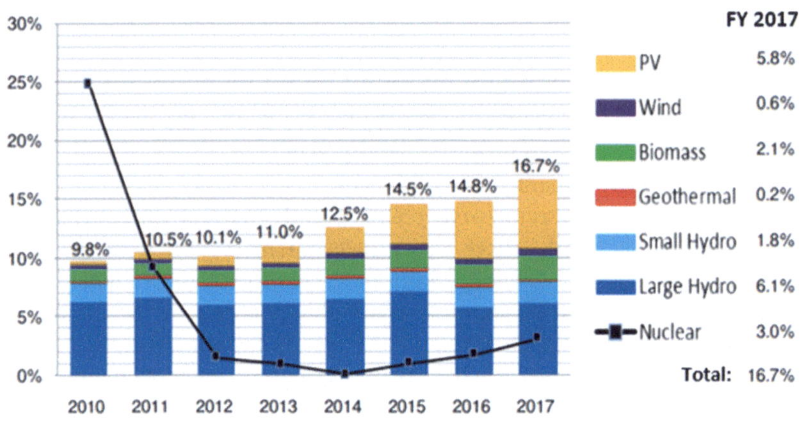

Fig. 2.6 Trends in nuclear energy and renewable energy generation in Japan (*Note* Author created figure based on data from ISEP [2018])

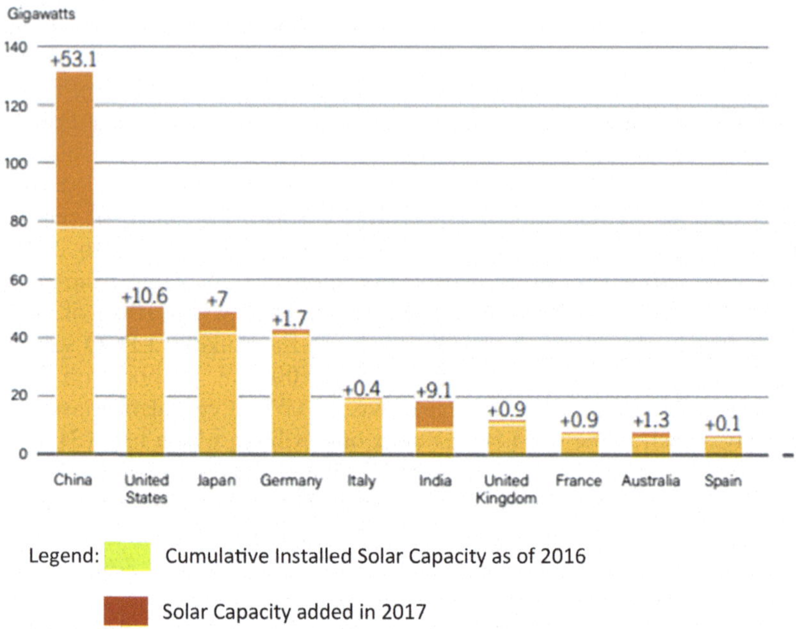

Fig. 2.7 Solar PV installed capacity of top 10 countries in 2016–2017 (*Note* Author created figure, with assistance from Mathias Shabanaj Janklila, using data from REN21 [2018])

This fall in Japan's global wind power ranking is due to the government's and utilities' reluctance to promote wind power.

The central government prefers the name "new energy (*shin enerugi*)" and avoids the term "renewable energy" and "natural energy (*shizen enerugi*)." Consequently, there is no government section with the title of "renewable energy." This is additional evidence of the reluctance of the central government to promote renewable energy. One of the political reasons for this is that before and after the Fukushima accident, most renewable energy supporters have strongly criticized nuclear energy.

The Community Power Movement in Japan

In order to grasp the specific situation regarding the promotion of wind turbines in Japan, where the central government and power companies are reluctant to support their expansion, we should distinguish between four types of wind-turbines initiatives: (1) local governments' initiatives called "municipal wind power generators"; (2) NGO initiatives called "citizens' communal wind power generators"; (3) electric utilities' initiatives; and (4) commercial wind power companies initiatives. Commercial wind power companies' initiatives are currently the mainstream for wind power development in Japan as well as worldwide. In Japan local government's initiatives were prominent during the 1990s, and NGO initiatives were impressive during the first decade of the twentieth century. They can be regarded as Japan's model of "community power" or community wind, which stresses local initiatives, local funding, local ownership, and local benefits (Hasegawa 2015, pp. 118–124). Both local government and NGO initiatives substituted for central government and EPCO initiatives, due to the reluctance of the latter two actors to promote wind turbines.

Community power is a new concept that the World Wind Energy Association launched in 2010, stressing local ownership, local initiatives, and local benefit. The emphasis is on localism. It is the antithesis of large-scale projects like gigantic offshore wind turbines. Wind power generation has become a major growth industry in the twenty-first century, motivated by the desire to tackle the climate change issue and its increasing economic profitability. The entry of giant energy and oil corporations such as GE and BP (British Petroleum) into the wind generation business has been increasingly noticeable in recent years as more and more sizeable projects are planned. However, are we right to simply praise the expansion of large-scale of projects? Are there any problems with large-scale projects? One of the advantages of renewable energy is supposed to be the sense of affinity experienced by ordinary citizens and communities that is at the opposite end of the spectrum from nuclear power stations, or large-scale thermal or hydropower generation facilities. Large-scale wind farms inevitably feel more remote in the minds of local people and encourage a Not-in-my-Backyard (NIMBY) response. This is the reason for the recent stress placed on promoting community power worldwide.

Community Wind Power Cases in Japan

In this section, I will introduce several cases of community wind power in Japan (Hasegawa 2015, pp. 115–129). These cases illustrate the advantages of community wind power.

Tachikawa: From the windiest town to a wind power pioneer, in May 1993, the town of Tachikawa (renamed the Shōnai since 2005), Yamagata Prefecture, installed three town-operated 100 kW wind turbines as the first municipal wind power project to sell electricity to a commercial utility. In August 1994, this town hosted the inaugural *National Conference for the Promotion of Wind Power*. In January 1996, Japan's first private power generation business began selling electricity generated by two 400 kW turbines located in this town. In July 1996, this town initiated *the National Association of Municipalities for the Promotion of Wind Power Generation* and hosted the secretariat until 2001 (Hasegawa 2015, pp. 115–117). In November 2015, the 18th *National Conference for the Promotion of Wind Power* was held in Shōnai town. In August 2003, this town launched the Electricity Saving Town campaign in summer, aiming to becoming a "100 percent renewable-energy (zero-fossil-fuel-consumption) town." Especially in 2011, 439 households joined the program and saved an average of 10.4% in reduced electricity use through conservation in August and September compared to the previous year.

Kuzumaki: A town of milk, wine, and clean energy, in June 1999 Kuzumaki, Iwate Prefecture, launched Japan's first commercial wind farm located on a high-altitude site, at 1000 meters above sea level. It began operating with three 400 kW wind turbines. The town owns a stake in this wind farm. Since December 2003, another twelve 1.75 MW wind turbines have been operating at another wind farm within the town that is wholly owned by a private company. Total electricity generation from these two wind farms is equivalent to two times the total amount of Kuzumaki's electricity consumption. The private company expanded its wind farm by adding an additional 44.6 MW of capacity in the form of 22 wind turbines, with this additional capacity starting operations in March 2019. Two other projects in this town are under environmental impact assessment: 72.32 MW of capacity with 32 wind turbines of 2.3 MW and 110. 4 MW projects with 60 turbines. These new projects are commercial wind power company initiatives, but the original trigger was three turbines started by the town's initiative in 1999 that demonstrated the good performance of producing electricity at the high-altitude site, 1000 meters above sea level. This initial project also demonstrated the town's support of hosting wind power (Hasegawa 2015, pp. 119–120; J Power 2017).

Citizens' Communal Wind Generator Campaign from *Sapporo*

In December 2000, an NGO based on the *Seikatsu Club Hokkaido*, a non-store consumer cooperative with 13,667 members as of the end of 2010, launched the first citizens' communal wind power generation campaign. There were three triggers. The first was a citizens' campaign, active since 1990, that has opposed the placement of a deep geological repository for high-level nuclear waste in Horonobe town, Hokkaido. The second trigger was another citizens' campaign opposing construction of the Tomari number 3 nuclear reactor, announced by Hokkaido EPCO in July 1996 (Hasegawa 2015, pp. 122–123).

The final trigger was my lecture at a workshop on October 15, 1996, at which I recommended a donation of ten percent of one's electricity bill toward a fund for promoting renewable energy in conjunction with a power-saving campaign to offset this ten percent by reducing electricity consumption. "You already have the Seikatsu club organization in place. As you have developed milk and other safe food products, you can develop your own safe electricity by pooling small donations. You can easily cut your electricity bills by ten percent by switching off [appliances and lights] frequently at home. All you need to do is to contribute that ten percent or an average of about 800 yen saved on your electricity bills" (Hasegawa 2015, p. 123).

After two and a half years of preparation, the green electricity levy program began in March 1999 through a contribution of five percent of power bills. The group felt that five percent was easier to save and would allow them to recruit more people. *The Hokkaido Green Fund* was established. The initial membership target was 1000 for the fiscal first year, and over 800 people joined in the first thirteen months to the end of April 2000. The membership revenue of the Green Fund for the year 2000 was about four million yen. The Hokkaido Green Fund decided it would operate the venture as a community business. The first turbine of 990 kW was built and started to operate in Hokkaido, after the Fund had collected 249 shares of 500 thousand Yen, total 1.4 billion Yen ($1.2 million). Almost of all the investors were local people. Many of them were nuclear energy opponents. Since then, a total of 21, communally owned wind turbines have been built and are operating nationwide. See Fig. 2.8 (Hasegawa 2015, pp. 124–126).

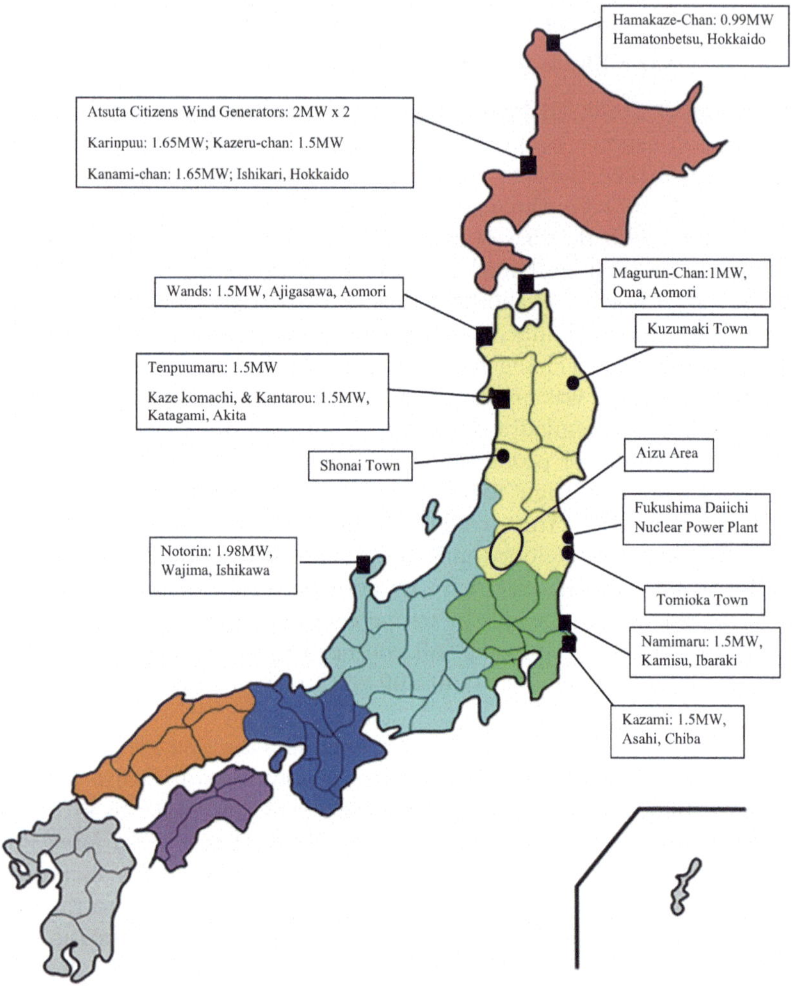

Fig. 2.8 Communally owned wind turbines (*Note* Author created figure with assistance from Mathias Shabanaj Janklila)

The *Fukushima* Recovery Projects of Solar PV in *Fukushima* Prefecture

After the Fukushima nuclear accident, Fukushima recovery projects promoting solar energy started in Fukushima Prefecture, especially in the Aizu region in the western part of this prefecture, more than 90 km away from the Fukushima nuclear power plant. The Aizu Electric Power Company (Ai-power) was established in August 2013, following meetings on recovery and renewable energy that had been held since November 2011. Ai-Power president, Satō Yaemon, is the owner of a local sake brewing company with a history of more than 320 years. Ai-power is aiming within the next ten years to widely disseminate knowledge about the existence of alternative energy forms that can replace nuclear generation of electricity. Keeping the safety of the public as the top priority, the company continually is striving for the best possible alternative energy sources for the Aizu community to improve its economy and culture for future generations.

Ai-power has 51 solar PV sites consisting of mainly 50 kW sites, with a total installed generating capacity of 4326 kW. It had received a total of 99.8 million yen in investments from citizens as of the end of December 2017. The largest facility is a 1000 kW site in Kitakata city. In order to show the younger generations how sustainable energy and a sustainable lifestyle can lead to a peaceful community, this company is stressing the role of visible small-size sites located in many places and built in collaboration with municipalities. Ai-power has enjoyed nationwide media coverage many times.

The International Community Power Conference was held in Kitakata, Aizu area, and other cities in Fukushima Prefecture from January 31 to February 1, 2012. The Fukushima Community Power Declaration was published on February 2, 2012, to launch the "community power alliance for 21st century" and the "Fukushima community power fund." The First World Community Power Conference was held in Fukushima City in November 2016, in conjunction with the 15th World Wind Energy Conference WWEC2016, Tokyo (WWEA 2016). Both conferences were managed by Iida Tetsunari, the founder of Institute for Sustainable Energy Policies (ISEP) and an advocator of community power in Japan.

Tomioka Recovery Solar Project in Fukushima Prefecture. A 33 MW PV solar project has been built on 34 hectares in Tomioka town, which is located only approximately 7 km from, and was severely contaminated by, the Fukushima Daiichi nuclear plant. This project cost 9.2 billion Yen and was financed by citizens and corporate investment. Following the construction period from November 2016 to March 2018, this plant started operating fully from April 2018. Income from this solar PV plant is being used to support elderly residents and young farmers (Kudou 2018). PV solar plants built on abandoned farming land are a symbol of recovery and the future of these local communities. Local citizens, citizens nationwide, companies, and media are supporting these projects. In 2013, Fukushima Prefecture established the goal of becoming a 100% renewable energy prefecture by 2040 and becoming a world leader in renewable energy.

Conclusions

Even after the Fukushima nuclear accident, the second severe nuclear accident after the 1986 Chernobyl accident, Japan's energy policy remains largely unchanged, with only minor modifications due to a relatively closed political-opportunity structure and the relative weakness of the counter-veiling power of opposition political parties and civil society. The central government and EPCOs companies are still reluctant to support the expansion of renewable energy sources. On the other hand, Japan's model of "community power" or community wind power in some areas, including Fukushima Prefecture, which stresses local initiatives, local funding, local ownership, and local benefits, has nonetheless aided the wider adoption of renewable energy in Japan. It reveals new ways to build public support toward realizing a real energy transition.

Notes

1. See Aldrich (2012) and Samuels (2013) for responses to these questions from non-Japanese scholars.
2. Based on information provided by Hirata Kimiko of *Kiko Network* on October 20, 2019.
3. This biomass plant is undergoing an environmental impact assessment process as of the end of December 2019.

References

Agency for Natural Resources and Energy. 2017. Kagakuteki tokusei mappu kōhyō saitto [The Website for Announcing the Nationwide Map of Scientific Features for Geological Disposal]. https://www.enecho.meti.go.jp/category/electricity_and_gas/nuclear/rw/kagakutekitokuseimap/. Accessed 10 Jan 2020.

Aldrich, Daniel P. 2012. Post-Crisis Japanese Nuclear Policy: From Top-Down Directives to Bottom-Up Activism. *Asia Pacific Issues* 103: 1–11.

BankTrack. 2018. Banks vs. The Paris Agreement: Who's Still Financing Coal Plant Development, December. https://www.banktrack.org/coaldevelopers/. Accessed 29 Feb 2020.

Chou, Kuei-Tien. 2018. Tri-Helix Energy Transition in Taiwan. In *Energy Transition in East Asia: A Social Science Perspective*, ed. Kuei-Tien Chou, 45–73. New York: Routledge.

Citizens' Commission on Nuclear Energy (ed.). 2015. *Our Path to a Nuclear-Free Japan: Policy Outline for a Nuclear Phaseout.* Tōkyō: Citizens' Commission on Nuclear Energy. http://www.ccnejapan.com/?page_id=2048. Accessed 31 Dec 2017.

Gorbachev, Mikhail S. 1996. *Memoirs*. New York: Doubleday.

Hasegawa, Kōichi. 2015. *Beyond Fukushima: Toward a Post-Nuclear Society.* Melbourne: Trans Pacific Press.

Hasegawa, Kōichi. 2017. Kakunenryō saikuru to 'Rokkasho mura' [Japan's Nuclear Fuel Cycle Facilities and Rokkasho Village]. In *Kettei-ban genpatsu no kyōkasho* [Nuclear Energy Handbook], ed. Tsuda Daisuke and Kojima Yuichi, 213–226. Tokyo: Shinyo sha.

Hasegawa, Kōichi. 2018a. Hisaichi Sendai-ko no sekitan karyoku wo sashitomeru [Can an Injunction Stop Coal-Fired Power at Disaster-Stricken Sendai Port]. *Kankyō to Kogai* 47 (4): 44–47.

Hasegawa, Kōichi. 2018b. Risk Culture, Risk Framing, and Nuclear Energy Dispute in Japan Before and After the Fukushima Nuclear Accident. In *Energy Transition in East Asia: A Social Science Perspective*, ed. Kuei-Tien Chou, 9–27. New York: Routledge.

Institute for Sustainable Energy Policies (ISEP). 2018. *Renewables 2017 Japan Status Report (Summary).* http://www.isep.or.jp/jsr2017. Accessed 20 Jan 2018.

ISEP. 2019. *Renewables 2018/2019 Japan Status Report (Summary).* https://www.isep.or.jp/archives/library/category/japan-renewables-status-report. Accessed 19 Jan 2020.

Ishibashi, Katsuhiko. 2011. Masani "genpatsu shinsai" da [A 'Combined Earthquake and Nuclear Disaster' Has Really Happened]. *Sekai* 817: 126–133.

Iwai, Noriko, and Kuniaki Shishido. 2015. The Impact of the Great East Japan Earthquake and Fukushima Daiichi Nuclear Accident on People's Perception of Disaster Risks and Attitudes Toward Nuclear Energy Policy. *Asian Journal for Public Opinion Research* 2 (3): 172–195. https://doi.org/10.15206/ajpor.2015.2.3.172. Accessed 20 Dec 2017.

JAERO. ud. Graphical Flip-Chart of Nuclear & Energy Related Topics. Accessed at: http://www.ene100.jp/www/wp-content/uploads/zumen/el-2–7.jpg.

JAERO. 2014. 2013 nendo genshiryoku riyō no chishiki fukyū keihatsu ni kansuru yoron chōsa, February. https://www.jaero.or.jp/data/01jigyou/tyousakenkyu25.html. Accessed 29 Feb 2020.

J Power. 2017. Kuzumaki dai ni fūryoku hatsudensho no kensetsu kōji kaisho ni tsuite, June 8. https://www.jpower.co.jp/news_release/pdf/news170608-1.pdf. Accessed 29 Feb 2020.

Kudou, Sousuke. 2018. 30 MW Solar Plant Completed in Restricted Residential Area of Fukushima. Nikkei Technology, January 26. https://tech.nikkeibp.co.jp/dm/atclen/news_en/15mk/012601. Accessed 11 Feb 2019.

Mainichi. 2019. Editorial: Is Japan Turning Its Back on World Trend to Phase Out Coal? December 13. https://mainichi.jp/english/articles/20191213/p2a/00m/0dm/011000c. Accessed 20 Jan 2020.

Ministry of Foreign Affairs of Japan. 2014. Sankō: hikaku san gensoku ni kansuru kokkai ketsugi [Reference Sources: Three Non-Nuclear Principles Adopted by the Diet]. https://www.mofa.go.jp/mofaj/gaiko/kaku/gensoku/ketsugi.html. Accessed 10 Jan 2020.

Office of Atomic Energy Policy of Cabinet Office. 2019. *The Status Report of Plutonium Management in Japan 2018.*

Reconstruction Agency. 2019. Zenkoku no hinanshasū [Number of Evacuees Nationwide]. https://www.reconstruction.go.jp/topics/main-cat2/sub-cat2-1/hinanshasuu.html. Accessed 10 Jan 2020.

REN21. 2018. *Renewables 2018 Global Status Report.* http://www.ren21.net/gsr-2018/chapters/chapter_01/chapter_01/. Accessed 10 Jan 2020.

Samuels, Richard J. 2013. *3.11: Disaster and Change in Japan.* Ithaca: Cornell University Press.

Vaughan, Adam. 2019. Germany Agrees to End Reliance on Coal Stations by 2038. *The Guardian*, January 26. https://www.theguardian.com/world/2019/jan/26/germany-agrees-to-end-reliance-on-coal-stations-by-2038. Accessed 10 Jan 2020.

Windpower Monthly. 2017. Connected Global Capacity Tops 500GW, Say WPI, October 18. https://www.windpowermonthly.com/article/1447740/connected-global-capacity-tops-500gw-says-wpi. Accessed 29 Feb 2020.

WWEA (World Wind Energy Association). 2016. *WWEC2016 Conference Resolution.* https://wwindea.org/blog/2016/11/01/wwec2016-conference-resolution/. Accessed 10 Jan 2020.

Yomiuri Shinbun. 2015. Naikaku shiji bizō 53% honsha zenkoku yoron chōsa kekka (morning edition), January 12, p. 16.

CHAPTER 3

Why Japan Is No-Longer a Front-Runner: Domestic Politics, Renewable Energy, and Climate Change Policy

Hiroshi Ohta

Introduction

Japan overcame severe environmental pollution problems in the 1960s and 1970s and weathered two oil crises in the 1970s. In surmounting these problems and crises, Japan has developed pollution-abatement and energy-efficient technologies. Regarding mitigation of climate change, it has maintained a technological lead, for instance, in developing and commercializing hybrid cars and solar panels. In tandem with these efforts, Japan has had a strong domestic political incentive to play a leading role in non-military issues like global climate change due to the "no-war clause" in its constitution, which does not recognize military force as a legitimate instrument of foreign policy. The general public has been supportive of the Japanese government playing a leading role in international negotiations for the mitigation of climate change.[1] As

H. Ohta (✉)
Waseda University, Tokyo, Japan
e-mail: h-ohta@waseda.jp

P. Midford and E. Moe (eds.), *New Challenges and Solutions for Renewable Energy*, International Political Economy Series,
https://doi.org/10.1007/978-3-030-54514-7_3

expected by many observers in the world, Japan, jointly with European countries, took the lead during negotiations for the United Nations Framework Convention on Climate Change (UNFCCC), it hosted the Kyoto Conference (COP3) in 1997 and contributed to the adoption of the Kyoto Protocol. However, since that time, it has been reluctant to grasp the initiative once more. Instead, while assigning itself the role of a mediator between the United States and the European Union or between developed countries and developing countries, it has recently become a laggard in taking bold steps to mitigate climate change.

After the historic electoral victory of the Democratic Party of Japan (DPJ) over the Liberal Democratic Party (LDP) in September (2009), Prime Minister Hatoyama Yukio pledged that Japan would reduce greenhouse gas (GHG) emissions by 25% by 2020 below 1990 levels at the UN Summit on Climate Change and in December at the 15th Conference of the Parties (COP15) to the UNFCCC. However, there was a condition attached to his pledge that all large emitters also commit to ambitious reduction targets. Hatoyama's successor Kan Naoto, who faced the Great East Japan Earthquake of March 11, 2011, and the Tokyo Electronic Power Company's (TEPCO's) *Fukushima Daiichi* (No. 1) nuclear power plants' triple disaster, did not have enough time to reshape Japan's climate policy. Nonetheless, he left the Japanese version of the feed-in-tariff system for the promotion of renewable energies in exchange for his resignation. However, the Japanese government led by Prime Minister Noda Yoshihiko refused to take any responsibility for the second commitment period under the framework of the Kyoto Protocol at COP17 in 2012.

The second Abe administration has not shown any enthusiasm for tackling this global problem either, although the first Abe administration in May 2007 showed some interest by announcing with much fanfare a new climate change initiative "Cool Earth 50" (to halve GHG emissions from the current level by 2050). The government in June even proudly declared that Japan would strive to "Becoming a Leading Environmental Nation Strategy in the 21st Century" (MOE 2007). Why has Japan been so reluctant to take a leadership role in climate change diplomacy despite the technological lead it holds that could generate an economic gain, vigorous public support, and the political incentive for non-military diplomacy?

This chapter argues that the source of Japan's inaction in international climate negotiations arises from its energy policy, which has been discouraging the extensive development of renewable energy. The root cause of this energy policy is Japan's energy security concerns that have prevailed in Japanese energy policy since the two oil crises in the 1970s. Since then, the Japanese government has promoted nuclear energy as the primary energy alternative, not renewable energy sources, to reduce overseas oil dependency while pursuing energy conservation through the development of nuclear power plants led by the private sector (Samuels 1987). In February 1974, the government established the Advisory Committee for Natural Resources and Energy (ANRE) under the auspices of the Ministry of International Trade and Industry (MITI) (since 2001, Ministry of Economy, Trade, and Industry: METI) to adopt a long-term energy supply and demand policy every five years (Kikkawa 2011; Yoshioka 2011). Since then, the government has prioritized the development of nuclear energy over renewable energy, which led to constructing 54 nuclear power plants, becoming the third largest nuclear-energy power in the world, just after the United States and France. As the construction of new nuclear power plants almost came to a halt in the late 1990s to the early 2000s,[2] the government began to advocate nuclear energy as one of the main policy measures to mitigate climate change, claiming that it does not emit any greenhouse gases (Yoshioka 2011). Thus, the government could continuously justify upholding the policy of prioritizing nuclear energy over renewable energy. However, the long-term energy policy of discouraging the full-fledged development of renewable energy while steadfastly maintaining the system of local monopolies over electricity generation and distribution by relying on nuclear power plants revealed themselves as major systemic flaws in both energy policy and climate change policy when the Fukushima Daiichi nuclear accident occurred. Japan had to restart old coal and gas power plants and even build new thermal plants, to import more fossil fuels such as natural gas and coal, and then to continue to emit a large volume of greenhouse gases, which render Japan's position in international climate negotiations laggard or passive, at best.

In sum, the reason why Japan has been reluctant since COP3 in Kyoto to take leadership in international negotiations is mainly that the policy coalition consisted of the MITI/METI and Japan Business Federation (*Keidanren*) under the strong influence of energy-intensive industries (Moe 2012, 2015; Tsunekawa 2010) that are very reluctant to support any substantial CO_2 emission reductions. This pressure was reflected even in Prime Minister Hatoyama's initiative of setting a mid-term target for reducing GHG emissions by 25% below the level of 1990 by 2020, which was premised on "the establishment of a fair and effective international framework in which all major economies participate and on the agreement of ambitious targets by all major economies." The United States and China were unwilling to commit to any substantial GHG reductions, enabling Japan to make an excuse for not committing to ambitious reduction targets. Besides, this mid-term target itself contained a major flaw, since it heavily relied on nuclear energy, whose prospects were already not promising due to the low-operating rate of nuclear power plants, their old age, and the difficulties of constructing new power plants in Japan. Then, the final blow for both Japan's energy strategy and climate policy was the Fukushima Daiichi nuclear accident. Japan is now edging away from taking the initiative in international efforts to mitigate climate change.

This chapter employs the process-tracing method (George and Bennett 2005). It does so by providing a detailed account of the critical decision-making processes and describing the core elements of energy policy closely related to climate change policy.

National Energy Strategy and Climate Change Policy in Japan

Both the Ministry of the Environment (MOE) and METI regard a stable global climate system as an international public good. Both ministries recognize Japan's responsibility for mitigating climate change and are willing to help the most vulnerable developing countries to adapt to adverse consequences of climate change. However, while the MOE has emphasized the importance of the principle of "common but differentiated responsibilities," METI has emphasized the fair and just distribution of costs among major emitters and efficient ways of alleviating global climate change. The MOE has attempted to take the initiative in introducing a global warming tax as a fair and effective means of sharing costs. METI's preference has been voluntary efforts by industry and is reluctant to promote both the introduction of a new environmental taxation and a "cap-and-trade" emissions trading system at home, reflecting Keidanren's policy preferences (Watanabe 2015; Kameyama 2017).

Against this backdrop, the most influential policy that determines Japan's domestic climate policy and climate diplomacy has been Japan's energy policy. The Fukushima Daiichi nuclear accident exposed its contradictions and ineffectiveness, and consequently turned Japan in on itself in international negotiations for the mitigation of climate change. Japan's energy policy has been heavily biased in favor of nuclear energy while turning a cold shoulder toward renewable energy sources. Japan has maintained a strong commitment to the development of nuclear power as the national energy strategy for reducing its dependence on imported oil. This policy has been funded exceptionally well through a complicated system of government subsidies based on the three laws for nuclear power development: namely, the Electric Power Development Taxation Law, the Special Budget Law for the Development of Electric Power, and the Law for the Adjustment for Areas Adjacent to Power Generating Facilities (Oshima 2010; Yoshioka 2011). This so-called Dengen sampō (the three laws for power development) was introduced in June 1974 by Tanaka Kakuei, one of the most dynamic postwar prime ministers. Since Tanaka's initiative, Japan has completed 54 nuclear reactors with an installed capacity of about 45 GW, which was the third largest in the world after the United States and France. According to a study (Oshima 2010, pp. 30–50), since the 1970s, 70% of the account for the smooth siting of power plants, whose source of revenue had been the electric power development tax, and 97% of the energy expenditure in the national general account budget have been spent for the development of nuclear power.

In conjunction with both energy and climate change policy, in 2006, METI's *New National Energy Strategy* set ambitious targets (METI 2006). Regarding energy conservation, it set a target of 30% improvement in energy efficiency by 2030. Japan's oil dependence as the primary energy sources should be reduced to lower than 40% from 52% (Toichi 2002, p. 1) by 2030. The transportation sector had to reduce oil dependency by 80%, while Japanese developers should develop 40% of total oil import by 2030. Nuclear power generation should increase by 30 to 40% or more by around 2030 (METI 2006, p. 14). Despite the future expansion of new energy technology under this *New National Energy Strategy*, the role assigned for renewable energy sources was very limited as the report failed to mention them as one of the primary sources in the Japanese energy strategy. Several years later, in June 2010, the cabinet meeting adopted the *Basic Energy Plan*. Accordingly, 70% of electricity generation would stem from non-fossil fuels or "zero emissions" energy sources by 2030, of which the share of renewable energy and nuclear energy would be 20% and 50%, respectively. Toward this end, Japan had to build five additional nuclear power plants by 2020 and another nine by 2030 (METI 2010).

This energy strategy, however, had to be fundamentally revised due to the Fukushima Daiichi nuclear accident that occurred on March 11, 2011, when a 9.0 magnitude earthquake struck off the coast of northeast Honshū not far Sendai. Colossal tsunamis swept away and destroyed essential facilities and equipment, including the emergency core cooling system (ECCS) of the plant. The damages done by the earthquake and ensuing tsunamis forced Tokyo Electric Power Co. (TEPCO) to decide to decommission four nuclear reactors (reactors numbers 1 to 4) of the Fukushima Daiichi nuclear power complex. On April 12, according to the Nuclear and Industrial Safety Agency (NISA), the level of the Fukushima Daiichi nuclear accident was categorized as scale seven, that is, the same level as the Chernobyl accident that occurred in 1986. Despite Prime Minister Noda's declaration of attaining the state of a "cold shutdown" of nuclear reactors in mid-December 2011, the accidents were not yet entirely under control, and more than 90,000 people at the time remained displaced away from the evacuation zone around the nuclear power plant (Tabuchi 2011). The sense of distrust of nuclear safety has now spread among the Japanese along with a shared sense of "risk society" à la Ulrich Beck (Beck 1992).

Compared with its nuclear energy policy, Japan's renewable energy policy is meager. As to METI's policy relating to renewable energy, the New Energy Law of 1997 and the Special Measures Law on Promoting Use of New Energy by Electric Enterprises (RPS law) of 2003 are notable. The former law defines "new energy" (or renewable energy excluding hydro and geothermal energy sources) and encourages measures to promote it. The RPS law required the power utilities to generate a certain amount of electricity from new energy sources such as solar PV, wind power, and biomass. However, this piece of law did not encourage the full-scale introduction of "new energy" by subsidizing with a fixed price for electricity generated from these sources as the equivalent German policy did. Thus, the 2010 target under the RPS law aimed at generating 12.2 billion kWh, which comprised only 1.2% of the total national electric power supply (Oshima 2010, p. 21).

As of 2005, nearly 50% of the world's photovoltaic (PV) solar cell production was manufactured in Japan. Japan led the world in thin-film PV cells with the highest capacity of operational manufacturing plants. Kyocera, Kaneka, Matsushita Battery, Sanyo, and Sharp dominated the market (Kimura and Suzuki 2006, p. 7). Helped by government subsidies, more than 80% of PV cells were installed on the rooftops of private households in Japan. However, the Japanese government ended the PV subsidy policy in 2005. As a result, Japan, a one-time top-runner in solar energy generation lost its lead to Germany, which in February 2000 enacted the Renewable Energy Act (or Erneuerbare Energien Gesetz: EEG). The objective of this law was to increase the amount of renewable electricity in the power supply to 12.5% by 2010, as one of the central elements of Germany's climate mitigation policy. The main feature of the EEG was to oblige grid operators in Germany to purchase electricity generated by renewable energy sources at a fixed tariff. This is commonly known as a feed-in-tariff (FIT) system, which aims to promote the investment into renewal energy technologies and to ensure investment security.

As for wind power generation, the Japanese market for wind power was even shrinking. Since 1990, the number of windmills had rapidly increased, and it reached 1050 in 2005 with the generation of more than 1000 MW. It appeared that this trend would continue to grow until the saturation of the wind power market. However, according to critics, there was no real "market" under the new RPS law since 99.5% of the certificate "market" was covered by the ten regional electric power companies (EPCOs) and the target for electricity generation by wind power set under the RPS was also small.

Another impediment to the full-fledged growth of wind (and other forms of renewable energy) market was grid connection issue. The EPCOs regionally monopolized power lines so that other electric generators under the RPS law needed their cooperation to transmit electricity. Since EPCOs insisted that "stability of electricity supply" was their highest goal, and fluctuating wind power was unreliable, they were unwilling to significantly increase their purchase of electricity generated by wind power. As a result, the market for wind power never grew much. In 2003, the power companies received connection applications for 2000 megawatts (MW) of wind power generation, but only 330 MW was approved for connection to the grid, according to the Institute of Sustainable Energy Policies (ISEP). The figure further diminished, in 2004, when only 50 MW (out of applications totaling 700 MW of capacity) was approved for grid connection (Iida et al. 2006). In sum, despite the plan for expanding the use of new energy technology, the role assigned to renewable energy sources was minimal. They altogether occupied a mere 3% of the total primary energy supply (excluding large-scale hydropower) in the year 2010 (Yoneda 2008).

However, in the wake of the Fukushima Daiichi nuclear accident and sensing a shift in public opinion on nuclear energy policy, Prime Minister Kan, who had barely avoided a non-confidence motion by his last-minute promise of resignation after the enactment of legislation on renewable energy, began advocating denuclearization. However, later he toned it down to a policy of pursuing a "nuclear power reduction." He pushed a bill to promote renewable energy through the Diet. On August 26, 2011, the Special Measures Law on Procurement of Renewable Energy Sources Electricity by Electric Utilities (the FIT law), a Japanese version of Germany's Renewable Energy Law, was passed through Diet and went into effect on July 1, 2012. According to this law, the power companies have to purchase electricity generated by solar for 10 to 20 years at a fixed price, from geothermal for 15 years and wind, small-scale hydro, and biomass for 20 years. According to the Renewable Energy Institute, renewable energy sources of electricity generation have been steadily increasing under the FIT law. For instance, as of November 25, 2019, the generation of electricity by renewable energy occupied 16.9% of the total (hydro 7.7%, solar 6.0%, wind 0.7%, bioenergy 2.3%, and geothermal 0.2%). In passing, natural gas was 38.4%, coal 31.2%, petroleum and waste 7.3%, and nuclear 6.2% (REI 2019).

We now can identify a set of plausible explanations for why the Japanese government has recently been inflexible and tenacious in maintaining its fixed position in international negotiations on climate change by refusing to accept a second commitment period to the Kyoto Protocol. This Japanese governmental position stems from the institutional inertia or the vested interests that cling to nuclear energy, its half-heartedness about the promotion of renewable energy (Moe 2012, 2015; Tsunekawa 2010), and the strong opposition of *Keidanren* to the introduction of a "cap-and-trade" system. However, neither the government nor economic interest groups are dominant in policymaking decisions: Rather, the "reciprocal consent" between them is still a norm in Japanese public policymaking, particularly in the field of energy policy (Samuels 1987). Without a dynamic political leadership, we cannot expect Japan's proactive role in international collective action to mitigate global climate change, even though Japan could contribute to bilateral mitigation schemes and aid adaptation efforts.

Japan's Domestic Politics and Diplomacy on Climate Change

Short-lived Governments, Energy Strategy, and Climate Change Policy. Japan's climate change diplomacy became reactive and lackluster, as the domestic political climate was stormy after the 65-month-long coalition government led by Prime Minister Koizumi Jun'ichirō ended in September 2006. Subsequent administrations led by Abe Shinzō, Fukuda Yasuo, Asō Tarō, Hatoyama Yukio, and Kan were all short-lived and unstable, plagued by political scandals, parliamentary impasses, particularly after the loss of the DPJ's Upper House majority in the July 2010 election, and intraparty conflicts. Except Hatoyama, the remaining prime ministers, who came to power without a lower house election, also suffered from a lack of legitimacy and low approval ratings. Noda was the sixth prime minister during the five years following Koizumi's resignation.

After the formation of the Abe Cabinet on September 26, 2006, the government pursued a conservative political agenda, inheriting from the previous Koizumi Cabinet a more than two-thirds majority in the House of Representatives (the Lower House) and a majority in the House of Councilors (the Upper House). However, this administration also showed interest in the issues of climate change. Abe was called on to create a new international framework beyond the Kyoto process in which the entire world would participate in emissions reductions. In May 2007, he announced a new climate change initiative, called "Cool Earth 50," which proposed a long-term target for halving global GHG emissions by 2050 as a common global goal. The proposed solutions included the establishment of innovative technologies, the creation of a "low carbon society," and a national campaign for energy conservation (Kantei 2007).

Despite the role that Abe played in getting the statement of "considering seriously . . . at least a halving of global emissions by 2050" into the communiqué of the Heiligendamm G8 summit, his sudden resignation in September 2007 left that policy line to drift without a leader. Various cases of political corruption by Abe's Cabinet appointees lowered its approval ratings, and this contributed to the crushing defeat the LDP suffered in the Upper House election on July 29, 2007. The ruling coalition parties, namely the LDP and the Clean Government Party (Komeito), lost their Upper House majority, whereas the DPJ significantly increased its number of seats to make it the largest party in the Upper House. Although Abe did not immediately resign, taking responsibility for this historic electoral defeat, he later gave up his post after reshuffling his cabinet.

After a brief political vacuum, the LDP chose Fukuda as the new party leader (and prime minister), a politician who had not previously placed the issue of climate change at the top of his political agenda. Thus, under the Fukuda government Japan could hardly be expected to play a significant role in the negotiations at COP13 in Bali in 2007. During the Bali Conference, one of the Japanese government side events demonstrated the apparent influence of established and energy-intensive Japanese industries such as steel, cement, transportation, and electric power, the leading industries of *Keidanren*.[3]

The predominance of the policy alliance between METI and energy-intensive industries also permeated Japanese policy at the G8 Tōyako Summit in July 2008. The Japanese government did not press G8 leaders to negotiate numerical targets for a mid-term scheme, while arguing for a sectoral approach to emission reductions as a viable policy option. Critics and environmental NGOs considered the G8 Tōyako Summit a failure since it did not result in agreement on establishing mid- and long-term numerical targets. The outcome of the G8 Tōyako Summit did not generate the desired tailwind for the international community, especially for COP15 in December 2009 in Copenhagen, where the parties were to devise an international framework beyond the Kyoto process.

Within a year, on September 1, 2008, Prime Minister Fukuda, who suffered from low approval ratings, gridlock in the Diet, and economic recession, abruptly resigned his position. Then, Asō filled this power vacuum by becoming prime minister on September 24, 2008. Asō's primary mission was to win the next general election within a year. However, soon after taking office, mainly because of various political gaffes, inconsistency, as well as halting and incoherent responses to the global economic recession, Asō also suffered from chronically low approval ratings. He did not manage to dissolve the Diet while the LDP had the upper hand over the DPJ before his term as prime minister expired in September 2009. He was forced instead to plunge into a "suicidal" Lower House election on August 30, 2009, in the wake of a major LDP defeat in July 12, 2009, Tokyo Metropolitan Assembly election, which often functions as a barometer for predicting the outcome of national elections.

Political change finally came to Japan and with it a new dynamic on climate policy. Following the DPJ's historic victory over the LDP in the

2009 Lower House election, Japan began to play a leading role in international climate change negotiations. Prime Minister Hatoyama, at the UN Summit on Climate Change on September 22, 2009, pledged that Japan would reduce GHG emissions by 25% by 2020 from 1990 levels if all significant emitters also commit to ambitious reduction targets. In order to promote broader cooperation, he proposed the "Hatoyama Initiative," an international scheme for the provision of additional technological and financial support to developing countries while ensuring developing countries' "measurable, reportable, and verifiable" emission reduction targets (MOFA 2009). However, there were already loud voices of dissent raised within Japan against Hatoyama's 25% reduction target. One such voice attacked the premise of DPJ leadership in promoting international cooperation by proclaiming an ambitious mid-term target, since developing countries were demanding that developed countries assume a 40% mid-term reduction, and that the prospect for passing legislation similar to the Waxman-Markey Bill in the US Senate on climate change was not bright (Sawa 2009).

On March 12, 2010, the Hatoyama Cabinet approved "Chikyu ondanka-taisaku kihon hōan" (the Basic Global Warming Bill). The bill set a mid-term target of reducing GHG emissions by 25% below the level of 1990 by 2020. The policy proposals to achieve this goal included: (1) establishing a domestic emissions trading system; (2) implementing a global warming tax in fiscal year 2011; (3) introducing a system for purchasing all electricity generated by renewable energy sources at a fixed price; (4) nuclear energy promotion with due consideration for safety; (5) further improvements in energy efficiency; and (6) promotion of R&D for innovative technologies (Shugiin 2010). Although the Lower House passed this bill, the Upper House did not pass it during the ordinary Diet session through June 2010, and the bill was dropped.

The Hatoyama government stumbled over a controversy surrounding a US military base and the money scandals that struck him and Ozawa Ichirō, the DPJ Secretary-General. Above all, Hatoyama's failure to fulfill a campaign promise to relocate the US Marine Corps Air Station at Futenma outside Okinawa prefecture was a fatal blow (Allen and Sumida 2010; *The Japan Times* 2010). After the resignation of Hatoyama, Kan became the new prime minister on June 8, 2010. The Kan Cabinet soon adopted the 2010 Basic Energy Plan, which specified that 70% of electricity would be generated from non-fossil-fuel or "zero emission" sources by 2030, of which the share of renewable energy and nuclear energy would be 20% and 50%, respectively. Toward this end, five new nuclear power plants were to be built by 2020 and another nine by 2030 (METI 2010).

This energy strategy had to be fundamentally revised due to the Fukushima Daiichi nuclear accident on March 11, 2011. As already mentioned in the previous section, after Prime Minister Kan managed to enact the renewable energy bill, he stepped down as prime minister. Then, Noda became prime minister on August 30, 2011. His political style resonated with the country's prime-ministerial norm of seeking compromise within his party and with opposition parties. However, Noda soon faced severe criticism from both pro-nuclear interest groups and anti-nuclear groups. Still, he maintained the "zero-nuclear reliance goal" by upholding a new rule limiting the operational life of reactors to 40 years, despite heavy criticism by pro-nuclear interest groups. Pro-nuclear interest lobbies consisting of nuclear industry groups, *Denjiren* (Federation of Electric Power Companies of Japan: FEPC), *Denryokusōren* (Federation of Electric Power Related Industry Worker's Unions of Japan), and a host of local governments, were concerned about the "zero-nuclear" goal's adverse economic impact.[4]

On June 8, 2012, Noda approved the restart of the number 3 and 4 reactors at the Ōi nuclear power plant on the justification that a stable supply of electricity was imperative and that the operator ensured the safety of these nuclear reactors. Since the government made this decision before the Fukushima Nuclear Accident Independent Investigation Commission of the National Diet of Japan (NAIIC 2012) published its official report, Noda's decision incited furious public opposition. Massive anti-nuclear rallies, which at their peak reached more than 200,000,[5] were held every Friday in front of the prime minister's residence (Aoki 2016).[6]

With the hope that citizens would understand the necessity of nuclear energy use, on June 29, 2012, the Energy and Environment Council (EEC) (established in June 2011) announced "Options for Energy and the Environment," based on proposals from the Japan Atomic Energy Commission (JAEC), the Advisory Committee for Natural Resources and Energy (ACNRE), and the Central Environment Council (CEC). The EEC presented three scenarios to reduce nuclear dependence while achieving various policy objectives for ensuring nuclear safety, energy efficiency, cost reduction, as well as reducing CO_2 emissions. The three nuclear options for the energy mix by the 2030s and CO_2 emission reduction targets below the 1990 level included: (1) a zero nuclear energy option (65% for thermal energy and 35% for natural renewable energy) with a 23% CO_2 reduction, (2) a 15% nuclear option (55% thermal and 30% natural renewable) with a 23% CO_2 reduction, and (3) a 20–25% nuclear option (50% for thermal and 25–30% for renewable) with a 25% CO_2 reduction (Enerugi-kankyo kaigi [EEC] 2012, pp. 8–12).

The government solicited public input through collecting public comments, holding public hearings, conducting opinion polls, and holding deliberative polls from 2 July to 12 August in 2012 (Asahi Shimbun 2012). During this period, the Noda government received 89,124 public comments and it made notes regarding 7000 of these comments. About 90% of respondents endorsed the zero nuclear option. Public hearings were held in 11 cities from 14 July and 4 August with the participation of 1447 people in total, 68% of whom supported the zero nuclear option. The deliberative poll included three separate surveys: a phone survey and written questionnaire surveys before and after the debate forum. In the phone survey, 32.6% of the respondents supported the zero nuclear option, and 285 of the respondents to the phone survey attended the debate forum. In the questionnaire survey before the debate, 41% of the participants endorsed the zero option, but following the debate, 46.7% of them supported the zero option.[7] This result differed from the government's preference for the 15% nuclear option (Nihon Keizai Shimbun 2012). In short, after having received information about nuclear power plants, listening to experts' opinions, and having a debate, more people became concerned about nuclear safety.

The Second Abe Administration's Energy and Climate Change Policies. The LDP won a landslide victory in the Lower House election of December 16, 2012, and formed a coalition government with Komeito. The LDP's electoral strategists were aware that the election was more a vote against the DPJ's mismanagement of economic affairs, security policies (e.g., the Senkaku/Diaoyu islands territorial dispute and the relocation of the US airbase from Futenma), and the triple disasters of the earthquake, tsunami, and meltdown of three nuclear reactors at the Fukushima nuclear power plant than a positive support for the LDP. Abe himself learned from the failure of his first administration (2006–2007), which attempted to promote his conservative and historical revisionist policies, such as promoting constitutional reform by insisting that the current constitution was imposed by the US occupation.

This time his policy focus was economic recovery from two decades of deflation, or so-called Abenomics, consisting of fiscal stimulus, an easy-money policy, and structural reform of the economy. Unlike his previous cabinet, whose members were mostly from his ideological clique, he formed his second administration by including moderate/liberal politicians from different factions. In this way not only could he ensure stable support from within the LDP, but he could also dilute his blatant conservative political inclinations (Nilsson-Wright and Fujiwara 2015). However, by the time Abe reshuffled his cabinet in December 2014, he was gradually putting forward his conservative political agenda. The Abe Cabinet established the "National Security Council (NSC)" on December 4, 2013, and changed the interpretation of Article 9 of Japan's Constitution to enable the exercise of the right of collective self-defense under the condition of a clear threat to the country's survival and the undermining of citizens' rights (MOFA 2014).

Contrary to Abe's forceful leadership in taking initiatives in security policy, both energy policy and climate change policy are lagging far behind. The mode of policymaking on energy and climate policy is becoming similar to that of the pre-Fukushima era. Abe's LDP-Komeito coalition government adopted a new "Strategic Energy Plan" (Basic Energy Plan in Japanese) at the cabinet meeting on April 11, 2014. The new energy strategy ranked nuclear energy as an essential baseload power. It allowed the restart of nuclear power plants without any conditions if the Nuclear Regulatory Authority (NRA) concludes that they have taken necessary measures to conform to the new safety regulations introduced after the Fukushima Daiichi nuclear accident. The new energy plan suggested the reversal of the nuclear policy of the previous administration by implying the construction of new nuclear power plants. The DPJ's "Innovative Strategy for Energy and the Environment" (ISEE) established the principle of a forty-year limitation of nuclear power plant operations and the policy that no construction of new nuclear power plants and no installations of additional nuclear reactors at existing nuclear power plant sites would occur. The new strategy also upheld the continuation of the nuclear fuel cycle policy by continuing to operate the trouble-plagued Monju prototype fast breeder reactor,[8] which was supposed to be decommissioned according to the DPJ government. The Abe government finally decided to decommission Monju on December 21, 2016. As to the target for renewable energy, while the ISEE set up the concrete target of more than 3000 billion kWh (3 times of the current level) by 2030, the LDP-Komeito government aims at only about 20% increase from the current level (METI 2014).

On July 16, 2015, METI officially set a goal for the power generation mix in 2030. It ranked nuclear energy as a stable and cost-effective source and allotted 20–22% for nuclear generation, while the allocation for renewable energy was 22–24% due to relatively high costs of generation according to METI (2015). METI's subcommittee of Long-term Energy Supply and Demand Outlook received 2060 public comments for about one month until July 1. Unlike previous treatment, which made public all the comments received from approximately 89,000 people, rating pro-and-con responses, and the results of their analysis of these comments, METI did not disclose these details this time (Asahi Shimbun 2015c). Facing difficulties to build new nuclear reactors, the alternative means for meeting the nuclear target in its energy policy is extending the lifespan of existing reactors.[9] It is unknown how many existing power plants can restart, but it is unlikely to meet the stipulated goal of the 20–22% nuclear option (see Koppenborg's chapter in this book).

Similar setbacks are discernable in the FIT policy and concerning electricity deregulation. In September 2014, two years after the implementation of the FIT system, Kyushu EPCO stopped making new contracts to connect renewable electricity generated by large private suppliers (over 10kWh) to its power grid. The reason was the lack of grid capacity to receive electricity generated by renewable energy (Asahi Shimbun 2014). Since other utilities faced similar problems, METI decided to provide subsidies to reinforce the power grid and to improve large-sized storage batteries, and at the same time, reviewed the FIT policy aiming at correcting the imbalance in different renewable energy sources. The supply of solar power electricity grew 8.3 times between 2012 and 2017. By contrast, the increase of other renewables between 2012 and 2017 was meager: wind power generation increased by about 35%, bioenergy by about 28%, small-scale hydro 11%, and geothermal generation even decreased slightly in 2017 (REI 2018a).

METI certainly needs to correct this imbalance in the supply of electricity generated by renewable sources. However, the fundamental problem is the way electric power companies decide the volume of different sources of electricity that they connect to the electricity grid, or the allocation of "possible amounts of connection." This problem and other problems relating to the policy of electricity deregulation stem from the existence of 10 vertically-integrated regional electricity companies, EPCOs, which still mostly monopolize generation, transmission and distribution, and retail sales of electricity with the support of economic and energy agencies and concerned industrial interest groups.

METI prioritizes nuclear, thermal, hydro, and geothermal power generation as baseload electricity sources and power companies likewise prioritize "possible amounts of connection" in the power grid to these sources of electricity. Then, remaining amounts are assigned to solar, wind, and biomass (Asahi Shimbun 2015a). This policy has two significant problems. The first problem is that this goes against the principle of the FIT system, which is supposed to prioritize the connection of electricity generated from renewable energy sources as the German system does. The other problem is that a power company includes the generation capacity of nuclear power plants in the "possible amounts of connection," even though they are to be decommissioned in the future, or idle due to safety inspections or the retrofitting of additional safety measures. This fact further reduces the allocation of renewable energy generation.[10] One of the reasons why Germany could rapidly increase its renewable electricity generation is the priority given to connecting renewable energy facilities to electricity grids. What is happening in Japan is just the opposite. On October 13, 2018, Kyushu EPCO, which faced electricity generation in excess of what its grid could handle, curtailed electricity from solar generators in order to keep four nuclear power plants connected to the grid (Asahi Shimbun 2018a). One month later, Tohoku, Chugoku, and Okinawa EPCOs announced that they were also preparing for restraining renewable energy generation (Asahi Shimbun 2018b).

Moreover, there are two other factors for promoting renewable energy that have yet to develop in Japan: electricity deregulation and the separation of electric power production from power distribution and transmission. The liberalization of the Japanese electricity market began on April 1, 2016. The market size is said to be USD150. Thus, successful liberalization can bring tremendous benefits to both the Japanese economy in general and consumers, especially through efficient electricity generation, lower electricity charges, and various retail services. However, these benefits are likely to emerge if the vertical integration of generation, transmission and distribution, and retail by regional monopoly EPCOs comes to an end, and fair competition between the old and new entrants is guaranteed. As mentioned in the previous paragraph, unfair practices already appear to exist regarding independent generators' access to transmission and distribution lines.

As of March 24, 2020, according to METI, 646 retailers, including major municipal gas companies, have registered and are providing services (ANRE/METI 2020). However, an NGO, Power Shift (powershift.org), reports that there are only 28 entities (as of June 13, 2019) that can be considered environmentally friendly (Powershift 2020).[11] It is fair to say that the liberalization of the electricity market alone will not promote enough renewable energy. That requires a more fundamental systemic change with proper procedures. A general procedure of electricity deregulation starts with the separation of transmission and distribution from a monopolistic EPCO, and at the same time, the deregulation of the power generation sector itself. As the final step, the retail sector should be deregulated (Ito 2016, p. 3).

In the case of Japan, there are three phases of electric deregulation, and the order of the phases is not normal compared with other countries' approaches. The first step in Japan was establishing a grid regulatory organization: the Organization for Cross-regional Coordination Operation (OCCTO), which was established on April 1, 2015, to "enhance the function of controlling the supply-demand balance of electricity in both normal and emergencies on a nationwide basis" (OCCTO 2019).[12] The second step was the deregulation of the retail sector, which started on April 1, 2016. However, the final step, namely the separation of electricity generation from power distribution and transmission, does not occur until 2020. It is desirable to establish a public or private entity to own and operate electricity grids independent of 10 EPCO regional monopolies. Alternatively, at least, as EU countries do, while leaving the ownership rights of transmission lines to electric power companies, the operation of transmission lines should be independently managed by a public entity such as an independent system operator (ISO). However, the Japanese governmental plan establishes "legally unbundling" of a transmission and distribution company, which is still likely to maintain a financial relationship with the EPCO parent. As late as March 2019, ANRE continues to promote a legal unbundling scheme by pointing out that it can maintain the unity of a group, which is necessary to meet the needs of financing and stability supply of electricity during disasters (ANRE/METI 2019, p. 6). Thus, it may leave room for a transmission/distribution company not to provide fair access to new entrants under the influence of the EPCO. For a full-scale spread of renewable energy, it is imperative to establish a fair and independent system for transmission and distribution of electricity by establishing a real separation of ownership and management between a power company and a transmission/distribution company (ISEP 2017, Chapter 2). Under the legally unbundling scheme, it is crucial how closely the OCCTO, along with the Electricity Market Surveillance Commission (EMSC), or another regulatory organization, can monitor fair access to power grids (Ito 2016, p. 4; ISEP 2014, pp. 28–29), but it is questionable whether these regulatory bodies can ensure open access to grids.

Against the backdrop of all the above, the Abe administration adopted Japan's climate policy in the run-up to the Paris climate conference (COP21) held in December 2015. In July 2015, Japan announced its "intended nationally determined contributions" (INDCs) of reducing its emissions 26% below 2013 levels by 2030. This target appears similar to those of European countries, but actually, it is equivalent to an 18% reduction below 1990 levels. Thus, it is much less ambitious than the EU target of 40% reductions below 1990 levels.

In July 2018, the Abe Cabinet approved a new "Strategic Energy Plan" drafted by ANRE/METI. Even though it states in its introduction by referring to the Paris Agreement that Japan leads the challenge to achieve decarbonization and energy transition, this basic energy plan still substantially relies on coal and nuclear power plants for electricity generation. The estimated share of coal-fired electricity generation is 26% by 2030, that of nuclear energy 20–22%, and that of renewables 22–24%, which is lower than coal and slightly higher than nuclear power, though renewable energy is supposed to be "a major power source" according to the new energy plan (METI 2018). Even the most advanced coal-fired plant (IGFC)[13] emits almost twice as much CO_2 as an LNG plant does (MOFA 2018, p. 27).

Furthermore, the promotion of ultra-supercritical (USC) power plants to replace conventional coal-fired power plants in China, India, and the US can, according to ANRE's estimate, only reduce around 18% of CO_2 emissions. Therefore, the remaining 82% of CO_2 emissions will be kept unchecked for 40 years (REI 2018b, p. 1). Moreover, the cost of nuclear power generation has been on the rise globally since 2012, while the costs of renewable energy sources are on the decline, with the costs of utility-scale (mega) solar PV and wind power declining dramatically from 2009 to 2017: declines of 86% and 68%, respectively (MOFA 2018, p. 28). Nonetheless, the cost of generating electricity is merely a tiny aspect of the entire nuclear problem Japan faces today. Other aspects include the costs of compensation for the victims of the Fukushima Daiichi nuclear accident, uncertainties regarding technological development and the economic costs of taking out spent fuel rods and decommissioning severely damaged nuclear reactors, finding ways and places to dispose of highly radioactive nuclear wastes, dealing with the excessive accumulation of plutonium, and more.

Conclusion

In Paris in December 2015, 195 countries completed the task set in Durban in 2011 and adopted the Paris Agreement. They also agreed that through five-year reviews, each nationally determined contribution (NDC) would be more ambitious than the last to keep global temperature rise "well below 2°C above pre-industrial levels and pursue efforts to limit the temperature increase to 1.5°C" (UNFCCC 2015). Japan received the Fossil-of-the-Day Award twice during COP25 held in Madrid in December 2019 because it could not commit to more ambitious reduction targets while sticking to coal power plants.

Climate change mitigation policy and energy policy, over which METI has jurisdiction, are closely related to each other. The lack of reliable and stable political leadership on the issue of energy policy and climate policy has allowed organized vested economic interests and METI to solidify their policy coalition against the substantial introduction of renewables that would enable fulfilling ambitious GHG reduction targets. Thus, despite its tremendous potential to become a vital force in climate change negotiations, Japan has relinquished both its leadership and its initiative. Instead, it has withdrawn into the role of an intermediate or supportive state and has even become a laggard. It is crystal clear that Japan should choose a renewable energy path, and as other chapters in this book make clear, there are lots of private and local communities' initiatives for developing renewable energy. What Japan lacks now is political will and visionary political leadership to embrace this path wholeheartedly.

Notes

1. According to an opinion poll carried out in 2005 by the Ministry of Foreign Affairs (MOFA), 72.2% of the respondents regard the issue of global warming as a grave problem in their daily life and a majority of them, 52.2%, consider that Japan should take the lead in the international community to tackle this problem (Gaimusho Kokusaishakai-kyoku Seisaku-ka 2005). Likewise, the Environmental Ministry's 2004 opinion poll conducted during the run-up to the implementation of the Kyoto Protocol showed overwhelming public support for stronger action by Japan. More than 85% of respondents wanted Japan to enhance further its efforts to stop global warming (85.4%) (Kankyosho 2004).
2. The construction of new nuclear plants slowed partially because of the Chernobyl nuclear accident in 1986, a massive leak of sodium coolant at the fast breeder reactor *Monju* in 1995, the accident at the Japan Nuclear Fuel Conversion Company (JCO) in 1999 in Tokaimura (Shimizu 2014, p. 7), and the ever-escalating costs of constructing new nuclear plants as well as growing societal opposition to further construction of nuclear plants.
3. This Japanese governmental side event was composed of panelists representing all energy-intensive industries endorsed by MOFA's Ambassador for Global Environmental Problems and MOE's Vice Minister of Global Environmental Affairs. The proposed policy outlined a bottom-up approach under a voluntary reduction scheme varying according to each industrial sector (Kawamata 2007).
4. The FEPC and nuclear industries are the core of nuclear interest groups. Local governments that host nuclear power plants also have high economic stakes in nuclear power since they can receive subsidies and create jobs for nuclear facilities and their operation. Besides, the DPJ relied on electoral support from labor unions, especially from *Denryokusōren*.
5. Manabe scrutinized this phenomenon by arguing that music played a central role in expressing anti-nuclear sentiments and mobilizing political resistance in Japan (Manabe 2015).
6. However, by July 2014, the massive anti-nuclear rallies had been eclipsed by another series of opposition movements against changing the interpretation of Article 9 of the Japanese Constitution concerning the exercise of collective self-defense, and a package of new security legislation introduced by the second Abe administration.
7. Eleven cities where public hearings were held include Fukushima, Sendai, Saitama, Osaka, Nagoya, Takamatsu, and Fukuoka (*Asahi Shimbun* 2012).
8. *Monju* was shut down in December 1995 following a sodium coolant leak and fire and subsequent cover-up attempt. It went online again in May

2010, but in August, it again was shut down because a fuel-loading device fell into the reactor vessel (Aoki 2015). It has been idle since then.

9. Two 40-year-old nuclear reactors, the No. 1 and 2 reactors at the Takahama plant, were approved by the NRA to extend their operation for another 20 years by completing all necessary safety measures (*Mainichi* 2016).
10. Moreover, when power companies have surplus electricity, as an operational rule, they can curtail without compensation electricity from thermal, biomass, large-scale solar and wind, and household solar generation in this order (Asahi Shimbun 2015b).
11. The criteria for their environmentally friendly electricity label include: (1) the disclosure of information about energy mix and environmental loads in a way easy for consumers to understand; (2) the main electricity supply comes from renewable energy power plants (including from the FIT); (3) no procurement from both nuclear and thermal power plants (except for 24-hour backup electricity); (4) prioritize the supply from renewable power facilities of local communities and citizens; and (5) no capital ties with a major power company (see http://power-shift.org/choice-2/ [April 29, 2020]).
12. From the web site of OCCTO: https://www.occto.or.jp/en/about_occto/about_occto.html.
13. IGFC stands for Integrated Coal Gasification Fuel Cell cycle.

References

Allen, David, and Sumida, Chiyomi. 2010. Hatoyama: Moving Futenma Operations Off Okinawa 'Not Possible.' *Stars and Stripes*, May 6.

ANRE/METI (Agency of Natural Resources and Energy, METI). 2019. Denryoku nettowāku kaikaku no kihonteki kangaekata nitsuite (The Basic Ideas About the Reform of the Electric Power Network). ANRE, March. Available at: https://www.meti.go.jp/shingikai/enecho/denryoku_gas/datsu_tansoka/pdf/002_01_00.pdf. Accessed 29 Apr 2020.

ANRE/METI (Agency of Natural Resources and Energy, METI). 2020. Denkijigyo seidono gaiyō—Toroku kouridenki-jigyosha ichiran. Available at: https://www.enecho.meti.go.jp/category/electricity_and_gas/electric/summary/retailers_list/. Accessed 29 Apr 2020.

Aoki, Mizuho. 2015. Fate of Troubled *Monju* Reactor Hangs in the Balance. *Japan Times*, November 23. Available at http://www.japantimes.co.jp/news/2015/11/23/reference/fate-of-troubled-monju-reactor-hangs-in-the-balance/#.VxNQqmO7dJM. Accessed 29 Apr 2020.

Aoki, Mizuho. 2016. Down But Not Out: Japan's Anti-Nuclear Movement Fights to Regain Momentum. *The Japan Times*, March 11. Available

at: https://www.japantimes.co.jp/news/2016/03/11/national/not-japans-anti-nuclear-movement-fights-regain-momentum/#.Xqj-5hO2E0o. Accessed 29 Apr 2020.
Asahi Shimbun. 2012. Genpatsu 0%' Shijisaita—Tōrongata yoronchosa, chōshukai, paburikku-komento, August 23.
Asahi Shimbun. 2014. Kyuden, saisei-ene kaitori jijitsujyo chushi-he, September 20.
Asahi Shimbun. 2015a. Keisanshō genpatsu to karyoku bi'iki, January 17.
Asahi Shimbun. 2015b. Saisei-ene yokusei, saisan fuan, February 5.
Asahi Shimbun. 2015c. 'Genpatsu niwari' pabukome sudōri, shōsai/sanpi shimesazu: 2030 nen dengenkōsei kettei, July 17.
Asahi Shimbun. 2018a. Taiyoko-ha yokusei, ugokashitsuzukeru genpatsu, Kushu-igaidemo okoriuru, October 14.
Asahi Shimbun. 2018b. Sai ene shutsuryoku yokusei, Kyushu-igaidemo junbi, Tohoku, Chugoku-de, November 13.
Beck, Ulrich. 1992. *Risk Society: Towards New Modernity*. London: Sage.
Democratic Party of Japan (DPJ). 2009. *The 2009 Manifesto, The Democratic Party of Japan's Platform for Government, Putting People's Lives First*, DPJ.
Enerugie-kankyo kaigi (Energy and Environment Council: EEC). 2012. Enerugi-kankyo ni kansuru sentaku (Choice About Energy and the Environment), June 29. Available at: http://www.env.go.jp/council/06earth/y060-110/mat01_2.pdf. Accessed 28 Apr 2020.
Gaimusho Kokusaishakai-kyoku Seisaku-ka. 2005. Chikyukibo mondai ni kansuru ishikichosa, May 11. Available at: http://www.mofa.go.jp/mofaj/press/release/17/rls_0511b.html. Accessed 28 Apr 2020.
George, Alexander. L., and Bennett, Andrew. 2005. *Case Studies and Theory Development in the Social Sciences*. Cambridge, MA: MIT Press.
Iida, Testsunari (ISEP) et al. 2006. Renewable Energy and Social Innovation in Japan, 18 February. Available at: http://www.isep.or.jp/images/press/060218swisspresen-1.pdf. Accessed 28 Apr 2020.
ISEP (Kankyo enerugie kenkyujyo). 2014. *Shizen enerugi hakusho 2014*. Available at: http://www.isep.or.jp/jsr2014. Accessed 28 Apr 2020.
ISEP (Kankyo enerugie kenkyujyo). 2017. *Shizen enerugie hakusho 2017*. Available at: http://www.isep.or.jp/jsr/2017report. Accessed 28 Apr 2020.
Ito, Koichiro. 2016. Deregulation of Japan's Electricity Market: Key Factors Needed for Success. Energy Policy Institute at the University of Chicago (EPIC), March 1. Available at: https://epic.uchicago.edu/news-events/news/deregulation-japan%E2%80%99s-electricity-market-key-factors-needed-success. Accessed 28 Apr 2020.

Japan Times. 2010. Hatoyama Offers an Apology to Okinawa, May 24. Available at: http://www.japantimes.co.jp/news/2010/05/24/national/hatoyama-offers-an-apology-to-okinawa/#.WTtxQjNI2SM. Accessed 29 Apr 2020.

Kameyama, Yasuko. 2017. *Climate Change Policy in Japan: From the 1980s to 2015*. London: Routledge.

Kankyosho. 2004. Chikyu ondanka mondai ni kansuru yoron chōsa (Kekka no gaiyo), October 16–17. Available at: https://www.env.go.jp/policy/report/h17-03/ref01-2.pdf. Accessed 28 Apr 2020.

Kantei. 2007. Innovative Technology Development, May 24. Available at: https://japan.kantei.go.jp/policy/ondanka/2007/0524inv/siryo1.pdf. Accessed 28 Apr 2020.

Kawamata, Kotaro. 2007. Road to Hokkaido Toyako Summit: Message from Japan as a G8 Host Country in 2008. Side Events/Exhibits Archive, UNFCCC COP13, December 10. Available at: https://seors.unfccc.int/seors/reports/archive.html?session_id=COP13. Accessed 28 Apr 2020.

Kikkawa, Takeo. 2011. *Tsusansangyo seisakushi 10: Shigen enerugie seisaku 1980–2000* (*The History of Trade and Industry Policy 10: Natural Resources and Energy Policy 1980–2000*). Tokyo: Keizaisangyo chōsakai.

Kimura, Osamu, and Tatsujirō Suzuki. 2006. 30 Years of Solar Energy Development in Japan: Co-evolution Process of Technology, Policies, and the Market. Paper Prepared for the 2006 Berlin Conference on the Human Dimensions of Global Environmental Change: "Resource Policies: Effectiveness, Efficiency, and Equity," Berlin, November 17–18.

Mainichi. 2016. NRA's Takahama Reactor Approval a Blow to 40-Year Lifespan Rule, June 21. Available at: https://mainichi.jp/english/articles/20160621/p2a/00m/0na/019000c. Accessed 29 Apr 2020.

Manabe, Noriko. 2015. *The Revolution Will Not Be Televised: Protest Music After Fukushima*. Oxford: Oxford University Press.

METI (Minister of Economy, Industry, and Trade). 2006. *New National Energy Strategy: Digest*, May. Available at: http://warp.da.ndl.go.jp/info:ndljp/pid/11245844/www.enecho.meti.go.jp/en/reports/pdf/newnationalenergystrategy2006.pdf. Accessed 29 Apr 2020.

METI. 2010. The Strategic Energy Plan of Japan: Meeting Global Challenges and Securing Energy Futures (Summary). Revised June 2010. Available at: http://warp.da.ndl.go.jp/info:ndljp/pid/3487098/www.meti.go.jp/english/press/data/pdf/20100618_08a.pdf. Accessed 29 Apr 2020.

METI. 2014. Strategic Energy Plan, April. Available at: http://www.enecho.meti.go.jp/en/category/others/basic_plan/pdf/4th_strategic_energy_plan.pdf. Accessed 29 Apr 2020.

METI. 2015. Long-Term Energy Supply and Demand Outlook, July. Available at: http://www.meti.go.jp/english/press/2015/0716_01.html. Accessed 29 Apr 2020.

METI. 2018. Strategic Energy Plan, July. Available at: http://www.enecho.meti.go.jp/en/category/others/basic_plan/5th/pdf/strategic_energyplan.pdf. Accessed 29 Apr 2020.

MOE (Ministry of the Environment). 2007. Becoming a Leading Environmental Nation Strategy in the 21st Century, June 6. Available at: https://www.env.go.jp/en/focus/070606.html. Accessed 11 Mar 2016.

Moe, Espen. 2012. Vested Interests, Energy Efficiency and Renewable in Japan. *Energy Policy* 40 (C): 260–273.

Moe, Espen. 2015. Japan: No Structural Change, Save for a Structural Shock? Vested Interests Pre- and Post-Fukushima. In *Renewable Energy Transformation or Fossil Fuel Backlash*, ed. Espen Moe, 38–71. New York: Palgrave Macmillan.

MOFA (Minister of Foreign Affairs of Japan). 2009. Statement by Prime Minister Yukio Hatoyama at the United Nations Summit on Climate Change. New York, September 22. Available at: https://japan.kantei.go.jp/hatoyama/statement/200909/ehat_0922_e.html. Accessed 29 Apr 2020.

MOFA. 2014. Cabinet Decision on Development of Seamless Security Legislation to Ensure Japan's Survival and Protect Its People, July 1. Available at: https://www.mofa.go.jp/fp/nsp/page23e_000273.html. Accessed 29 Apr 2020.

MOFA. 2018. Report of Advisory Panel to the Foreign Minister on Climate Change: Recommendations on Energy Promote New Diplomacy on Energy Through Leading Global Efforts Against Climate Change, February 2. Available at: https://www.mofa.go.jp/files/000337158.pdf. Accessed 29 Apr 2020.

National Diet of Japan Fukushima Nuclear Accident Independent Investigation Commission (NAIIC). 2012. *The Official Report of the Fukushima Nuclear Accident Independent Investigation Commission*, Executive Summary. Available at: http://warp.da.ndl.go.jp/info:ndljp/pid/3856371/naiic.go.jp/en/report/. Accessed 29 Apr 2020.

Nihon Keizai Shimbun. 2012. Fenpatsu zero shiji, sankago 47% ni zoka—Tōrongata yoronchosa, August 22. Available at: http://www.nikkei.com/article/DGXNASGC22005_S2A820C1MM0000/. Accessed 29 Apr 2020.

Nilsson-Wright, John, and Fujiwara Kiichi. 2015. Japan's Abe Administration Steering a Course Between Pragmatism and Extremism. Research Paper, Chatham House: The Royal Institute of International Affairs, September. Available at: https://www.chathamhouse.org/sites/default/files/publications/research/20150914JapanAbeAdministrationNilssonWrightFujiwara.pdf. Accessed 29 Apr 2020.

OCCTO (Organization for Cross-regional Coordination of Transmission Operators, Japan). 2019. Role of OCCTO in the Electricity System Reform, March 18. Available at: https://www.occto.or.jp/en/about_occto/about_occto.html. Accessed 29 Apr 2020.

Oshima, Kenichi. 2010. *Saiseikano enerugie no seijikeizai gaku: enerugie seisakuno gureen kaikakuni mukete (The Political Economy of Renewable Energy: Toward the Green Reform of Energy Policy)*. Tokyo: Tōyōkeizai shimpōsha.

Powershift. 2020. Pawā sifutona denryokugaisha ichiran, June 13. Available at: http://power-shift.org/choice-2/. Accessed 29 Apr 2020.

REI (Renewable Energy Institute). 2018a. Comment on the Draft Basic Energy Plan: Japan Should Join the Global Trend for Renewables Toward Decarbonization, Persisting to Coal and Nuclear Will Put Japan's Future at Risk, June 28. Available at: https://www.renewable-ei.org/en/activities/reports/img/pdf/20180628_02/20180628_CommentBasicEnergyPlan_EN.pdf. Accessed 28 Apr 2020.

REI. 2018b. Electricity Generation Mix FY 2017 (The Source: METI/ANRE "Total Energy Statistics"). Available at: https://www.renewable-ei.org/en/statistics/electricity/. Accessed 5 Apr 2019.

REI. 2019. Electricity Generation Mix FY 2018 (Preliminary) (Based on the Data from METI/ANRE "Total Energy Statistics"), November 25. Available at: https://www.renewable-ei.org/en/statistics/energy/?cat=electricity. Accessed 28 Apr 2020.

Samuels, Richard J. 1987. *The Business of the Japanese State: Energy Markets in Comparative and Historical Perspective*. Ithaca: Cornell University Press.

Sawa, Akihiro. 2009. "The Fragility of Hatoyama's 25% Reduction Initiative: The Risks of Losing International Leadership," Circulated Through the "Climate Change Info Mailing List," the Speech at the 21st Century Public Policy Institute on September 14. Available at: http://www.21ppi.org/english/pdf/090924.pdf. Accessed 29 Apr 2020.

Shimizu, Tetsuro. 2014. Genshiryoku hatsuden-to nihon-no enerugie jyukyu (Nuclear Generation and Japan's Energy Supply and Demand). *Norinkinyu* (*Co-operative Finance for Agriculture, Forestry and Fishery*), 10: 2 (622), 14 (634).

Shugiin (House of Representatives). 2010. Chikyu ondanka-taisaku kihon hōan. Dai 171 kai, San. Dai 19 gō. Available at: http://www.shugiin.go.jp/internet/itdb_gian.nsf/html/gian/honbun/houan/g17106019.htm. Accessed 29 Apr 2020.

Tabuchi, Hiroko. 2011. Japan's Prime Minister Declares Fukushima Plant Stable. *The New York Times*, December 16.

Toichi, Tsutomu. 2002. Japan's Energy Policy and Its Implications for the Economy. *IEEJ*, April. Available at: https://eneken.ieej.or.jp/en/data/pdf/110.pdf. Accessed 29 Apr 2020.

Tsunekawa, Keiichi. 2010. Kiseikanwa-no seijikatei: Naniga kawattanoka (The Political Processes of Deregulation: What Did Change?). In *Kōzomondai to kiseikanwa*, ed. Jyuro Teranishi, 77–147. Baburu/defureki-no Nihonkeizai to keizaiseisaku Dai 7 kan (The 7th Volume of *Japanese Economy and Economic Policies During the Period of the Bubble Economy and Deflation*) Under the Supervision of the Economic and Social Research Institute of the Cabinet Office (ESRI). Tokyo: Keio University Press. Also Available at: http://www.esri.go.jp/jp/others/kanko_sbubble/analysis_07_03.pdf. Accessed 29 Apr 2020.

UN Framework Convention on Climate Change (UNFCCC). 2015. The Paris Agreement. FCCC/CP/2015/L.9/Rev.1, December 12.

Watanabe, Rie. 2015. *Nihon to Doitsu-no kikōenerugie seisakutenkan: paradaimutenkan-no mekanizumu (Climate and Energy Policy Changes in Japan and Germany: A Path to Paradigmatic Policy Change)*. Tokyo: Yushindo.

Yoneda, Yuriko. 2008. The Spread of Solar Power Generation in Japan. *Japan for Sustainability, JFS Newsletter*, 70, June. Available at: https://www.japanfs.org/en/news/archives/news_id027851.html. Accessed 28 Apr 2020.

Yoshioka, Hitoshi. 2011. *Shinpan Genshiryoku-no shakaishi: Sono nihonteki tenkai*. Tokyo: Asahi Shimbun Publications Inc.

CHAPTER 4

Japan's Nuclear Safety Regulation Policy

Florentine Koppenborg

The Fukushima Daiichi Nuclear Accident[1] and Its Aftermath

On 11 March 2011, the so-called triple disaster (3/11) shook Japan's energy policy to its core. It consisted of a magnitude nine earthquake and a resulting tsunami, which, together, caused a meltdown in three of six reactors at the Fukushima Daiichi nuclear power plant. The nuclear accident directly impacted Japan's energy policy and, by extension, Japan's climate policy. It also caused large anti-nuclear demonstrations[2] (Hasegawa 2014) and a shift in public opinion towards a more critical stance on nuclear power[3] (Shibata and Tomokiyo 2014). Around the

I would like to express my gratitude to the editors Paul Midford and Espen Moe for the invitation to a book workshop in Trondheim in 2018. I am immensely grateful for the comments I received from the editors and workshop participants' comments on an earlier version of the manuscript, although any errors are my own. I thank Fiona Kinniburgh (Technical University Munich) for assistance with English language editing.

F. Koppenborg (✉)
Technical University Munich, Munich, Germany

P. Midford and E. Moe (eds.), *New Challenges and Solutions for Renewable Energy*, International Political Economy Series,
https://doi.org/10.1007/978-3-030-54514-7_4

same time, the government embarked on a reform of the nuclear safety administration in order to improve nuclear safety. Another measure was to shut down all nuclear power plants for a stress test, causing a considerable electricity gap. Nuclear power plants provided almost 30% of Japan's electricity supply just before the nuclear accident. To address this electricity gap, immediate energy savings measures were introduced and the use of fossil fuels such as oil, gas, and coal increased. The move to fossil fuels led to higher greenhouse gas emissions (ANRE 2018), posing a challenge to Japan's emissions reductions pledge under the United Nations Framework Convention on Climate Change (UNFCCC) (Koppenborg 2017). Against this background, energy policy deliberations commenced early in 2011.

The role of nuclear power took centre stage in discussions about Japan's future energy policy. It was an essential means to achieve the triad of energy policy goals, namely energy security, environmental friendliness, and economic efficiency, the "3Es" of nuclear energy policy. While the 2014 *Strategic Energy Plan* resolved to retain nuclear power as a "baseload" power source, it also pledged to reduce dependence on nuclear power "as much as possible". At the same time, it added safety as a fourth aspect to the existing three energy policy goals and pledged to mitigate the risks posed by nuclear power plants. It also stressed expanding renewable energy sources (METI 2014). Thus, the 2014 *Strategic Energy Plan* added "S" for safety to the existing three "3Es" in order to enable the continued use of nuclear power. An energy mix with numerical targets was published one year later in Japan's 2015 *Long-term Energy Supply and Demand Outlook*. It envisioned a share of 20–22% for nuclear power and 22–24% for renewable energy (METI 2015). These targets were confirmed once again in the most recent *Strategic Energy Plan* published in 2018 (METI 2018).

Powerful pro-nuclear actors were pushing for swift restarts to achieve the share of 20–22% as soon as possible. For example, Prime Minister Abe Shinzō called for a restart of all reactors within three years in his first New Year's address in 2013 (Cabinet Office Japan 2013). However, despite pressure for a swift return to nuclear power as an important "baseload" power source and a means to lower Japan's greenhouse gas emission, in 2016, nuclear power accounted for only 2% of Japan's electricity supply (ANRE 2018). Meanwhile, it is becoming increasingly clear that nuclear power will provide, at the most, 15% of the electricity generated by 2030 (Koppenborg 2016), and perhaps as little as 10% (Izadi-Najafabadi 2015).

This raises the question: Why is Japan unable to meet the nuclear power targets it set for itself?

This chapter argues that this development is due to the creation of an independent nuclear safety agency after 3/11, called the Nuclear Regulation Authority (NRA). The findings here challenge earlier assessments of the NRA, published not long after the nuclear safety reforms following the March 2011 nuclear accident. The question many scholars asked was whether the new safety agency would be independent and able to impose stricter safety standards.

This question emerged from Japan's history of collusion between regulator, the regulated, and the government, which was at the root of the problems leading to nuclear disaster. In its July 2012 report, the Nuclear Accident Independent Investigation Committee (NAIIC) stated:

> The TEPCO Fukushima Nuclear Power Plant accident was the result of collusion between the government, the regulators and TEPCO, and the lack of governance by said parties. ... We believe the root causes were the organisational and regulatory systems that supported faulty rationales for decisions and actions ... (NAIIC 2012, p. 16)

Such collusion is also referred to as "regulatory capture", a situation where the regulator decides in favour of the industry it was supposed to regulate (Wilson 1980) as opposed to imposing the safety measures necessary to protect the people from harm.

Since the power of the NRA's predecessor was severely constrained by regulatory capture, many scholars expected the same fate to befall the NRA—particularly in the face of pro-nuclear actors dominating energy policy. Centred on the Ministry of Economy, Trade and Industry (METI), the nuclear industry, and the Liberal Democratic Party (LDP), the group of pro-nuclear actors was characterised by vested interests, wielded considerable power, and had a propensity for keeping critics out of the decision-making process (Cotton 2014; Kingston 2013; Vivoda and Graetz 2014). Critics termed these pro-nuclear actors the "nuclear village"[4] to express the proximity of relations between actors within. Defying expectations, the NRA began to exhibit signs of independent regulatory behaviour (Hymans 2015; Shiroyama 2015). The premise of this book chapter is that more attention needs to be paid to the NRA as an independent regulatory agency.

The rest of this chapter is divided into three parts. First, it introduces the NRA as an independent regulatory agency. Second, it outlines the nuclear safety standards the NRA introduced and studies how they are implemented as part of the safety review process. The third part looks at how the nuclear industry and the Japanese public responded to the new safety agency's work.

The NRA

Reform Process. The March 2011 nuclear accident exposed flaws in the administrative system governing nuclear safety regulation and emergency response in three ways: the NRA's predecessor, the Nuclear and Industrial Safety Agency (NISA), failed to prevent the accident (The New York Times 2011), to gauge its severity correctly (IAEA 2011a, b, c), and to play its assigned role in crisis management. As a result, Prime Minister Kan Naoto questioned the competencies of the safety agency and perceived it to be under the influence of METI (Kan 2012), the ministry most actively promoting nuclear power. The Kan government initiated a reform of the nuclear safety administration that aimed to separate nuclear safety regulators from those promoting nuclear power. Concretely, the Kan Cabinet decided to create a regulatory body which would have autonomy from METI in order to achieve regulatory independence (Cabinet Secretariat 2011). Accordingly, the idea of establishing an independent agency played a crucial role throughout the legislative reform process.

Legal Framework and Purpose of the NRA. Debates about the appropriate legal status for the new nuclear safety agency to achieve regulatory independence were at the heart of the legislative process. They boiled down to two choices, Articles 3 and 8 of the National Government Organisation Act. Most bodies in the Japanese government are established based on one of the two. Whereas Article 3 serves to establish ministries and other comparatively autonomous entities, Article 8 is used to set up divisions or agencies within Article 3 organisations. One major difference pertains to the staff and budget of an administrative body. While Article 3 organisations hold staff and budget rights, meaning they control human resource policies and formulate their own budget requests, Article 8 organisations are dependent on a superordinate organisation for both. Eventually, a decision was made in favour of an Article 3 independent regulatory commission that is loosely affiliated with the Ministry of Environment.

The Nuclear Regulation Authority Establishment Act (NRA Act), passed by the Japanese parliament in June 2012, contains provisions to ensure organisational autonomy from both the remainder of the administration and the nuclear industry. Making the NRA an Article 3 commission organisation granted it many of the same rights as a ministry. It also gave the NRA necessary competencies vis-à-vis the industry to regulate nuclear power plants. Article 3 grants the NRA the sole right to exercise authority within its area of jurisdiction. This makes courts the only bodies with the authority to override NRA decisions on nuclear safety. The NRA Act also gave the NRA board autonomy to decide all aspects of the new safety agency not specified by the law. The board took the opportunity to implement rules reinforcing the NRA's organisational autonomy by adopting guidelines for transparency, neutrality of experts, neutrality of commissioners, and rules against regular staff transfers with the remainder of the administration, especially METI, and the nuclear industry. The legal framework and operational guidelines served as important cornerstones for the NRA's independence.[5]

Shortly after its inauguration, the NRA determined its purpose to consist of translating the lessons learned from 3/11 into organisational practices with three goals in mind: to prevent a large-scale accident from happening again, to restore public trust in nuclear safety administration, and to rebuild nuclear safety management by putting a genuine safety culture in place. Its declared mission was "to protect the general public and the environment through rigorous and reliable regulation of nuclear activities" (NRA 2014b). To achieve this, the NRA board adopted a number of guidelines, including transparency guidelines for NRA Commission meetings, guidelines for managing administrative documents, and a code of conduct for NRA Commission members (NRA 2013b). Soon after, guidelines followed for ensuring transparency and neutrality when consulting external experts (NRA Commission 2012). Basically, the NRA vowed to regain public trust in nuclear safety administration that was lost after the March 2011 nuclear accident by implementing strict safety standards in an independent, transparent, and neutral manner.

An Independent Regulatory Agency. By 2016, the NRA consisted of a board of five scientists and nuclear experts, as well as a secretariat[6] of almost 1000 full-time employees. As such, it reached about half of the size of the Ministry of Environment with which it was affiliated. According to the literature on regulatory agencies, the NRA should meet certain criteria to live up to the ideal of an independent safety agency. These include competent staff and in-house expertise, sufficient financial resources, independent decision-making, as well as transparency as a prerequisite for public scrutiny (Gilardi and Maggetti 2011; Wilson 1980). The progress the NRA has made towards becoming an independent regulatory agency, according to these criteria, differs from area to area.

Regarding in-house expertise, the NRA implemented different measures, such as hiring fresh graduates and mid-career officials, limiting staff rotation with those parts of the administration promoting nuclear power, restricting retirement to the nuclear industry, and creating a career in nuclear regulation, including comprehensive in-house training. Without a doubt an ongoing effort, the NRA embarked on a good path to achieving sufficient in-house expertise. Turning to the financial resources, the NRA's predecessor depended on METI for its budget, making it vulnerable to external influence. In contrast, the NRA became as financially independent as possible within the administration with seemingly sufficient financial resources.

All decisions, including regulatory judgements, are made within what I term a "council system". Essentially, a two-step process, it ensures the board, the "NRA Commission", has the last word in every decision. First, NRA officials from different divisions, or utility representatives in the case of safety review meetings, provide information and answer questions about their activities to the respective council. In a second step, one or two representatives from this council present the matter to the NRA Commission, which makes all final decisions. As part of the council system, new regulatory standards were developed without influence from the regulated parties. Industry representatives were only consulted regarding practical aspects of conducting the safety review. The "council system" allows the board of five scientists to control decision-making processes and to minimise outside influence.

Transparency is an integral part of how administrative agencies should conduct themselves in a democratic system. By inviting public scrutiny, it can also limit the extent to which others can openly try to exert influence. The NRA achieved a high degree of transparency by taking a proactive approach to being open to the public. Guidelines for transparency and administrative documents interlink to form an information release system not subject to disclosure requests. This includes streaming all meetings[7] live on its YouTube channel "NRAJapan" as well as publishing reference materials on its website. The NRA also opened press conferences to all media and not just those associated with journalist clubs ("***kisha clubs***"). While a high degree of transparency allowed the public to watch the NRA at work, there was no citizen participation in the decision-making process.

In sum, the NRA was on a good path towards becoming an independent regulatory agency. It enjoyed considerable independence as a regulatory agency due to control over its human resources and budget, rules for keeping a distance from the industry applying to both the NRA board and consulted external experts, and transparency towards the public as a prerequisite for public scrutiny.

However, there was considerable political pressure on the NRA to return to pre-3/11 business as usual. For example, the energy policy strategy adopted by the Abe government in 2014 and 2015 was contingent on the NRA swiftly approving reactors and on relaxing the newly introduced safety requirements, discussed in the next section. Another avenue for political influence was through the nomination of board members with close ties to the industry.

A controversial choice by the Abe government was Tanaka Satoru due to his closeness to the nuclear industry. As a professor at the Tokyo University's Department of Nuclear Engineering, he had long-standing ties with the nuclear industry and had actively promoted and supported nuclear power expansion in the past. Once appointed, NRA transparency guidelines required Tanaka to disclose industry funding. He received 6 million yen in research funds from the nuclear power plant operator J-Power and the plant manufacturers Hitachi and GE between 2004 and 2010. He also received funding in the three years prior to his nomination, for example from TEPCO Memorial Foundation, Hitachi GE, and Taiheiyo Consultant, an engineering firm. In the light of this information, critical voices warned that someone like Tanaka joining the NRA board could undermine the independence of the nuclear safety regulator (Asahi Shinbun 2014; The Japan Times 2014b). Regardless of criticism and all seven opposition parties opposing Tanaka Satoru's appointment, the Abe government pushed the decision through the Diet.

Had Tanaka Satoru been appointed as NRA Chairman when the term of the first chairman, Tanaka Shunichi, ended in 2017, this would have been an instance of political influence. However, Tanaka Shunichi was followed by Fuketa Toyoshi, an old colleague of Tanaka Shunichi and the second board member present since the NRA's creation in 2012. There clearly was pressure on the NRA, but it passed the litmus test for political influence when changing chairman.[8]

Nuclear Safety Regulation

Looking at the NRA's regulatory authority, one can speak of a medley of competencies, reflecting the reform goal to unify tasks related to nuclear safety, security, and safeguards in one new agency. An area where the NRA shares competencies with the Cabinet Office is that of emergency response measures. The NRA Act specifies that the NRA is responsible for the technical aspects of emergency response measures and gives the prime minister a role in handling operational aspects such as coordinating relevant ministries. In practice, the NRA formulates guidelines while the Cabinet Office monitors their implementation. In the area of nuclear security and safeguards, the safety agency translated provisions of international treaties Japan had signed, such as the Nuclear-Non-Proliferation Treaty, into domestic regulation and oversaw the implementation. Regarding radioactive materials and radioisotopes, a division of labour between the NRA and the Ministry of Health, Labor and Welfare emerged with the NRA devising rules for research and medical treatments and the ministry setting standards regarding human health. Another task the NRA took on was the supervision of the process of decommissioning the crippled Fukushima Daiichi nuclear power plant.

The area where the NRA holds the most regulatory authority is nuclear safety. Article 4 of the NRA Act put rulemaking for nuclear facilities and the right to grant permits for taking up operation into the hands of the NRA. When the Japan Nuclear Safety Organization was incorporated into the NRA in 2014, it integrated the task of monitoring the implementation of safety measures on the ground. Since the right to grant permits can be used as a sanctioning mechanism by threatening to withholding permits, the NRA is in charge of formulating safety standards, implementing them, and imposing sanctions in case of violations. The next section takes a closer look at the nuclear safety regulation for commercial nuclear power plants adopted by the NRA as well as their implementation.

New Safety Standards

When the NRA was established, a major task was to formulate new safety guidelines. It established four councils to handle the matter, two discussing safety requirements for commercial nuclear power plants, one dealing with nuclear fuel cycle facilities, and one more in charge of a safety review procedure. Each council consisted of at least one NRA

Commissioner, external experts selected in line with the neutrality criteria, NRA Secretariat officials, and sometimes members of so-called incorporated administrative agencies or semi-private research institutes, such as the Japan Atomic Energy Agency (JAEA). New safety requirements for commercial nuclear power plants were adopted by the NRA Commission, issued in June 2013 and went into force on July 8. For nuclear fuel cycle facilities, the NRA decided to take a graded approach, meaning that new regulatory requirements will be developed for each facility due to vast differences in construction design (NRA 2014a).

The new safety regulation introduced several innovations. These can be subsumed under the following umbrella terms, each of which will be explained in detail hereafter: a forty-year rule, a back-fit system, defence-in-depth, and severe accident countermeasures (NRA 2013a). First, the NRA incorporated a *forty-year rule* limiting a reactor's lifespan into the new safety regulations, meaning that an operating licence is only granted for 40 years. If a plant operator wishes to operate a reactor for longer, they can apply for a one-time extension of 20 years. Such an extension is tied to a renewed safety review to determine whether the reactor is fit for another 20 years in operation. This is the area where the Abe government exerted pressure on the NRA to relax the forty-year rule and to let reactors operate for 60 years. This is presumably because the 20–22% target for nuclear power in the 2030 energy mix, adopted by the Abe government, is only possible if reactors generally have their lifespan extended.

The *back-fit system* means that updating reactors in line with new scientific developments is not at the discretion of electricity utilities anymore. Now, plant operators not only have to conform to a given set of safety standards when they commission the plant, but also need to update them in line with adjustments in regulation over time. This forms the basis for the currently undertaken reactor review. It is a safety review to make sure that reactors which were already in operation are back-fitted in accordance with NRA regulation. The back-fit system and the 40-year limit provide the NRA with leverage over the industry. In stark contrast to its predecessor, which had to rely on the industry following its recommendations, the NRA can ensure compliance by threatening to resolve the operating licence of a nuclear reactor in case of non-compliance.

The concept of *defence-in-depth* is related to predefined threats that a nuclear power plant needs to be able to withstand. Previous safety requirements assumed that a single function or component within a reactor might fail (e.g., that an earthquake might cut off external electricity supply). Omitted were measures against severe accidents where more than one component within the reactor might fail, as was the case with the Fukushima nuclear power plant where off-site power supply was cut off by the earthquake and back-up generators were destroyed by the tsunami. One way the NRA has chosen to implement this is by strengthening the reactor design basis, which designates the construction design a new reactor should adhere to in order to withstand predefined threats. New standards require a seismic ground motion assessment for each nuclear power plant in order to determine the specific level of earthquake resistance required for the plant. Volcanoes within a 160-km radius are to be surveyed to assess the risk and that appropriate measures are to be taken. Mandatory tsunami protection walls need to exceed the largest one ever recorded for the area. Strengthened design basis requirements are meant to prevent the nuclear power plant from being damaged by a natural disaster in the first place. In case they prove insufficient, severe accident countermeasures are in place to prevent a nuclear disaster like the one at the Fukushima Daiichi power plant.

The idea of defence-in-depth is also represented in *severe accident countermeasures*. Such countermeasures address both natural phenomena and other events. Natural phenomena are defined to include earthquakes, tsunamis, volcanic eruptions, tornadoes, and forest fires. Events other than natural phenomena refer to events such as a fire inside a reactor, internal flooding, power supply failure, and the like. Severe accident countermeasures include the installation of waterproof doors and fire-proof cables, and the preparation of switchboards for operating the plant in diverse locations. Measures to increase the reliability of off-site power sources prescribe connecting two or more power substations with at least two transmission lines to ensure they remain functional even if some of the stations and some of the power lines are damaged. This off-site power supply is further supplemented by mobile power units placed on a hill nearby. Measures to strengthen capacities to cool parts of the nuclear power plant such as a spent fuel pool or a reactor itself focus on the use of mobile water injection systems as well as permanently installed water injection systems. To prevent hydrogen explosions at boiling water reactors, there needs to be filtered venting system to let out hydrogen if needed to reduce pressure within a reactor. Another novelty is the requirement of an extra control room that is located on higher ground and is radiation-proof to a certain degree. Next to introducing a back-fit system, accident countermeasures are the second area not at the discretion of electricity utilities anymore but prescribed by NRA regulation.

Looking at safety requirements in detail, some are relative criteria, whereas others are mandatory requirements. Relative ones pertain to volcanic and earthquake-related risks. Their implementation depends on underlying risk assessments and on the interpretation of how severe the risk of an earthquake or a volcanic eruption is for each nuclear power plant. Necessary earthquake resistance levels, for example, are determined by looking at possible sources of earthquakes around the plant and the characteristics of the surface structure to determine ground motion. Hence, they depend on how many and which possible sources for an earthquake are considered.

Mandatory requirements were modelled after the types of interventions which could have prevented the events of the March 2011 nuclear accident. They focus on preventing a loss of power, leading to a loss of cooling capacity, followed by an increase in pressure inside the reactor, ultimately erupting in a hydrogen explosion. For example, any plant operator is required to build a tsunami protection wall, to install filter vents and waterproof doors, and to provide backup electricity generators and reactor cooling systems. Absent from the new safety regulations were new threats such as a cyber-attack on a nuclear reactor or other IT-related safety requirements. While measures aimed at preventing an accident like the one at Fukushima Daiichi were the strictest, the 2013 safety requirements failed to address new threats.

Safety Reviews

In contrast to what one might assume about a nuclear safety agency, the NRA itself avoids expressions like "ensuring safety". The term was used by PM Abe himself when describing the NRA's work. Asked about this at a press conference, former Chairman Tanaka Shunichi laid out his understanding of what the NRA does:

> Safety is a vague term and says little about the degree of safety. As I have mentioned many times before, we are conducting different compliance investigations from the point of view of lowering the risk as much as possible. However, the understanding of what safety means differs from person to person. Since it creates misunderstandings when we speak of

> certifying safety, we use the term to conduct compliance investigations. (NRA 2014c)

Despite the long-windedness, his response made it clear that "safety" is hard to achieve in nuclear power. While the NRA aims to reduce the risks, it simply cannot guarantee absolute safety, if defined as an absence of any risks. Reducing risks as much as possible crucially depends on what one understands to be a risk.

Risk assessments are part of the safety review process. Given that safety requirements for volcanic and earthquake-related risks are of relative nature, the extent of required safety measures depends on the severity of these risks assumed for a given nuclear reactor. For the earthquake risk assessment, it is necessary to estimate precisely the maximum amount of ground movement and, together with that, the maximum amount of shocks the base of the plant's facilities must be able to withstand. For such an estimate, it is important to have a clear grasp of the characteristics of fault lines where the epicentre of a possible earthquake could be. Safety measures related to volcanic eruptions depend on the risk posed by volcanoes in the vicinity of the nuclear power plant. This requires a judgement call about the likelihood of a volcanic eruption. This is not limited to full-blown eruptions including lava flows, but spewing ash that accumulates on reactor buildings could also be a threat due to the accumulated weight. One might assume that in a country like Japan, which faces both earthquakes and active volcanoes regularly, both risks are assessed to be relatively high.

However, the safety review conducted for the Sendai nuclear plant in Satsumasendai city, Kagoshima prefecture, suggested otherwise. According to a critical Japanese expert in seismology, the NRA could have been stricter in this regard. One problem in seismology as a research field is that developing precise assessments of the impact of earthquakes on reactor buildings is an ongoing effort. With no established scientific consensus, it is possible to portray some fault lines as trivial in estimating the impact on nuclear power facilities (Tateishi 2015). In the report presented to the NRA, Kyushu EPCO included the most obvious earthquake sources around the Sendai nuclear power plant. However, it chose not to include potential others, in particular a fault line off the coast whose inclusion would have led to stricter earthquake countermeasure requirements. The NRA accepted this conservative interpretation of the current state of knowledge in the field. Regarding volcanic eruptions, in case of the Sendai nuclear power plant, the NRA judged the risk to be negligible. This was despite Kyushu island being known for its volcanic activity (The Japan Times 2014c). Rather than taking a precautionary approach by addressing all possible, including disputed, earthquake sources, it agreed to focus on the undisputed ones and took a conservative stance on the risk of a volcanic eruption.

In contrast, the NRA took decisive action in the face of an undisputable earthquake risk. In case of the Tsuruga nuclear plant in Fukui Prefecture, the NRA decided to conduct so-called fracture zone investigation. The aim was to make sure that the fault line the reactor was sitting on was not an active one, thus not posing an earthquake risk. The NRA's definition of an active fault line states that surrounding geological layers that are approximately 120,000–130,000 years old must show signs of displacement or deformation as a result of fault line activity. While the operator, the Japan Atomic Power Co., claimed the fault line was inactive and even applied for a restart permit, the NRA conducted an investigation with a strong element of on-the-ground activity. Instead of considering information provided by the utility, the NRA sent experts to conduct excavations around the fault line in question and thus collected their own data. Ultimately, the NRA came to the conclusion that the fault line underneath reactor number 1 was in fact active (NRA Commission 2014). After confirming that the fault line was active, the NRA refused a restart permit, leading to the permanent shut down of reactor number 1 at the Tsuruga nuclear plant. Reactor number 2 was still under review at the time of writing.

The NRA also proved relentless in the face of apparent mismanagement. In 2013, the NRA criticised how the Japan Atomic Energy Agency (JAEA) maintained the Monju fast breeder nuclear reactor. During two on-site safety inspections, irregularities surfaced. In response, the NRA ordered the JAEA to revise Monju's operational safety programme. However, the NRA later found out that the JAEA submitted a report about the requested revisions before completing them. During another on-site inspection, more irregularities surfaced (NRA 2014a). With problems persisting, the NRA acted in November 2015. It issued its first recommendation (***kankoku***), in which it concluded that the "JAEA does not have the capacity to operate Monju safely". The recommendation indicated that a failure to find a new operator would lead to the NRA decommissioning the facility (NRA Commission 2015). The recommendation itself was not binding, but the NRA lent substance to it by publicly threatening to shut down the facility at the heart of METI's fuel cycle ambitions. In December 2016, the government made the decision to decommission Monju as it could not find a suitable operator (The Japan Times 2017).

Slow Reactor Restarts and a Shrinking Reactor Fleet

Examining the number of restarts granted by the NRA as an indicator for measuring its independence intuitively might seem like an obvious choice. Taking the difference between submitted applications for security screenings and granted restart permits as representative of the degree of regulatory independence makes the rather far-reaching assumption that the first screenings are representative of all safety checks. Looking at the strategy pursued in conducting security screenings, though, it becomes clear that the assumption of representativeness is unsustainable in this case. As the reactors which were screened first are the ones with the least amount of required construction work and changes to fit the new standards, it can be expected that the ratio of reactors passing the screening decreases as time goes on. Thus, the ratio of granted permits to applications is only telling once all screenings have been finished.

Looking at the bigger picture of safety screenings, how much time the NRA took and how many reactors were decommissioned serve as better indicators for regulatory independence. By 2018, the NRA concluded safety reviews and approved restarts of 14 nuclear reactors (JAIF 2018). This stands in stark contrast to PM Abe Shinzō's demand to review almost 50 reactors by 2016. Thus, the NRA resisted pressure from the pro-nuclear group to speed up the safety review process. With every additional year the NRA takes for approvals, the older Japan's fleet of nuclear reactors get. Reactors getting closer to the 40-year limit on operations means that utilities must carefully assess whether additional safety investments are worth it despite limited time in operation left to generate enough revenue to make refurbishments a worthwhile investment.

The first step in the review process is electricity utilities submitting a request for a safety review of a specific reactor. By 2019, requests for a total of 26 nuclear reactors were submitted. Adding up the electricity generation capacity of these 26 nuclear reactors, it amounts to about half of the installed capacity available prior to the Fukushima accident. In 2010, nuclear power covered 29% of Japan's electricity demand. Assuming a stable electricity demand, the 26 reactors could provide close to 15% of Japan's electricity demand by 2030—far from enough to reach the government's goal of 20–22% by 2030. Even if the NRA approved restarts of all reactors utilities are currently trying to bring back into service, it would not be enough to reach the target of providing 20–22% of Japan's electricity supply.

The strategy pursued by electricity utilities was to submit safety review applications not for all reactors, but the ones with the highest chances of improved safety measure investments paying off eventually. Accordingly, utilities mainly submitted review requests for 26 newer reactors with large electricity generation capacity. Meeting safety requirements even for those 26 reactors forced utilities to make costly investments. Asked about the estimated costs of investments for reactors they submitted a review request for, utilities estimated total costs of 1.6 trillion yen in January 2014. By July 2015, that figure rose to 2.4 trillion yen (The Tokyo Shinbun 2015b) and even reached 5 trillion yen by July 2019 (Asahi Shinbun 2019b). With higher than anticipated refurbishing costs, electricity utilities chose to invest only in the most profitable nuclear reactors and have begun shutting down older and smaller ones.

Almost ten years after 3/11, Japan's once large nuclear reactor fleet, which at the time ranked as the third largest in the world after the US and France, has shrunk considerably. As of September 2019, a total of 35 reactors remained. Even with two additional reactors under construction, that is a far cry from the 54 commercial nuclear reactors Japan before 3/11. Thus, the margin between reactors under review and additional reactors utilities can apply for has decreased rapidly.

Looking at the time taken by the NRA to assess safety and the additional safety costs imposed on electric utilities shows regulatory independence on the part of the NRA. It took much longer than politicians asked for and imposed much higher than anticipated safety investments on utilities—possibly high enough to undermine economic viability of the business of operating nuclear reactors. Further adding to difficulties in achieving nuclear policy targets are lawsuits against the restarting of nuclear power plants based on safety concerns. With courts as the only bodies with the authority to override NRA decisions, the outcome of these lawsuits has strong implications regarding the future of nuclear power in Japan.

Critical Public Opinion and Lawsuits Challenging Safety Assessments

One of the NRA's goals was to regain public trust in Japan's nuclear safety administration by conducting independent and neutral safety investigations in a transparent manner. Looking at a recent public opinion poll about whether idled nuclear power plants should be restarted, a majority of 56% opposed, while 32% of respondents were in favour (Asahi Shinbun 2019a). Thus, many citizens still had reservations about operating nuclear power plants in Japan. Whether citizens trust the NRA's safety assessment can be inferred from the extent to which they challenge NRA decisions. With efforts picking up to restart nuclear power plants, several lawsuits were brought before courts all over Japan. Court cases reflect the plaintiff's safety-related concerns and, as such, their stance on whether NRA safety checks were sufficiently strict and trustworthy.

Ōi nuclear plant was the first where reactors were set for restarts. They were also the first to face legal action. The plaintiffs, a group of almost 200 people from the plant's host prefecture Fukui as well as from Tokyo and other prefectures, contended that the basic earthquake ground motion figures used as the basis for the reactor's anti-seismic design were insufficient. The court's ruling echoed that stance and granted the injunction (The Japan Times 2014a). Ultimately, the injunction was overruled.

In April 2015, the Fukui District Court under judge Higuchi Hideaki ordered an injunction against bringing the reactors numbers 3 and 4 of the Takahama nuclear plant in the Fukui Prefecture back online. The decision came following a group of citizens seeking an injunction after the reactors numbers 3 and 4 cleared the NRA safety review and received green light for a restart. In the ruling, Judge Higuchi criticised the data the operator KEPCO used for the earthquake simulation as unreliable and thus not suitable to show the plant's earthquake resilience. The verdict, furthermore, criticised the NRA safety standards as "not strict enough" (The Japan Times 2015; The Tokyo Shinbun 2015a). Ultimately, the injunction was overruled.

After only a few days in operation, the two reactors at Takahama nuclear power plant had to be shut down again following an injunction ordered by the Ōtsu District Court in March 2016. This court is located in the adjoining Shiga Prefecture, where a group of almost thirty citizens sought an injunction. The Ōtsu court ruled that KEPCO failed to fulfil its obligation to provide information on safety and evacuation-related issues in case of an emergency. Furthermore, the court ruled that the NRA's approach to safety inspections violated Japanese's citizens constitutionally protected human rights by not adequately taking risks into account (Asahi Shinbun 2016). In June 2016, the Ōtsu District Court rejected an appeal by KEPCO asking the court to lift the injunction (The Japan Times 2016). After KEPCO filed a lawsuit to the Osaka High Court challenging the Ōtsu District Court's ruling, the earlier ruling was overturned.

There are several lawsuits against the Ikata nuclear plant, brought before the Hiroshima and the Matsuyama District Court. The filing of these lawsuits took place on the grounds of insufficient basic earthquake ground motion figures used for reactor's anti-seismic design. While the Hiroshima District Court rejected a request for an injunction of reactor 3 at the Ikata nuclear power plant (Asahi Shinbun 2017), two more lawsuits seeking a permanent halt of all three reactors were still pending.

Most lawsuits filed against nuclear power plants were rooted in concerns about safety measures addressing natural hazards, particularly earthquakes. Plaintiffs criticised risk assessments that used low earthquake ground motion figures as insufficient due to their exclusion of possible earthquake sources. Regardless of the outcome, these time-consuming lawsuits stalled reactor restarts, causing electricity utilities to lose valuable time to generate revenues in order to offset the investments necessary to pass the NRA safety review in the first place. Whether or not these lawsuits ultimately lead to a shutdown, they certainly stall restarts and add additional costs for electricity utilities already burdened with high refurbishing costs.

Conclusions

The aim of this chapter was to understand why the process of restarting power plants was not progressing in accordance with nuclear power targets. To that end, first, it introduced the NRA as an independent safety agency and, second, took a closer look at revised nuclear safety standards and their implementation during reactor safety reviews. Finally, it illuminated the response of the nuclear industry and the Japanese public to the unexpected emergence of an independent regulatory agency.

To begin with, the NRA as an independent regulatory agency was able to ward off pressure from others, such as the Abe government. Looking at safety standards the NRA adopted in 2013, they were wider in scope, covering not only reactor design basis but also accident countermeasures. More importantly, NRA safety regulations were binding for the electricity utilities. The introduction of a back-fit system prevented situations where electricity utilities failed to follow recommendations for better safety measures by the NRA's predecessor. These two factors alone, binding regulation and the capacity to force utilities to back-fit, amounted to a revolution in nuclear safety regulation in Japan.

Nonetheless, independent safety regulation did not necessarily translate into the strictest possible safety standards. For example, the 2013 safety requirements failed to address new threats, such as cyber security. In its risk assessments, too, the NRA was not as strict as it could be. In areas of scientific debate, such as which geographic fault lines will affect a given nuclear power plant or how likely a future volcanic eruption will be, decisions fell on the more conservative side. The NRA forced utilities to include the scientifically undisputed sources of an earthquake, but not all theoretically possible ones. On the other hand, in the face of an undeniable risk, such as an active fault line running underneath a reactor, and apparent mismanagement, as in the case of Monju, the NRA took decisive action by refusing an operating licence. Thus, there was decisive regulatory action. At the same time, there was room for improvement regarding new threats and, in areas of scientific debate, it turned out to be less strict than it could have been.

The NRA enforcing new safety standards led to a massive increase in the additional safety investments electric utilities must make. Thus far, electric utilities have not moved to refurbish enough nuclear reactors to reach the Abe government's current energy policy goals of generating 20–22% of electricity from nuclear power plants by 2030. Rather, the reactors currently undergoing refurbishments and safety reviews will amount to a maximum of 15% in Japan's 2030 electricity mix. This presents an opportunity for renewable energy development in Japan. If Japan wants to achieve the climate change mitigation targets it has set for itself, it is crucial that the gap in nuclear power generation is replaced with renewable energy as opposed to fossil fuels.

While the NRA's safety standards were strict enough to render some nuclear reactors economically unviable, they were not strict enough to convince the majority of Japanese people that nuclear reactors should be restarted. In particular, the NRA not taking the toughest stance possible in scientific grey areas leaves part of the public unsatisfied with safety assessments. A closer look at lawsuits filed by citizens revealed that a common reason for legal action has been the NRA's relative lenience regarding the risk posed by earthquakes and volcanoes. Lawsuits challenging the NRA's safety assessments can add to the already high additional costs of nuclear safety by prolonging the restart process. Even if only successful in a first instance, they shorten the remaining operating time, and thereby possible revenues to be earned by electric utilities.

In conclusion, three aspects coming together explains why restarting power plants is not progressing in accordance with nuclear power targets. First, the NRA as an independent regulatory agency has defied industry influence and made independent decisions. Second, increased technical refurbishing costs mean that the maximum feasible share of nuclear power by 2030 will be about 15%. Third, the NRA had yet to regain public trust in nuclear safety as evident from the lawsuits against its restart decision. These lawsuits have the potential to lower the share of feasible nuclear power generation even further.

The findings here point to three areas where further research seems in order. These are the role of scientific debates in safety regulation, how the addition of the NRA as an independent safety regulator has altered the political economy of nuclear power in Japan, and how remaining public safety concerns can be an additional hurdle for nuclear reactor restarts.

Notes

1. This chapter avoids the term "Fukushima nuclear accident" due to the negative image this creates for the prefecture Fukushima. Rather this chapter speaks of "March 2011 nuclear accident" or simply "3/11".
2. Public demonstrations began in April 2011 with about 15,000 people demonstrating in Koenji, Tokyo, and reached their zenith on 19 September 2011 with 60,000 people gathering in Meiji Park, Tokyo, according to the organisers. Protests targeted the government and TEPCO for the way they handled the crisis response and for not preventing a severe nuclear accident in the first place.
3. A poll conducted by the *Yomiuri Shimbun*, Japan's daily newspaper with the widest circulation, from April 2011 (before the full scale of the nuclear disaster had become apparent to the wider public) through November 2011, showed a shift in public opinion towards a more critical stance on nuclear power. In the poll, participants were asked about the future role of nuclear energy for Japan's energy supply. They were given four possible answers: (1) increase, (2) maintain, (3) decrease, and (4) phase out. In April, 10% favoured an increase of nuclear power, a number that went

down to 1% by November. The number of people wishing to maintain the ratio of nuclear energy dropped from 46 to 23%. Whereas only 29% favoured decreasing the role of nuclear power at first, the number went up to 51% in November. Finally, the share of those supporting a nuclear phase out increased from 12% in April to 22% in November.

4. The term nuclear village was coined by Iida Tetsunari in the late 1990s. He left the nuclear industry to become an advocate of renewable energy development. He is the executive director of the Institute for Sustainable Energy Policies (ISEP) in Japan.
5. For the full analysis of the NRA's legal framework and how it established the foundation for organisational autonomy, see Koppenborg (forthcoming).
6. The official name is '原子力規制庁' (Genshiryoku-kiseichō). The term 'secretariat' was introduced during the reform process in 2011 and 2012. It has survived even though the current secretariat is far larger than originally anticipated.
7. An exception to this rule is meetings and documents discussing the safeguarding of nuclear materials and other safety-related aspects.
8. The results presented here are part of a detailed analysis the author conducted for a book manuscript (in progress) on "Japan's Nuclear Disaster and the Politics of Safety Governance".

References

ANRE. 2018. Japan's Energy 2017 Edition: 20 Questions to Understand the Current Energy Situation. Available at: http://www.enecho.meti.go.jp/en/category/brochures/pdf/japan_energy_2017.pdf. Accessed 25 Sept 2019.

Asahi Shinbun. 2014. Kiseii no dokuritsu ayabumu koe. *Asahi Shinbun* (morning edition), June 11, p. 3.

Asahi Shinbun. 2016. Takahama genpatsu san, yongoki, untensashitome Ōtsuchisai ga karishobun. *Asahi Shinbun* (online), March 9. Available at: http://www.asahi.com/articles/ASJ37454PJ37PTJB00C.html?iref=comtop_list_nat_n02. Accessed 23 Oct 2016.

Asahi Shinbun. 2017. District Court Rejects Request to Temporarily Halt Ikata Nuke Plant. *Asahi Shinbun* (online), March 30. Available at: http://www.asahi.com/ajw/articles/AJ201703300062.html. Accessed 28 Apr 2017.

Asahi Shinbun. 2019a. Yoronchōsa-shitsumon to kotae, Fukushimaken, 2 gatsu 23, 24 nichi jitsushi. *Asahi Shinbun* (online), February 27. Available at: www.asahi.com/articles/ASM2T4TW2M2TUZPS007.html. Accessed 29 Oct 2019.

Asahi Shinbun. 2019b. Genpatsu anzen taisakuhi, 5chō en chōni seifu no "saiyasu" hyōka yuragu. *Asahi Shinbun* (online), August 12. Available at:

www.asahi.com/articles/ASM7R6KNCM7RULBJ00S.html. Accessed 29 Oct 2019.

Cabinet Office Japan. 2013. Press Conference by Prime Minister Shinzo Abe: Friday, January 4, 2013. Available at: http://japan.kantei.go.jp/96_abe/statement/201301/04kaiken_e.html. Accessed 10 Jan 2020.

Cabinet Secretariat. 2011. Review on an Organization in Charge of Nuclear Safety Regulation. Available at: http://www.cas.go.jp/jp/genpatsujiko/info/shian_en_110805.html. Accessed 23 Oct 2016.

Cotton, Matthew. 2014. Structure, Agency and Post-Fukushima Nuclear Policy: An Alliance-Context-Actantiality Model of Political Change. *Journal of Risk Research* 18 (3): 1–16.

Gilardi, Fabrizio, and Martino Maggetti. 2011. The Independence of Regulatory Agencies. In *Handbook on the Politics of Regulation*, ed. D. Levi-Faur, 201–214. Cheltenhan, UK: Edward Elgar.

Hasegawa, Koichi. 2014. The Fukushima Nuclear Accident and Japan's Civil Society: Context, Reactions, and Policy Impacts. *International Sociology* 29 (4): 283–301. https://doi.org/10.1177/0268580914536413.

Hymans, Jacques E.C. 2015. After Fukushima: Veto Players and Japanese Nuclear Policy. In *Japan: The Precarious Future*, ed. F. Baldwin and A. Allison, 110–138. New York: NYU Press.

IAEA. 2011a. Fukushima Nuclear Accident Update Log: Updates of 12 April 2011. Available at: https://www.iaea.org/newscenter/news/fukushima-nuclear-accident-update-log-15. Accessed 23 Oct 2016.

IAEA. 2011b. Fukushima Nuclear Accident Update Log: Updates of 12 March 2011. Available at: https://www.iaea.org/newscenter/news/fukushima-nuclear-accident-update-log-14. Accessed 23 Oct 2016.

IAEA. 2011c. Fukushima Nuclear Accident Update Log: Updates of 18 March 2011. Available at: https://www.iaea.org/newscenter/news/fukushima-nuclear-accident-update-log-25. Accessed 23 Oct 2016.

Izadi-Najafabadi, Ali. 2015. Japan's Likely 2030 Energy Mix: More Gas and Solar. Available at: https://about.bnef.com/blog/japans-likely-2030-energy-mix-gas-solar/. Accessed 10 Oct 2020.

JAIF. 2018. Current Status of Nuclear Power Plants in Japan. Available at: https://www.jaif.or.jp/cms_admin/wp-content/uploads/2019/09/jp-npps-operation20190905_en.pdf. Accessed 26 Sept 2019.

Kan, Naoto. 2012. *Tōden Fukushima Daiichi genpatsu jiko: Sōridaijin toshite kangaeta koto* (Gentōsha Shinsho, 283). Tokyo: Gentōsha.

Kingston, Jeff. 2013. Japan's Nuclear Village: Power and Resilience. In *Critical Issues in Contemporary Japan*, ed. J. Kingston, 107–119. London: Routledge.

Koppenborg, Florentine. Forthcoming. Legal Fallout from the Fukushima Nuclear Accident: Fundamental Changes in Nuclear Safety Administration.

In *Fukushima and the Law*, ed. J. Weitzdörfer and K.C. Lauta. Cambridge: Cambridge University Press.

Koppenborg, Florentine. 2016. Japan's Nuclear Power Plants Don't Add Up. Available at: http://www.eastasiaforum.org/2016/03/03/japans-nuclear-power-plans-dont-add-up/. Accessed 10 Jan 2020.

Koppenborg, Florentine. 2017. Will the Silent Comeback of Coal Threaten Japan's Climate Goals? Available at: www.eastasiaforum.org/2017/04/06/will-the-silent-comeback-of-coal-threaten-japans-climate-goals/. Accessed 10 Jan 2020.

Ministry of Economics, Trade and Industry (METI). 2014. Strategic Energy Plan. Available at: http://www.enecho.meti.go.jp/en/category/others/basic_plan/pdf/4th_strategic_energy_plan.pdf. Accessed 23 Oct 2016.

METI. 2015. Long-Term Energy Supply and Demand Outlook. Available at: https://www.meti.go.jp/english/press/2015/pdf/0716_01a.pdf. Accessed 8 Oct 2018.

METI. 2018. Strategic Energy Plan. Available at: https://www.meti.go.jp/english/press/2018/pdf/0703_002c.pdf. Accessed 25 Sept 2019.

NAIIC. 2012. The Official Report of the Fukushima Nuclear Accident Independent Investigation Commission (NAIIC). Executive Summary. Available at: http://www.nirs.org/fukushima/naiic_report.pdf. Accessed 23 Oct 2016.

NRA. 2013a. Enforcement of the New Regulatory Requirements for Commercial Nuclear Power Reactors: July 8, 2013. Available at: https://www.nsr.go.jp/english/regulatory/. Accessed 23 Oct 2016.

NRA. 2013b. FY 2012 Annual Report. Available at: https://www.nsr.go.jp/data/000067053.pdf. Accessed 23 Oct 2016.

NRA. 2014a. FY 2013 Annual Report. Available at: https://www.nsr.go.jp/data/000067054.pdf. Accessed 23 Oct 2016.

NRA. 2014b. Nuclear Regulation Authority, Japan: Nuclear Regulation for People and Environment. Available at: http://www.nsr.go.jp/english/e_nra/nsr_leaflet_English.pdf. Accessed 23 Oct 2016.

NRA. 2014c. Genshiryoku-kisei inkai-kishakaikenroku Heisei 26 nen 7 gatsu 16 nichi. Available at: https://www.nsr.go.jp/data/000068796.pdf. Accessed 23 Oct 2016.

NRA Commission. 2012. Genshiryokukiseiinkai-kettei. Available at: http://www.nsr.go.jp/data/000047083.pdf. Accessed 23 Oct 2016.

NRA Commission. 2014. Tsuruga hatsudensho shikichinai hasaitai no chōsa nikansuru yū shikishakaigō Heisei 26 nen 12 gatsu 10 nichi. Available at: http://www.nsr.go.jp/data/000090805.pdf. Accessed 23 Oct 2016.

NRA Commission. 2015. Kankokubun: Recommendation Text. Available at: https://www.nsr.go.jp/disclosure/law/RAR/20151113monju.html. Accessed 23 Oct 2016.

Shibata, Tetsuji, and Hiroaki Tomokiyo. 2014. *Fukushima genpatsu jiko to yoronchōsa*. Tokyo: Energy Review Center (ERC).

Shiroyama, Hideaki. 2015. Nuclear Safety Regulation in Japan and Impacts of the Fukushima Daiichi Accident. In *Reflections on the Fukushima Daiichi Nuclear Accident: Toward Social-Scientific Literacy and Engineering Resilience*, ed. J. Ahn et al., 283–296. SpringerOpen: Cham.

Tateishi, Masata. 2015. Genpatsu no taishinsei anzen mondai to shinkisei kijun. In *Hyōryū suru genshiryoku to saikadō mondai: nihon kagakusha kaigi dai 35 kai genshiryoku hatsuden mondai zenkoku shinpojiumu (Kanazawa) yori*, ed. Nihon kagakusha kaigi genshiryoku mondai kenkyū iinkai, 95–104. Tokyo: Hon no izumisha.

The Japan Times. 2014a. Fukui Court Blocks Oi Nuclear Reactor Restart, in Landmark Ruling: Operations Halted Pending Verdict of Ongoing NRA Safety Probe. *The Japan Times* (online), May 21. Available at: http://www.japantimes.co.jp/news/2014/05/21/national/fukui-court-blocks-oi-nuclear-reactor-restart-landmark-ruling/. Accessed 23 Oct 2016.

The Japan Times. 2014b. Abe Picks for NRA 'Undermine' Nuclear Watchdog's Independence. *The Japan Times* (online), June 11. Available at: https://www.japantimes.co.jp/news/2014/06/11/national/abe-picks-nra-undermine-nuclear-watchdogs-independence/. Accessed 10 Oct 2020.

The Japan Times. 2014c. Volcano Near Sendai Nuclear Plant Is Shaking and May Erupt: Japan Weather Agency. *The Japan Times* (online edition), October 24. Available at: http://www.japantimes.co.jp/news/2014/10/24/national/volcano-near-sendai-nuclear-plant-shaking-may-erupt-japan-weather-agency/. Accessed 23 Oct 2016.

The Japan Times. 2015. Fukui court halts reactor restarts at Takahama. *The Japan Times* (morning edition), April 15, p. 1.

The Japan Times. 2016. Kepco Loses Challenge to Takahama Nuclear Injunction. *The Japan Times* (online), June 17. Available at: http://www.japantimes.co.jp/news/2016/06/17/national/crime-legal/kepco-fails-suspend-injunction-takahama-nuclear-plant/. Accessed 23 Oct 2016.

The Japan Times. 2017. Fukui Prefecture Oks Decommissioning of Monju Reactor. *The Japan Times*, June 7. Available at: www.japantimes.co.jp/news/2017/06/07/national/fukui-prefecture-oks-decommissioning-monju-reactor/WYRfKHrs44k. Accessed 8 Nov 2017.

The New York Times. 2011. Japan Extended Reactor's Life, Despite Warning. *The New York Times* (online), March 21. Available at: http://www.nytimes.com/2011/03/22/world/asia/22nuclear.html?pagewanted=all. Accessed 23 Oct 2016.

The Tokyo Shinbun. 2015a. Takahama saikadō mitomezu. *The Tokyo Shinbun* (morning edition), April 15, p. 1.

The Tokyo Shinbun. 2015b. Genpatsuanyenhi 2.3 chōenzō 13nen shinkijungo, yuragu keizaisei. *The Tōkyō Shimbun* (morning edition), May 17, p. 1.
Vivoda, Vlado, and Geordan Graetz. 2014. Nuclear Policy and Regulation in Japan After Fukushima: Navigating the Crisis. *Journal of Contemporary Asia* 45 (3): 1–20.
Wilson, James Q. 1980. The Politics of Regulation. In *The Politics of Regulation*, ed. J.Q. Wilson, 357–394. New York: Basic Books.

CHAPTER 5

The Politics of Nuclear Power Plant Restarts Versus Renewable Energy Promotion

Paul Midford

Introduction

This chapter examines the post 3-11 politics in Japan surrounding restarting nuclear power plants versus promoting renewable energy. This political debate has attracted international media attention and is often cast in terms of two irreconcilable and extreme options: go back to the pre-March 11, 2011 (3-11)[1] status quo of relying on, and even expanding the use of, nuclear power, or adopting an immediate zero nuclear strategy. The Abe administration's support for restarting some nuclear power plants, based on the safety authorizations given by the Nuclear Regulatory Authority (NRA), and Abe's personal pronuclear position (Koizumi 2018, pp. 105–106), are often assumed to mean that the Abe administration was hostile toward renewables and wants to return to the pre-3-11 goal of nuclear expansion (e.g., Aldrich 2016; Aldrich et al. 2019; Incerti

P. Midford (✉)
Norwegian University of Science and Technology (NTNU), Trondheim, Norway
e-mail: paul.midford@ntnu.no

P. Midford and E. Moe (eds.), *New Challenges and Solutions for Renewable Energy*, International Political Economy Series,
https://doi.org/10.1007/978-3-030-54514-7_5

and Lipscy 2018). This chapter asks whether this is an accurate assessment of the Abe administration's energy policy. It finds that the answer is no.

While the Abe administration supported restarts, it did not intervene politically to speed up the process of restarts, expand the relatively small number of nuclear reactors that are likely to qualify for restarts, nor did it authorize the building of new nuclear power plants. Rather, its policy legacy suggests that nuclear power will be largely phased out by the end of the 2050s. At the same time, the Abe administration continued the policies of its DPJ predecessors, especially by pursuing a road map laid out by Prime Minister Kan Naoto, and elaborated by Prime Minister Noda Yoshihiko. Abe followed the Kan-Noda road map by promoting renewables, both directly, and by investing in enabling infrastructure, including grid development, and developing the hydrogen economy, including as a medium for storing variable renewable energy (see Uriu's chapter in this volume). Finally, the Abe administration implemented another DPJ policy: electricity sector reform, arguably the leading example of structural economic reform in the third arrow of Abenomics. The weakening of the nuclear village by 3-11, METI's independence from these vested interests, and the pressure being exerted by public opinion largely explain these characteristics of the Abe administration's energy policies and offer insight into the energy policies that Abe's successor, prime minister Suga Yoshihide, is likely to pursue.

The rest of this chapter consists of eight sections. The next section outlines the Kan-Noda road map that emerged in the wake of 3-11 and the Fukushima Daiichi nuclear accident. The following section assesses the degree to which the Abe administration adhered to the Kan-Noda road map. The subsequent section analyzes public opinion regarding nuclear restarts. The section after that examines the establishment of a grid regulatory authority, and the temporary suspension of grid access for solar PV in several regions of Japan over several months in 2014–2015. The next section considers METI's responses, especially its storage subsidy system to aid renewable energy producers facing grid curtailment risks. The following section discusses the limited curtailment that occurred in Kyushu in late 2018 as Japan's first case. The section after that analyzes Japan's electricity retail market liberalization from April 2016 and the chance this provided to producers of renewable energy. The concluding section demonstrates that renewable energy thrived under the Abe administration and argues that the influence of public opinion and the strong-state perspective offer the best explanations.

The Kan-Noda Road Map

Before the Great East Japan earthquake, tsunami, and Fukushima Daiichi nuclear accident of March 11, 2011 (hereafter 3-11), Japan, under its 2010 *Energy Strategy* enacted by the DPJ, pursued a policy of expanding nuclear power generation from just under 30% of generation to 50% by 2030. This entailed building five new nuclear power reactors by 2020, and another nine by 2030 (METI 2010). The LDP and the DPJ both embraced this goal. Nonetheless, implementation of this policy was already largely deadlocked as rising public concerns about nuclear safety, concerns that had slowed the rapid expansion of nuclear power already from the late 1980s,[2] and brought expansion to a halt by the end of the 1990s, in the wake of the Tokaimura nuclear accident.[3]

Six weeks after 3-11 and the Fukushima Daiichi nuclear meltdowns, Prime Minister Kan Naoto scrapped the Strategic Energy Plan's goal of increasing nuclear power to over 50% of generation by 2030. Then, on July 13, just over four months after 3-11, Kan, at a press conference, issued a call for Japan to free itself from reliance on nuclear power. According to the prime minister, Japan "should...aim at a society where people can live without nuclear power plants." To achieve this non-nuclear future Kan called for reducing dependence on nuclear power "in a planned and gradual manner" while working "to actively secure new renewable energy sources" (Kantei 2011).[4]

A few weeks after this speech Kan succeeded in pushing through the Diet an ambitious feed-in tariff (FIT) to promote renewable power sources, including solar, wind, small-scale hydro, and geothermal generation of electricity. Shortly thereafter, in September, Kan resigned and turned over the prime minister's chair to fellow DPJ leader Noda Yoshihiko, leaving it to his successor to come up with the plan to denuclearize Japan. Noda largely supported his predecessor's position on nuclear power. He enacted Kan's proposal to strip METI of its authority over nuclear power plant safety regulation, placing this in the hands of a new Nuclear Regulatory Authority (NRA), and mandating that the NRA come up with new and more stringent safety standards for nuclear plants.

In October 2011, Noda instructed his cabinet's Energy and Environment Council to consider three options for Japan's energy mix by 2030, and to seek citizen input. In the first option, nuclear power would provide 20–25% of Japan's electricity, renewables would provide 20–30%, and fossil fuels the remainder. This option would require building additional nuclear reactors to supplement reactors retired due to age. The second option would see Japan generate approximately 15% of its energy from a reduced number of nuclear power plants, with no new builds, 30% from renewables, and the rest from fossil fuels. The third option would see Japan eliminate nuclear power, with renewables generating 35% of the country's electricity, and fossil fuels the rest (Oppenheim 2013, p. 95).[5]

Over several months, these options were debated nationally through several "town hall" meetings that solicited public input. Public input overwhelmingly favored the zero nuclear option, a conclusion supported by media polls. An *Asahi* poll taken in July 2012 showed that a plurality of 42% supported the zero option, 29% supported the 15% nuclear option, and 15% supported the 20–25% nuclear option. Another question in the same poll found that a total of 83% had great (29%) or moderate expectations (54%) about the potential of renewable energy such as wind and solar, versus 14% who had little (12%) or no (2%) expectations (*Asahi Shimbun* 2012a, b).

Following this debate, the Council presented its report to the Noda cabinet. The report set the goal of "realization of a society not dependent on nuclear power in earliest possible future," which was defined as the end of the 2030s (Pollitt et al. 2014, pp. 243–244). Zero reliance on nuclear power would mean replacing the approximately 29% of electricity provided by nuclear power before 3-11. In the meantime, some nuclear power plants were to be "restarted as an important energy source," which would generate up to 15% of electricity. Nuclear power was to be phased out as nuclear power plants reached their newly mandated legal limit of 40 years of operation, and that no new nuclear construction would be authorized. The report called for replacing nuclear power by the 2030s through "realization of a green energy revolution."[6] This entailed primarily developing the renewable energy sources of solar and wind power, with geothermal, biomass, and small-scale hydro serving as backup base-load sources. To realize this goal, the report called for "reform of the electricity power systems," through eliminating "monopolies and promote competition in the market, and separate electricity generation from transmission and distribution" (Energy and Environmental Council of the Government of Japan 2012, pp. 2, 4).

Although the cabinet did not formally adopt this plan, it pledged to use it when adopting future energy policies. This statement was similar to a prime minister's or Chief Cabinet Secretary's "*danwa*" or statement, in that it carried significant authority despite not being a formally adopted cabinet position (Kantei 2012).

Abe Follows the Kan-Noda Road Map

Two months after mandating nuclear phaseout, Noda dissolved the Diet and called new elections for the Lower House. The DPJ and Noda were both unpopular. This unpopularity did not stem from distance between the DPJ's policy positions and public opinion, but rather from public doubts about Noda and the DPJ's ability to implement and even formulate policies.[7] Contradictions in DPJ policies, including in energy policy, most notably their refusal to cancel the stalled fast-breeder reactor program even while simultaneously calling for gradual nuclear phaseout by the end of the 2030, damaged the DPJ's image, even in the eyes of nuclear power opponents.

In the December 2012 Lower House election, the LDP and its coalition partner Komeito won a huge victory. The LDP won almost as many seats as the DPJ had in 2009, in part because the LDP's opponents became divided as the DPJ found itself competing with new parties for the non-LDP vote, even in first-past-the-post single-member districts. Among the major political parties, the LDP was the only one not to promise phasing out nuclear power, whereas Komeito was the only party among those pledging to phase out nuclear power not specifying a date for doing so.

Although the LDP was the most pronuclear party, it noticeably did not criticize the Noda administration's policy of phasing out nuclear power by the end of the 2030s. Indeed, the LDP joined all other major political parties in its election manifesto by calling for "introducing renewable energy as much as possible." The LDP manifesto even contained a Kan-like statement when it advocated "creating economic and social structures that can be strong even without relying on nuclear power" (Jimintō 2012, pp. 23–24). In other words, in its election manifesto the LDP, under pressure from public opinion, was influenced by Kan's antinuclear declaration, and by the Noda administration's phaseout policy. Although the LDP did not endorse nuclear phaseout, it failed to present an alternative vision to the Kan-Noda road map.

Once in office Abe's coalition government acted largely along the lines laid out in the LDP's election manifesto and the coalition agreement it forged with Komeito. Even while the Abe administration interfered with the independence of Japan's ostensibly independent central bank (The Bank of Japan), it was careful to respect the independence of the NRA, even as the NRA devised stringent, time-consuming and costly new safety guidelines for nuclear power plants. Abe did not attempt to push for the speedy restart of nuclear power plants as the Noda administration had done with two reactors at Oi in Fukui prefecture in July 2012. Indeed, the Oi reactors remained the only reactors online from the time Abe took office in late December 2012 until they went off-line in fall 2013, and Japan again found itself in a state of zero nuclear power. This came despite Abe's enthusiasm at the beginning of 2013 for restarting all of Japan's nuclear reactors as soon as possible and desire to consider building new ones (Cabinet Office 2013). The Abe administration's caution about interfering with the NRA stemmed from strong public skepticism about nuclear power (see below) and the fear that this could lead to retrospective voting against Abe and the LDP so soon after they had clawed back into power from several historic election defeats.[8] This meant that Japan remained at zero nuclear power from September 2013 until August 2015. Even then, restarts were slow, with only nine reactors back on line by 2019 (and several subsequently ordered shutdown again by the NRA for failing to meet new counter-terrorism measures), and a large percentage of Japan's 48 operable reactors probably never to restart (see Koppenborg's chapter).

The Abe administration ended up following the Kan-Noda road map to break up the regional Electrical Power Companies (henceforth EPCOs), which were regional integrated electricity monopolies, with a plan to financially unbundle transmission grids from generation businesses. The Abe administration essentially picked up where the Noda administration had left off, finalizing a bill in spring 2013, inspired by Nordic models of electricity sector reform, consisting of three stages.

The first stage involved creating an Independent System Operator (ISO), established in 2015, to regulate electricity flows through all public grids and regulate access to those grids. The second entailed deregulating the consumer market by 2016, thereby allowing small-scale consumers to choose their electricity provider. The third stage, scheduled for 2020, called for unbundling power transmission from the current regional electricity monopolies.

The Abe administration succeeded in having the Diet enact this Kan-Noda road map inspired legislation in fall 2013. In June 2015, the Abe administration enacted two further bills specifying how liberalization of the retail market would occur in April 2016, and how to unbundle the EPCOs' generation businesses from grid ownership (*Japan Times* 2015). Specifically, this enacted legislation mandated the separation of grid ownership from ownership of generation assets through the creation of separate companies. However, the Abe administration did not require full divesture of the grid companies by the EPCOs, leading to criticism that the grid companies would favor the incumbent EPCOs and fail to actually provide grid parity to renewable energy generators (Hiranuma 2014, pp. 25–27).

At the same time, the Abe administration has continued to promote renewable energy. One way it has done so is by establishing a new and high-priced FIT for offshore wind power (36 Yen per KWh) which is producing large increases in installed wind power capacity, such as a 2 GW four-stage off-shore project in Akita prefecture (*Asahi Shimbun* 2020b; Matsubara 2018, pp. 23). The Abe administration is also promoting the development of the hydrogen economy, including using hydrogen as an energy carrier and storage medium (see Uriu's chapter for more on this) (METI 2014, p. 69), and smart grids, two key technologies necessary for expanding the use of variable renewable energy power sources such as solar and wind. Public support for the FIT made it harder for opponents of renewable energy within the Abe administration, the EPCOs, or the "nuclear village" more generally to push for any significant rollback.

Despite the high initial FIT rates, a poll conducted in March 2015 by *Yomiuri Shimbun* (a newspaper not generally friendly toward renewable energy) found that when respondents were told that the average extra cost resulting from the FIT per household in electricity rates averaged ¥225 (around USD $2 at 2020 rates) per month, 6% responded that they wanted this amount to go up, 64% that this amount was fine, and 29% that they wanted to see this amount reduced, and 1% who did not answer.[9] In short, at least 70% opposed reducing or eliminating the FIT (*Yomiuri Shimbun* 2015). This is broadly consistent with earlier polling by *Asahi* from June 2011, which asked respondents whether the share of renewable energy should be increased even if electricity prices increased as a result. In response, 65% favored increasing the share of renewable energy, versus 19% who opposed this (*Asahi Shimbun* 2011). Moreover, an overwhelming majority supported the gradual phaseout of nuclear power, as indicated by Fig. 5.1.

In 2014, the Abe administration adopted a new *Strategic Energy Plan*, replacing the 2010 *Strategic Energy Plan*. This plan abandoned both the 2010 Plan's call for increasing the use of nuclear power to 50% of electricity generation by 2030, and the Kan-Noda road map's plan to eliminate nuclear power by the end of the 2030s. Although the 2014

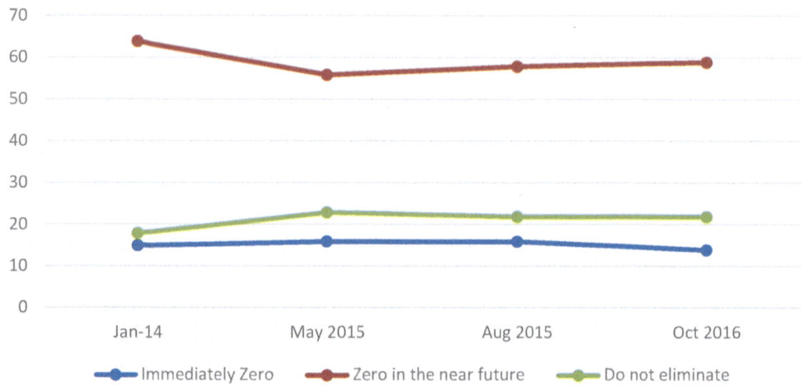

Fig. 5.1 *Asahi Shinbun*: The timing of nuclear phaseout? (*Source* Author created figure based on data from various issues of *Asahi Shinbun*)

Energy Plan was a formal repudiation of the Kan-Noda road map, in fact the plan reflected the influence of the Kan-Noda road map, and not only because it abandoned calls for building new nuclear power plants and expanding the use of nuclear power. It identified nuclear power as an important "base load" source of power, although this was only a slightly more descriptive concept than the Kan-Noda road map's description of nuclear power as an "important" power source. Moreover, Abe's Strategic Energy Plan called for reducing nuclear power "as much as possible," clearly echoing the language in the LDP's 2012 Lower House Election Manifesto. As such it effectively endorsed the policy of no new builds of nuclear reactors, meaning that nuclear power would continue to gradually phase out over time, a reality not explicitly mentioned by the 2014 Plan. In effect, the difference between the Kan-Noda road map and Abe's 2014 *Strategic Energy Plan* appeared to be about 15–20 years: phaseout by the end of the 2050s instead of the 2030s.

A year later, in 2015, METI issued its *Long-term Energy Supply and Demand Outlook*, a document that specified the best energy mix for 2030. Strikingly, this document claimed that renewables, in particular "geothermal, hydro and biomass, which can be operated stably despite weather conditions, are expected to replace nuclear power." However, no date for this expected replacement was stated. By 2030, this document foresaw nuclear power accounting for 20–22% of electricity generation, and renewable energy 22–24%. Within renewable energy, base-load sources hydropower, biomass, and geothermal were expected to provide 8.8–9.2%, 3.7–4.6%, and 1–1.1% of total electricity generation, respectively. Variable solar PV was expected to contribute 7% to total electricity generation, while wind power was expected to contribute 1.7% (METI 2015, pp. 7, 8). The 20–22% target for nuclear power in 2030 was not much more than the 15% share for nuclear power estimated by METI in 2012,[10] during the Noda administration (Shizen enerugi- kenkyūkai 2012, p. 90), suggesting a lack of drastic change in policy between the Noda and Abe administrations.

Japan's 2018 *Strategic Energy Plan*, its fifth, which updated the 2014 *Plan* and the 2015 *Outlook*, left the 2030 targets for renewable energy and nuclear energy unchanged, but had three notable policy changes. First, more focus was placed on upgrading grids, including for the first time discussing the role of smart grids, to promote the further integration of renewable energy, and second, there was more focus on further promoting the use of storage batteries. Finally, the 2018 *Plan* included the abandonment of the Monju fast breeder reactor (METI 2018, pp. 65, 92, 124). Overall, however, the 2018 *Strategic Energy Plan* mostly represented continuity with the 2018 Plan (which some critics claimed did not reflect rapid improvements in renewables technology), with continued promotion of renewable energy plus support for restarting some nuclear power plants (*Asahi Shimbun* 2018a).

Public Opinion on Nuclear Restarts

Moreover, the Abe administration followed the Kan-Noda road map policy of non-interference with the new stringent safety regulations put in place by the NRA, regulations that were expensive to implement, with recent estimates suggest the industry is spending as much as US$123 billion (Kyodo News 2020a), and which involve a long multi-year and uncertain path from application to restart. From the time Japan went zero nuclear in September 2013, it took 23 months until the first reactor, at the Sendai plant in Kyushu, was restarted in August 2015, ending Japan's zero nuclear state. A second reactor at the Sendai plant was restarted in November 2015. A third reactor, at Takahama in Fukui, was restarted in January 2016, but when the fourth reactor, also at Takahama was restarted in February, it suffered a system failure during start up, prompting an emergency shutdown. In March 2016, a judge ordered both Takahama reactors shut down for not meeting NRA earthquake safety standards (Johnson 2016). In 2020, four reactivated reactors in Kyushu Fukui were ordered shut down by the NRA for failing to meet new standards for withstanding terror attacks (Kyodo 2020). These developments have dealt major blows to efforts to restart Japan's aging nuclear power plants, and indicate that restarts will continue to move ahead only at a glacial pace. Indeed, with most boiler-type reactors in eastern Japan unlikely to meet NRA safety standards, observers believe that only around 20 of Japan's 48 intact reactors are likely to be restarted. Consequently, METI's goal of relying on nuclear power for 20–22% of nuclear power in 2030 appears unrealistically high; the Noda administration's target of 15% appears closer to what is achievable (see the Koppenborg chapter in this volume).

Behind this glacial pace of restarting nuclear reactors were not only the NRA's stringent safety standards, but strong public wariness about political interference in the process of restarts. Poll results from *Asahi Shimbun* and *Shimbun*, which respectively support and oppose nuclear power in their editorial pages, show that a stable public opinion majority opposes restarts, as depicted in Fig. 5.2.[11]

An *Asahi Shimbun* poll from July 2014 asked respondents whether they thought there would be negative economic consequences from not restarting nuclear power plants. In answer the public was evenly divided: 43% answered not restarting nuclear power plants would have no negative economic consequences, versus 42% who answered that not restarting reactors would have negative consequences. Another question in the same poll asked whether nuclear power could be made safe with proper management and technology, or whether due to human error nuclear power could never be made safe, 25% answered that nuclear power could be made safe, versus 63% who answered that it could never be made safe (*Asahi Shimbun* 2014). Clearly, a far larger portion of the public worried about the safety of nuclear power than worried about the economic consequences of not restarting nuclear power plants.

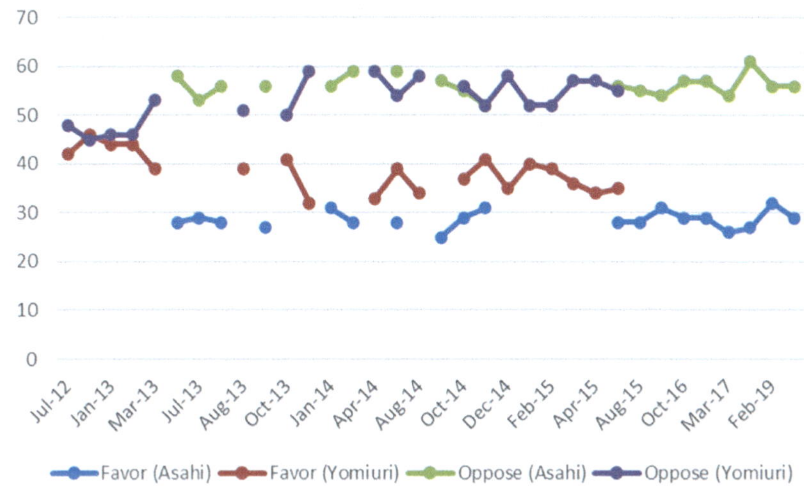

Fig. 5.2 Asahi Shimbun and Yomiuri Shimbun polls on nuclear restarts (*Source* Author created figure based on data from various issues of *Asahi Shinbun* and *Yomiuri Shimbun*)

Establishing a Grid Regulator and Grid Access Suspension

Implementation of electricity sector reform advanced in 2015 with the creation of the Organization for Cross-regional Coordination of Transmission Operators (OCCTO) in April 2015 (OCCTO 2019). OCCTO has a mandate to both ensure equal access to the grid for non-EPCO power suppliers and to ensure grid stability even as the share of variable solar and wind power increase. This includes the power to order the EPCOs, who own regional grids (directly until 2020, indirectly thereafter), to build greater grid capacity, including tie-lines to other EPCO areas outside their service regions, to increase the capacity of the grid to absorb more variable renewable energy. The OCCTO also developed an online system to allow retail electricity customers to switch electricity suppliers quickly and easily in support of the April 2016 retail liberalization (Yamazaki 2014, slides 11, 16).

Even before the OCCTO was launched the issue of grid access emerged as a key bottleneck for the promotion of renewable energy in late summer 2014, barely two years after the FIT had been launched. Five

EPCOs, Hokkaido, Tohoku, Shikoku, Kyushu, and Okinawa, suspended their acceptance of new grid connection applications from mega solar projects. The feed-in tariff that Kan had pushed through the Diet in summer 2011 came into effect in July 2012, and lead to spectacular growth in installed capacity of PV solar, both residential and mega solar, with an especially large amount of new capacity coming online in mid-2014. As of September 2014, projects totaling 72 GW of solar electricity generating capacity had been approved by METI, a capacity far exceeding that of Japan's surviving 48 nuclear reactors. As of June 2014, 11 GW of this total had already begun generating electricity, or about 15% of the approved total (Watanabe 2014). Despite the suspensions of new grid connections by these five EPCOs, PV solar continued to grow rapidly: by the end of 2014 installed PV capacity surpassed 20 GW, and approved projects exceeded 82 GW (Yabe et al. 2015, slides 3–4).[12]

By the time of these suspensions Japan's EPCOs had yet to implement any curtailment, compensated or uncompensated. Indeed, a comparison with China finds that with comparable rates of renewable energy penetration, Japan by 2014 had zero curtailment, while China in 2012 and 2013 had curtailment ratios ranging between 11 and 17% (Yasuda et al. 2015, p. 3).[13] An analysis of the EPCOs carrying capacity suggested that none of them had reached, or were even very close to reaching, their carrying capacity for intermittent PV solar power. However, three of the five that had suspended new grid hookups, Hokkaido, Shikoku, and Kyushu, exceeded 50% of their carrying capacity for connected PV solar plants. Nevertheless, in terms of approved, but not yet built or connected projects, these six EPCOs faced the prospect of more PV capacity than they could handle: just over 100% in Hokkaido, nearly 250% in Tohoku, just over 100% in Hokuriku, nearly 150% in Shikoku, over 200% in Kyushu, and over 125% in Okinawa (Kankyō Bijinesu 2016, p. 32, Figure 1). Likewise, a 2013 study by the New Energy and Industrial Technology Development Organization (NEDO) found that approved PV solar capacity in Hokkaido, Tohoku, and Kyushu exceeded off-peak electricity demand, with approved capacity almost double off-peak demand in Kyushu (Yabe et al. 2015). Although Japan's EPCOs had a looming capacity problem, the suspension of grid hookups by five EPCOs was arguably premature and was seen by industry critics as designed by the EPCOs to limit the spread of solar PV by rival generating companies, and perhaps even to protect their investments in nuclear power (*Japan Times* 2014).

In the end, this suspension served to modestly worsen grid access conditions for solar generators, albeit with significant regional variation.[14] METI did not oppose the suspensions and formulated modifications to the grid connection rules for new grid connections that went into effect in early 2015. The preexisting rules allowing EPCOs to implement uncompensated grid curtailment up to 30 days a year for PV generators with over 500 KW of capacity were expanded to include facilities between 10 and 500 KW of capacity, with permissible cut off times shortened from a per day to a per hour basis, but only up to 360 hours per year (or 15 days) before the EPCOs must compensate generators. However, specially designated EPCOs were given the right to unlimited curtailment of new generation capacity without compensation if their grid networks become overwhelmed by an influx of variable renewable electricity exceeding their capacity. METI designated the same five EPCOs that had originally suspended grid connection applications: Hokkaido, Tohoku, Shikoku, Kyushu, and Okinawa, plus Hokuriku and Chugoku. However, notably, the largest EPCOs, TEPCO, KEPCO, and Chubu EPCO (covering the Nagoya region) were not included and hence continued to operate under relatively favorable connection rules for renewable energy generators. Overall qualifications for applying for grid connection were also tightened. Nonetheless, these new conditions were not especially onerous and resolved the suspension of applications by the five EPCOs, suggesting a basic continuity of policy (Publicover 2016).

METI's Storage Subsidy Response to Growing Curtailment Risks

There is reason to suspect that the EPCOs' vested interests in the form of sunk costs in nuclear power and even fossil fuel generating facilities, and a more general desire to stifle competing generators, have been underlying motivations in pressing to limit grid connections by solar PV and other renewable energy producers, to underinvest in grid improvements needed to accommodate more renewable energy, and to prefer restarting their nuclear powers plant and curtailing solar PV supply to the grid rather than relying more on renewable energy. Nonetheless, while this book's vested-interest hypothesis may thus offer a good explanation for the behavior of the EPCOs, it does not offer a good explanation of METI's behavior, or that of the OCCTO, both of which have been pursuing policies that expand grid capacity to connect more renewable energy and setting and raising targets for how much renewable energy each EPCO can safely accept as their grid capacities improve. METI has been studying the EU's "Connect and Manage" system for expanding the amount of renewable energy connected to the grid through conditional access and the flexible use of spare capacity (METI 2018, p. 53). Rather the policies of METI and the OCCTO appear to be better explained by the strong-state hypothesis that sees government institutions in strong states as having bureaucratic capacity and sufficient distance from societal interests to make policy for the larger good of the country (see the Introduction chapter). Their behavior also arguably reflects the Abe administration's renewable energy policy that is more supportive and follows the Kan-Noda road map more than many observers recognize, and the influence of public opinion, which supports the expansion of renewable energy while remaining very skeptical of nuclear power.

One example of METI supporting renewable energy was its introduction of the "Emergency Subsidy for Suspension of Renewable Energy Connection (storage battery introduction support system for renewable energy producers)." Beyond the FIT, this program, launched shortly after the five EPCOs' grid application suspension, provided renewable energy producers with a subsidy to install megawatt-sized storage batteries, allowing mega-solar facilities to generate and store much of their power during the day and sell to the EPCOs at night and other times when solar generation is less and demand is higher (Kaneko 2017). Such batteries not only largely eliminate the problem of variable production, they even allow solar PV producers to earn additional revenue selling balancing services to the grid, as TESLA's megawatt battery storage system does in Australia (Deign 2018). For fiscal year 2015, METI budgeted 74.40 billion Yen for emergency responses to the suspension of grid connections for renewable energy, of which 26.5 billion Yen (approximately USD 240 million at 2020 rates) was allocated to this battery subsidy program. A total of 6.25 billion (approximately USD 58 million) was used between 2015 and 2017 to fund 137 MWh of battery storage capacity for 230 MW of installed solar and wind generating capacity (SII 2017, slide 3).[15] These funded 16 large projects, 13 small power company projects, and 254 individual/personal firm projects, with an acceptance rate ranging from 68 to 93% (SII 2017, slide 18)

Among METI's "Emergency Subsidy for Suspension of Renewable Energy Connection" funded storage battery projects were thirteen large ones in the Tohoku, Hokkaido, and Kyushu EPCO service regions.

In Tohoku, METI funded lead–acid batteries for three wind power projects: two storage batteries with a combined capacity of 24.4 MWh for a 36.1 MW capacity wind farm in Noshiro City, Akita prefecture; a 17.3 MWh battery for a 25.3 MW capacity wind farm in Ichinohe town, Iwate prefecture; and a 20.7 MWh battery for a 32.2 MW capacity wind farm in Rokkasho[16] and Yokohama villages in Aomori prefecture. METI subsidized lithium-ion batteries for seven solar PV projects in Hokkaido: a 13.8 MWh battery for a 28 MW solar PV plant in Chitose, that included funding from a large Korean electric power company; a 8.3 MWh battery for a 25 MW solar PV plant in Tomakomai; a 9 MWh battery for a 17 MW solar PV plant in Shinhidaka town; a 2.2 MWh battery for a 4 MW solar PV plant in Obihiro city; a 6.75 MWh battery for a 14.5 MW solar PV plant in Kushiro town; a 8.3 MWh battery for a 20 MW solar PV project in Akkeshi town; and a 7.2 MW battery for a 17.5 MW solar PV project in Shiriuchi town. In Kyushu, METI funded four battery projects for solar PV. It funded a 1.2 MWh lithium-ion battery for a 0.5 MW solar PV facility (part of a larger 1.99 MWh plant) in Kumamoto city, Kumamoto prefecture; a 7.1 MWh lead–acid battery for a 1.1 MWh solar PV plant in Miyazaki city, Miyazaki prefecture; and a 1 MWh lithium-ion battery for a 1.75 MWh solar PV plant in Amagi town on the remote island of Tokunoshima, Kagoshima prefecture (SII 2017, slides 23–36).

Perhaps the most ambitious METI storage battery grant under its emergency subsidy system was for a 1.2 MW solar PV plant that was built in Isa city, Kagoshima prefecture, Kyushu. The developer received a 240-million Yen subsidy from METI (approximately USD 2.23 million) that covered more than one-third of the total construction cost for the entire project of approximately 700 million Yen (USD 6.5 million). This subsidy was used toward the purchase of a 6.5 MWh lithium-ion battery from South Korea (Kaneko 2017), allowing nearly 6 hours of stored generation for release at night. In addition to receiving the subsidy, the decision to add a battery and use a DC connection allowed the developer to negotiate better connection terms with Kyushu EPCO and reduced land rent terms from Isa city (Kaneko 2017).

Although the Abe administration's energy policy has been dismissed as little more than industrial policy masquerading as an environmental policy (Incerti and Lipscy 2018), it is striking that this subsidy system lacked "buy Japanese" requirements, even though lithium-ion batteries are a strategic industry with major Japanese manufacturers such as Panasonic. The Isa-city project used an imported battery. Overall, 56% of the projects funded through this subsidy purchased imported batteries, which represented 42% of total added storage capacity. Foreign manufacturers supplied 76% of installed lithium-ion battery capacity, although only 10% of lead–acid battery capacity. In at least four METI subsidized projects, the main investors were from outside Japan (SII 2017, slides 15–17).[17]

While there was some slowdown in solar PV growth from the breakneck pace in 2014, healthy growth in solar PV connections continued through the rest of the decade. On the other hand, in sunny southern Kyushu the new rules, along with falling FIT rates for newly connected solar PV projects, may have contributed to developer decisions to delay or scrap up to 1.4 GW of new solar PV capacity there (Publicover 2016). Nonetheless, even in Kyushu solar PV growth continued to be robust through 2018 (Yamashita 2018). Thanks in part to the large-scale introduction of storage batteries through METI's initial subsidy, and a nearly 50% fall in the levelized cost of such batteries (to USD 150/MWh) between 2018 and 2020 (Eckhouse 2020), increasingly large storage batteries have been used at more and more renewable energy plants. Recent notable cases include a 6.5 MW solar PV project in Akita prefecture with a 24.4 MWh battery, a 3.7 MWh battery for a solar project in Fukuoka, Kyushu, an 11 MWh battery for a solar project in Nagano, and a 21 MWh battery for a solar project in Hokkaido. In total, 507 MWh worth of storage batteries were deployed to back up renewable energy in Japan in 2019, although over half of this total, 267 MWh, were batteries for household systems (PV Tech 2020).

Limited Curtailment Begins in Kyushu

Kyushu has proven to be an especially problematic region, not only because of the rapid buildup of solar PV facilities there in view of the abundant sunshine on this southern most of Japan's four major islands (by August 2018 Kyushu had 8 GW of installed solar PV capacity), but also because Kyushu EPCO, the fifth largest EPCO overall, became the leading EPCO for restarting nuclear reactors. In late August 2018, it restarted its fourth reactor, by which time KEPCO was the only other company to have restarted a nuclear reactor. In mid-October, Kyushu EPCO initiated Japan's first curtailment of renewable energy, restricting up to 4.2 GW of capacity (but not another 3.8 GW of capacity, which included facilities operating under the older rules that made curtailment more difficult). 8.2 GW of grid connected solar PV capacity had been the maximum that a government panel had estimated Kyushu EPCO's grid could handle without curtailment (*Asahi Shimbun* 2018a; Tsukimori 2018). Kyushu EPCO curtailed solar PV four times in October, curtailing 1.3% of solar PV generated electricity, and in four more instances in November (including for the first-time wind power), for a total of 8 times in 2018 (Yamashita 2018).

Although METI found these curtailments to comply with its regulations, Kyushu EPCO received protests from citizens' groups who argued that renewable energy should have priority over nuclear energy (Tanaka 2018), and criticism from the governor of Saga prefecture, who demanded Kyushu EPCO expand their investment in storage batteries and grid improvements to facilitate greater use of renewable energy (Sugiura and Kuroda 2019). Despite these first cases of curtailment, which had been announced months in advance, solar PV continued to grow at a healthy clip in Kyushu, with 50 MW of new capacity being added on average per month (Yamashita 2018). ANRE responded to the curtailments by formulating a plan with OCCTO to expand grid capacity for sending excess solar PV electricity from Kyushu to Chugoku and other service regions (*Asahi Shimbun* 2018b).[18] One option that ANRE and METI have not so far pursued is to respond to proposals from South Korea for building an interconnector between Kyushu and South Korea. Beyond enhancing energy security in western Japan, this could create a significant export market for "excess" solar PV and wind power from Kyushu.

In March 2019, curtailment of solar PV in Kyushu reached 6.8%, and curtailment of wind reached 1.4% of total generation. A year later, during the COVID19 pandemic, and before the shutdown of two reactors, when electricity demand dropped significantly, curtailment reached 12.3% for solar and 12.7% for wind. By contrast, neighboring Shikoku, with no operating nuclear power plants, had no curtailment during March 2019 or 2020, and during solar peaks on several days in April and May solar PV there supplied more than 80% (up to 88%) of electricity demand (REI 2020).

Retail Electricity Market Liberalization

Another major electricity sector reform was implemented in April 2016 with the opening of Japan's retail market to competition. Over 200 companies registered in advance as electricity suppliers to compete against the incumbent EPCOs; by October 2017 over 450 new market participants were supplying 12% of the market (Movellan 2016; METI 2018, pp. 82–83). Although most market participants, including new entrants, have been competing based on lower prices (including bundling electricity service with other services, such as cell phone and cable TV) (Nagata 2016), some new entrants compete on the type of power they provide. Among these companies, all offer to maximize renewable electricity supply, none promise to maximize nuclear power (even in the future), a fact that appears to reflect a lack of demand for nuclear power. Electricity liberalization thus gives consumers the ability to directly influence Japan's future energy mix by choosing between fossil-and-nuclear-fueled generation versus renewable energy generation.

Polling conducted by METI in 2014 showed that 54.4% of respondents were positive about being able to choose their electricity generator. By contrast, only 8.7% were negative about this prospect, 27.1% were neutral. Of the 35.8% who opposed or were neutral, a lack of information, uncertainty, and worry about difficult procedures were the main reasons cited. By contrast, only 1% expressed satisfaction with the EPCO serving them, and only 9% expressed no dissatisfaction with their EPCO supplier as their primary reason (see Figs. 5.3 and 5.4).

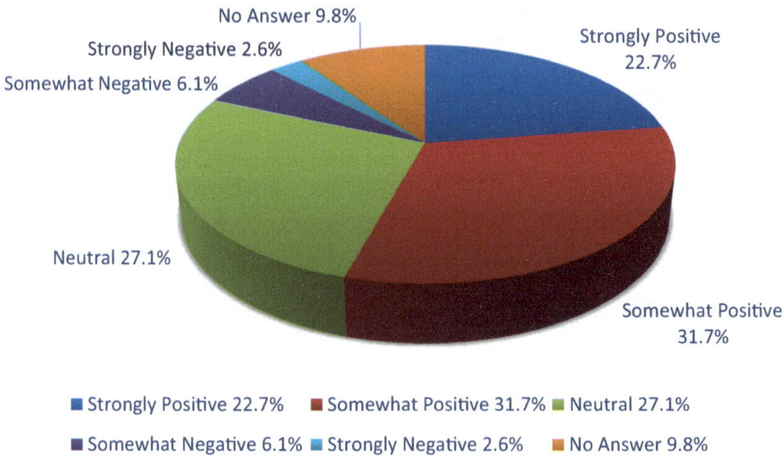

Fig. 5.3 Consumer attitudes about being able to choose an electricity supplier (*Source* Author created figure based on data from Yamazaki [2014])

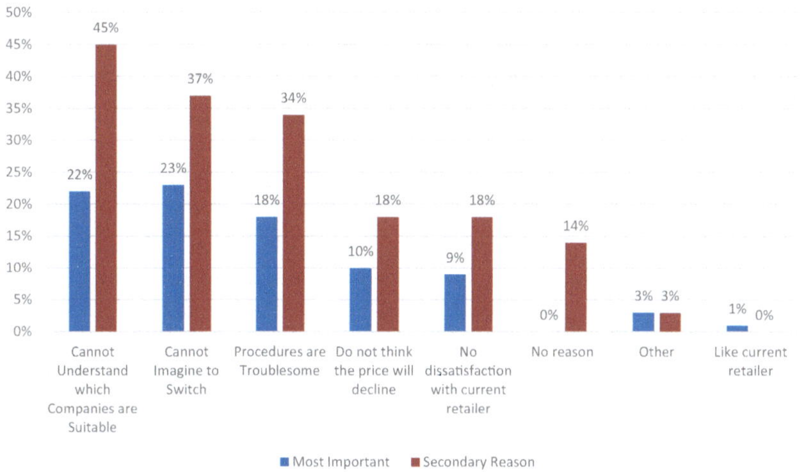

Fig. 5.4 Reasons respondents are negative or neutral about choosing their electricity supplier (*Note* Follow-up question asked of the 36% who answered: Neutral, Negative, or Strongly Negative in results depicted in Fig. 5.3. Respondents could only choose one reason as most important, but multiple secondary reasons. *Source* Author created figure based on data from Yamazaki [2014])

Nonetheless, an article by *Asahi Shimbun* bemoaned that as of the start of retail competition on April 1, 2016, only four suppliers not financially connected to the EPCOs would be able to supply electricity to customers with high-renewable content: Mito Denryoku, Loop, Mina Denryoku (see Yuki Takebuta's chapter in this volume about Mina Denryoku), and Miyama Smart Energy. In addition, members of the Izumi Shimin Coop in Sakai, Osaka, would have the option to subscribe to a renewable electricity plan (*Asahi Shimbun* 2016a, b).[19]

Several larger entrants into the electricity market, such as the cell phone service-provider Softbank, announced plans to offer renewable electricity plans by the beginning of retail competition but did not launch them in time (*Asahi Shimbun* 2016a, b). However, by 2018 Softbank was offering a renewable energy plan to consumers that was cost competitive with other plans and maximized the use of renewable energy through Softbank's national network of 34 solar PV plants, plus one affiliated wind farm and two affiliated solar PV facilities (Softbank 2019).

The large EPCOs initially did not offer renewable-energy-only options for consumers. A TEPCO employee interviewed by the author in March 2016 stated that the company did not offer consumers a renewable energy plan, and was not considering any such plan moving forward. A KEPCO employee was more positive: KEPCO did not offer such a plan, but was considering offering one in the future.[20] By 2020, the options for consumers had improved as four out of the ten EPCOs were offering retail renewable energy plans, including both KEPCO and TEPCO, along with Chubu EPCO (covering the Nagoya region), and somewhat unexpectedly the more rural Shikoku EPCO. On the other hand, Hokkaido, Tohoku, Hokuriku, Chugoku, Kyushu, and Okinawa EPCOs did not offer such plans. Chubu EPCO and TEPCO's plans offer consumers 100% hydro-generated electricity (TEPCO, n.d.; Chuden, n.d.). Shikoku EPCO distinguishes itself by offering the most diversified and transparent portfolio of renewable energy, with over 78% of its renewable energy being supplied by smaller-scale hydropower plants (under 30 MW), and small amounts being supplied by solar PV and biomass (under 0.5% each) (Yonden, n.d.), while KEPCO distinguishes itself by offering no transparency for the sources of its renewable energy plan on its Web site (Kansai Denryoku, n.d.).

Under the slogan "choose electricity change society," the Power Shift Campaign[21] is a small NGO helping consumers find retailers that meet five conditions. These are easy-to-understand information disclosure about the type of electricity and environmental footprint of the company, mainly procure electricity from renewable sources, no procurement of electricity from nuclear power or fossil fuels (except as a backup power source), emphasize regional and citizen sources of renewable energy, and have no financial connections with the EPCOs. This site identified only one company meeting these requirements that started selling power outside of a single region as of April 1, 2016: LOOP, with sales in the Tokyo, Nagoya, and Osaka regions. This company aimed to only sell renewable energy. However, in practice LOOP was procuring 6% of its energy from large-scale hydroelectric power, 20% from other renewable energy sources, mostly solar, covered by the FIT, and 74% from non-renewable sources through JEPEX (mostly conventional fossil-fuel sources). Four years later, LOOP was still obtaining only slightly more than a quarter of its electricity from renewable sources, but it was the only renewable-energy-promoting new retailer that was selling in every region of Japan except Okinawa (Pawa-shifuto 2020; Loop, n.d.).

Seikatsu Kurabu Eneji-, part of a nationwide consumer cooperative, began offering members renewable energy plans using solar, wind, and biomass in Tokyo from June 2016, and in Hokkaido, Tohoku, and the Nagoya and Osaka regions from October 2016. To date, it has not offered power sales in other regions (Seikatsu kurabu enaji-, n.d.). Another distinctive sub-category of new entrants was renewable-electricity-promoting retailers owned by municipal governments, with 31 such companies already established by mid-2017. Miyama Smart Energy, in Kyushu, is one example. It started selling power to residential customers from April 2016, and by September 2017, 20% of the city's residents were buying their electricity from this company, which was able to procure 30% of its electricity from renewable sources, compared with 12% for Kyushu EPCO (Bloomberg 2017). Another example is Izumisano Denryoku company in Osaka prefecture that was established by Izumisano city, and sells solar electricity from 2.4 MW of contracted capacity (Izumisano Denryoku 2019).

Retail electricity deregulation has produced significant competition, with many entrants, including EPCOs competing in each other's service areas, and price competition evident at least so far. The number of companies that initially offered consumers access to renewable power as a major power source remains limited, although less limited in number than in geographic scope and in the percentage of renewable energy in their power mix. During the first two months of liberalization, a total of 1,030,000 consumers switched electricity suppliers throughout Japan, but major urban centers were heavily overrepresented. Thus, while over 640,000 switched in the Tokyo region, and over 210,000 in the Osaka region, none switched in Okinawa, while in the regions of Hokuriku and Chugoku only 2300 and 2500 switched, respectively. On the fourth main island of Shikoku only 4200 switched. The paucity of new market entrants outside of large metropolitan regions was the most important reason, with companies offering renewable energy as a choice being even more scarce outside of big cities (*Asahi Shimbun* 2016b). By the end of 2016 2.6% of low-voltage retail consumers had switched suppliers, rising to 11.4% by the end of 2018. By March 2019, out of 88 million low-voltage customers, 10 million had switched suppliers (REI 2019b).

Renewable energy offerings to consumers have been expanding significantly since April 2016, albeit from a low base. By 2020, 26 renewable-energy-promoting retailers were selling power in every EPCO service region of Japan except Okinawa (it is perhaps no coincidence Okinawa

EPCO did not offer consumers a renewable energy option). Choices were limited in Hokkaido, with electricity only sold to members of two consumer cooperatives and by LOOP. In Hokuriku, LOOP was the only renewable choice, and in Shikoku only Minna Denryoku and LOOP were options, plus Shikoku EPCO, with two more renewable-energy promoting companies preparing to enter the market (Pawa-shifuto 2020; LOOP, n.d.).[22]

It remains to be seen how much influence consumers' "voting" for renewable energy (so far no companies are offering pure nuclear-power packages) will have on Japan's energy mix moving forward, but at least this choice (along with the choice of becoming a small-scale rooftop solar generator oneself, or a prosumer) has become available to retail consumers who own their own roof. Given that an overwhelming majority of Japanese favor nuclear phaseout, and that nearly as large majorities are willing to pay more for renewable energy, there is reason to expect that many more consumers will choose providers that offer renewable electricity plans if they are given reasonably competitive choices in their service region.

Conclusions: Resilient Renewables

The fundamental shift in energy policy following 3-11 that DPJ Prime Minister Kan launched with his abandonment of nuclear expansion and his zero nuclear declaration, which were fleshed by his successor through what this chapter calls the Kan-Noda road map, largely survived the shift in administration back to the LDP and an openly pronuclear prime minister.[23] As per the Kan-Noda road map, the Abe administration forced the EPCOs to accept major electricity sector reform and liberalization, unbundling their ownership of the grid, forcing them to buy electricity from hundreds of thousands of essentially rival electricity producers, including ordinary retail customers, ending their profitable monopoly on retail electricity sales that had accounted for approximately 90% of their profits (*Daily Yomiuri* 2013). Indeed, electricity sector reform was arguably the leading achievement of structural reform under Abenomics.

At the same time, nuclear restarts have been expensive and have proceeded at a glacial pace, even as installed solar PV capacity has grown dramatically since 2012. Although the Abe administration abandoned the Kan-Noda road map goal of phasing out nuclear power by the end of the 2030s, its actual policy trajectory suggests that phaseout will happen around fifteen years later. The Abe administration's de facto adherence, with some adjustments, to the Kan-Noda road map, reflected in large part strong public support for phasing out nuclear power. Although the public has some tolerance for restarts in the short run, an overwhelming majority supports phasing out nuclear power, a majority that precludes a return to nuclear expansion or even new reactor builds. At the same time, the Abe administration put in place the legal, regulatory, and technological infrastructure for renewables to gradually replace nuclear power as Japan's main domestic source of energy.

Despite pessimistic assessments of renewable energy development under the Abe administration, especially in relation to nuclear power, renewable energy thrived, with annual new capacity and generation outstripping restarted nuclear reactors. This was especially the case with solar PV: After over 9 GW of new capacity was added in 2014, growth in new capacity peaked at 10.8 GW in 2015,[24] and then declined to 8.6 GW in 2016, 7.5 GW in 2017, and 6.7 GW in 2018, before rebounding to 7.5 GW in 2019, with pre-COVID19 pandemic estimates of up to 8 GW of new capacity to be added in 2020. Cumulative installed solar PV capacity grew from 39 GW in 2016 to 63 GW in 2019, already exceeding the 7% target for solar PV (7.4% was achieved) set in METI's 2014 and 2018 energy strategies (PV Tech 2020; Matsubara 2018, p. 9; IEA 2019, p. 12; ISEP 2020). Japan climbed into the ranks of top-ten nations for solar PV capacity in 2013 with a fourth-place ranking, and since 2018 has had the second largest installed capacity behind only China (IRENA 2020).

Wind power has lagged the explosive growth of solar, with only 32% growth in installed capacity between 2012 and 2018. The amount of new capacity held up by Japan's rigorous environment impact assessments for wind power has reached five times the amount of total installed capacity, indicating that a spurt of new capacity is coming (REI 2019a).

Even with the reopening of some nuclear power plants solar PV generation alone outstrips nuclear power generation (ISEP 2020). In short, despite Abe's personal pronuclear proclivities, renewable energy, especially solar PV, continued growing robustly during his administration even while nuclear reactor restarts were fitful and far slower than what he had called for. Continued and new forms of policy support by METI for renewables have been a major cause for this robust growth in renewables, along with falling prices and rapid technological progress, especially in solar PV and storage-battery technologies.

In early 2020, the Abe administration introduced a new support scheme for renewable energy: the feed-in premium (or FIP), which provides payments to renewable energy producers if prices for their electricity fall below a set floor. The legislation also includes more funds for grid expansion and eases the rules for establishing local grids independent of the EPCO grids (*Asahi Shimbun* 2020a), which is important for some renewable energy producers, like Aizu Power in western Fukushima prefecture, who are trying to set up local grids.[25] The Abe administration's stimulus package in response to the COVID19 pandemic allocated 108 trillion Yen (almost USD 1 billion) to subsidize the development of onsite renewables in support of corporate Power Purchase Agreements (PPAs). This support was included under the rubric of "Development of Resilient Economic Structures" (Bhambhani 2020).

These new policies reflect a significant shift in how renewable energy was framed by the Abe administration after September 2019, when a powerful typhoon devastated Chiba prefecture just east of Tokyo, and left hundreds of thousands of customers there without electricity for days and even weeks, piercing TEPCO's (and the other EPCOs') track record of reliability (Kyodo News 2020b). This encouraged many Japanese households to purchase rooftop solar and storage battery sets to protect themselves against blackouts (PV Tech 2020). METI and the Abe administration responded by reframing renewable energy as a form of disaster preparedness and resilience, promoting distributed renewable energy generation and local grids to minimize electricity disruptions during natural disasters.

The findings of this chapter point to the influence of two variables identified in the introduction, namely the influence of a strong state and public opinion, and indicate that the influence of vested interests has been less than the literature and many observers have found. Abe's personal pronuclear opinions were less influential than often perceived. Clearly, the EPCOs have an interest in preserving their investments in nuclear power and restarting those assets while stopping retail market liberalization, the unbundling of their transmission business, and preventing rival renewable energy producers from having grid access. Other members of the so-called nuclear village have even clearer incentives to favor a return to nuclear-power expansion policies. Yet, EPCOs and these other actors have not been able to realize their preferences. 3-11 itself is a partial explanation, as it empowered METI to implement reforms it had long been advocating, as well as empowering advocates of renewable energy and opponents of nuclear power.

The vested interest factor is especially unpersuasive if it is understood as predicting no change, a position that probably does not even capture the EPCOs' long-term preferences. Rather, the real issue has not been a return to nuclear power or not, but how, and how slowly to proceed with nuclear and fossil fuel phaseout, and how, and how quickly, to ramp up renewable energy. Here, METI, if not the Abe administration, appears to have acted as a balancer among interest groups, which has helped them maintain significant independence from any interest group.[26]

The thrust of METI's policy appears to be the tried and true "convoy model," (Katz 2003, pp. 95–96) or what Richard Samuels calls "nurturance" (Samuels 1994, pp. 52–54), which is a policy of avoiding what METI has traditionally viewed as socially damaging "creative destruction" that produces bankruptcies, unemployment, and other social dislocations in favor of gradually managed change that facilitates the orderly movement of capital, labor, and other resources out of declining industries and into rising ones (Samuels 1994, pp. 53–54). In this sense, nuclear power may now be like Japan's domestic coal industry in the 1950s and 1960s, whose decline METI managed. More immediately, METI and the Abe administration have had an interest in preventing the sudden bankruptcies of most EPCOs, and the need to bail them out with taxpayer money (which already proved necessary with the largest of them, TEPCO, in the wake of the Fukushima Daiichi accident), which would almost certainly become necessary if Japan were to suddenly abandon nuclear power.

As it stands, many nuclear plants have been scrapped and others will probably never be restarted. A gradual phaseout of nuclear power from an already greatly diminished base over several decades while simultaneously making a transition to renewables as the main source of energy is the trajectory that Japan traveled during the Abe administration, and is continuing on under the Suga administration. Japan, specifically METI and OCCTO, is already devoting significant effort to tackling the second-phase challenges to the adoption of renewables that this book identifies, namely grid upgrading through smart grid and expansion investments, and investment in electricity storage. While Japan's promotion of a renewable energy transition and phaseout of nuclear power is too slow from a climate perspective, and from the perspective of nuclear opponents, it is nonetheless the direction Japan is heading.

Notes

1. The date of the Great East Japan Earthquake, tsunami, and the nuclear accident at the Fukushima Daiichi nuclear plant.
2. The Chernobyl accident in 1986 was the first accident that seriously challenged public trust in the safety of nuclear power. In response, Japanese regulators imposed new safety rules that required more frequent and longer reactor shutdowns for safety inspections, thereby reducing utilization rates of nuclear power plants.
3. Regarding the rise and fall of support for nuclear power before 2011 see Midford (2014), Shibata and Tomokiyo (2014).
4. 「原発に依存しない社会を目指すべき」. Kan claims he began thinking that Japan needed to abandon nuclear power a week after the Fukushima nuclear accident (Shizen enerugi- kenkyūkai 2012, pp. 68–69).
5. See Ohta's chapter for greater detail.
6. Some politicians in the ruling DPJ, notably former Prime Minister Kan, had a more ambitious plan. He formed a "Zero Nuclear Roadmap Study League" within the DPJ and came up with a proposal for phasing out nuclear power by 2025 (Shizen enerugi- kenkyūkai 2012, pp. 90–94).
7. This image for policy incompetence was created by Hatoyama Yukio, the DPJ's first prime minister, and Kan and Noda, despite their best efforts, were unable to rebuild the party's image (Midford 2013).
8. Abe allies, such as right-wing journalist Sakurai Yoshiko, have railed against the NRA's strict regulations as not being based on science and "erasing Japan's vitality." She also criticized "evil" solar PV for contributing to landslides on hillsides where they are deployed and falsely claiming that China and Germany are moving away from solar (Sakurai 2018, 2020). Her views are probably close to Abe's personal views, although not his policies.
9. By 2019, this amount had reached 767 Yen per average household per month (about USD 80 per year) (Mainichi 2019).
10. METI's actual 2015 estimate was 17–18% for nuclear power and 19–20% for renewables, but it raised the number for both based on assuming energy conservation (METI 2015, p. 8).
11. The figure shows essentially the same results regarding opposition to restarts, but the *Yomiuri* poll has a somewhat higher support rate for restarts than does the *Asahi* poll, although that higher rate is nonetheless a decided minority.
12. However, the distinction between installed capacity and electricity generation needs to be remembered. Solar produces almost no electricity at night, and at a significantly reduced level on a cloudy day. Depending on the average solar radiance for a particular location, and the type of PV facility (e.g., whether it has panels that rotate to follow the sun),

utilization rates can average as low as 12–13%. Thus, 70 GW of installed PV solar capacity will produce essentially this much electricity on a sunny day, but significantly less on a cloudy day, and no power at night, hence the need for storage and smart grids facilitating demand management. It is also important to consider the utilization rate of nuclear power. In 2014, the utilization rate for nuclear power in Japan was 0%, and by the end of 2015, this had increased to approximately 2%.

13. China in 2012 and 2013 had solar PV and wind penetration ratios (the percentage of electricity generated from these sources of 2.0 and 2.6%, respectively, and curtailment ratios of 17.12 and 10.74%, showing a high but falling curtailment rate even as renewable energy penetration increased. By comparison, in 2014 Hokkaido had a penetration ratio in of 2.7%, Tohoku 2.3%, and Kyushu of 1.8%, yet zero curtailment. These are three of the initial five EPCOs that suspended new grid connections. One note of caution regarding this comparison is that the figures for Japan are regional while those for China are nationwide, meaning that higher local penetration rates may be driving higher national as well as regional curtailment rates. Nonetheless, it seems unlikely that this difference alone explains the large variation. This study also includes future estimates of curtailment based on penetration, and only one scenario, Kyushu with a hypothesized penetration rate of 19.7%, produces an estimated curtailment rate that is comparable to the lower end of China's curtailment: 10.99% (Yasuda et al. 2015, p. 3).
14. For sensationalist accounts that exaggerate the change and its impact, see Sentaku (2014), and Sakakibara (2017).
15. This capacity was more than a third larger and earlier than the 100 MW storage battery Tesla completed in Australia in late 2017 that grabbed international headlines (Deign 2018).
16. Rokkasho is also known as the location for a nuclear-fuel reprocessing plant whose construction has stalled due to legal challenges.
17. Foreign manufacturers also supplied 20% of power conditioning systems (by capacity) for these projects (SII 2017).
18. Already in spring METI announced new grid-capacity rules, effective from September, expanding the amount of variable renewable energy that could be sent over the grid (Aida 2018).
19. Two independent companies not mentioned in the *Asahi* article are Kinki Denryoku and Aichi Denryoku. Kinki Denryoku only sold solar generated electricity from the Kansai (Kinki) region. Kinki Denryoku would actively try to procure other forms of renewable energy through JPEX (Japan Electricity Exchange), and aimed to achieve 100% reliance on renewable energy in the future. It boasted 20 generation facilities with a total capacity of 2.5 MW, all of them solar PV facilities. It was aiming for 10 MW of capacity by September 2016, and 200 MW by February 2018.

Aichi Denryoku aimed to only sell renewable energy. On the eve of retail liberalization, its generation was entirely based on solar, although it was exploring utilizing wind, hydro, geothermal, and biomass sources. Beginning in April 2016, more than 10% of its generation was to be supplied by solar. The largest of Aichi Denryoku's seven facilities was a 58 KW facility, the smallest was a 35 KW facility, and the average size was approximately 50 KW. Sources: brochures author obtained at Smart Energy Week on Electricity Market Liberalization, Tokyo Big Sight, March 4, 2016, and interviews with company representatives at the two companies' joint booth, and at Kinki Denryoku's headquarters, May 31, 2016. Since 2016 Kinki Denryoku has abandoned a focus on renewables in favor of focusing on price. www.kinki-epco.co.jp. Aichi Denyroku continues to promote renewables, but only sells to small- and medium-sized commercial customers, not yet to households. www.aichi-denryoku.jp.

20. Interview at Smart Energy Week on Electricity Market Liberalization, Tokyo Big Sight, March 4, 2016.
21. パワーシフト・キャンペーン.
22. For some reason, Power Shift Campaign no longer recognizes LOOP as meeting its criteria.
23. The LDP appears to be significantly less pronuclear than Abe personally, suggesting that the LDP may move further way from nuclear power under Suga or a subsequent LDP prime minister. For example, Kōno Tarō, who is one of the leading candidates to succeed Suga, is vehemently antinuclear and pro-renewables, while Koizumi Shinjirō, another leading candidate, is known to be skeptical of nuclear power, and his father, former Prime Minister Koizumi Jun'ichirō, has emerged as a leading nuclear opponent. Regarding the prospects for a decisive move away from nuclear power by the LDP after Abe, see Koizumi (2018, pp. 126–127).
24. By comparison, Germany's annual growth in solar capacity peaked at under 8 GW in 2013 (Matsubara 2019, p. 12).
25. Interview with the President of Aizu Denryoku, July 1, 2019.
26. While the range of vested interests is often described as big EPCOs versus small renewable energy producers (a description that understates how big renewable energy producers are becoming, and how many, like Softbank, are already large corporations), one important check on the EPCOs has been largely ignored: gas companies. Gas companies in major metropolitan areas, such as Tokyo Gas and Osaka Gas, were among the largest and fastest new entrants to the liberalized electricity market, and have been promoting renewable energy through Ene-farm systems that combine natural gas reforming to produce hydrogen with small-scale solar PV systems, allowing near if not complete independence from electricity grids (see Uriu's chapter for more on Ene-farms). The rivalry between gas and electric companies is a two-way street, and Japan appears distinctive

in that electricity retail market liberalization was quickly followed by retail gas market liberalization, allowing EPCOs to sell gas.

References

Aida, Satoshi. 2018. Sōdensen no 'aki yōryō' unyō minaoshi-sai ene dōnyū kakudaisaku. *Asahi Shimbun*, April 18 (chōkan), p. 9.

Aldrich, Daniel P. 2016. Anti-Nuclear Sentiment and Japan's Energy Choices. *Asia Unbound* (Council on Foreign Relations Blog), April 12, at http://blogs.cfr.org/asia/2016/04/12/anti-nuclear-sentiment-and-japans-energy-choices/. Accessed 12 Apr 2016.

Aldrich, Daniel, Phillip Y. Lipscy, and Mary M. McCarthy. 2019. Correspondence: Japan's Opportunity to Lead. *Nature Climate Change* 9 (July): 492.

Asahi Shimbun. 2011. Yoron chōsa – shitsumon to kaitō. June 13. http://www.asahi.com/national/update/0613/TKY201106130538.html. Accessed 25 November 2013.

Asahi Shimbun. 2012a. 2030 nen no genpatsu isondo "0%" shiji, saita 42%, Asahi Shimbunsha yoron chōsa. July 10 (chōkan), p. 5.

Asahi Shimbun. 2012b. Noda seiken, kiken suiki hairi. August 6 (chōkan), p. 3.

Asahi Shimbun. 2014. Asahi Shimbunsha yoron chōsa shitsumon to kaitō. July 29 (chōkan), p. 4.

Asahi Shimbun. 2016a. Saisei ene denryoku sentakushi wazuka. February 28 (chōkan), p. 1.

Asahi Shimbun. 2016b. Yasusa kyōsō ka sumu saisei ene. February 28 (chōkan), p. 35.

Asahi Shimbun. 2018a. Editorial: Commitment to Sustainable Energy Tested Like Never Before. October 17. www.asahi.com/ajw/articles/AJ201810170029.html. Accessed 18 Oct 2018.

Asahi Shimbun. 2018b. Shutsuryoku yokusei teigen he taisakuan, chiigai ikigai sōden saran ni. November 13 (chōkan, seibu), p. 8.

Asahi Shimbun. 2020a. Sai-ene jigyō ni shinshien. February 7 (morning edition), p. 7.

Asahi Shimbun. 2020b. Yōjō fūryoku akita ni tsuifuu? Kokunai saidai no hatsuden keikaku susumu. February 22 (chōkan), p. 4.

Bhambhani, Anu. 2020. Renewables Find Mention in COVID-19 Economic Stimulus Package of Japanese Government. *Taiyang News*, April 17. Accessed at: http://taiyangnews.info/markets/japan-corporate-re-ppas-in-covid-19-stimulus-package/.

Bloomberg. 2017. Utilities Put on Notice: Municipalities Looking to Sell Energy to Residents. *Japan Times*, September 15. https://www.japantimes.co.jp/

news/2017/09/15/business/utilities-put-notice-municipalities-looking-sell-energy-residents/#.Xt12TDr7TIU. Accessed 25 Nov 2017.

Cabinet Office Japan. 2013. *Press Conference by Prime Minister Shinzo Abe: Friday, January 4, 2013.* http://japan.kantei.go.jp/96_abe/statement/201301/04kaiken_e.html. Accessed 10 Jan 2020.

Chuden. n.d. CO2 Furi- menyu-. https://miraiz.chuden.co.jp/home/electric/menu/co2_free/index.html. Accessed Feb 2020.

Daily Yomiuri. 2013. 90% of TEPCO's Annual Average Profits Come from Households. May 24, p. 1.

Deign, Jason. 2018. Did Tesla's Big Australian Battery Kill the Business Case for More? *Greentech Media* (May 18). https://www.greentechmedia.com/articles/read/has-teslas-big-australian-battery-killed-the-business-case-for-more#gs.uml.AOfHlA8. Accessed 29 Feb 2020.

Eckhouse, Brian. 2020. Solar and Wind Cheapest Sources of Power in Most of the World. *Bloomberg Green*, April 28. https://www.bloomberg.com/news/articles/2020-04-28/solar-and-wind-cheapest-sources-of-power-in-most-of-the-world. Accessed 11 May 2020.

Energy and Environmental Council of the Government of Japan. 2012. Innovative Strategy for Energy and the Environment. September 14, Tokyo: Japan. http://japan.kantei.go.jp/noda/actions/201209/14kaigi_e.html. Accessed 3 Mar 2016.

Hiranuma, Hikaru. 2014. *Japan's Energy Policy in a Post 3/11 World: Juggling Safety, Sustainability and Economics*. Paris: IFRI Center for Asian Studies, Asie Visions 72 (September).

Incerti, Trevor, and Phillip Y. Lipscy. 2018. The Politics of Energy and Climate Change in Japan under Abe. *Asian Survey* 58 (4): 607–634.

IRENA (International Renewable Energy Agency). 2020. Country Rankings. https://www.irena.org/Statistics/View-Data-by-Topic/Capacity-and-Generation/Country-Rankings. Accessed 17 May 2020.

ISEP (Institute for Sustainable Energy Policies). 2020. *Share of Renewable Energy Electricity in Japan, 2019* (Preliminary Report). April 10. https://www.isep.or.jp/en/879/. Accessed 17 May 2020.

Izumisano Denryoku. 2019. Reiwa gan nendo jigyou keikaku. https://izumisano-pps.or.jp/wp-content/uploads/2019/12/a5405bc37746e7d94bd354004008e61f.pdf. Accessed 1 Mar 2020.

Japan Times. 2014. Planting a Green Grid. October 23. https://www.japantimes.co.jp/opinion/2014/10/23/editorials/planting-a-green-power-grid/#.XtamDm5uJhg. Accessed 25 Nov 2019.

Japan Times. 2015. Editorials: Electricity and Gas Liberalization. July 5. http://www.japantimes.co.jp/opinion/2015/07/05/editorials/electricityandgasliberalization/#.VZnXQPntmkp. Accessed 7 July 2015.

Jimintō. 2012. Jyuten seisaku 2012. pp. 23–24. http://www.jimin.jp/policy/manifest/. Accessed 3 Jan 2013.

Johnson, Eric. 2016. Takahama Injunction Delivers Body Blow to Japan's Nuclear Power Industry. *Japan Times*, March 10. http://www.japantimes.co.jp/news/2016/03/10/national/takahamainjunctiondeliversbodyblowtojapansnuclearpowerindustry/#. VuGiEfmLTIU. Accessed 29 Feb 2020.

Kaneko, Kenji. 2017. 1.2 MW Solar Plant Uses 6.5 MWh Battery to Sell Power at Night. *Nikkei Business Publications*, March 14. https://tech.nikkeibp.co.jp/dm/atclen/news_en/15mk/031401. Accessed 11 Feb 2019.

Kankyō Bijinesu. 201. PV shisutemu dōnyuu kosuto no teigen ha dōhanba. Smart Energy Week special issue, March 2016, p. 32.

Kantei. 2011. Press Conference by Prime Minister Naoto Kan. July 13, pp. 2, 3. www.kantei.go.jp/foreign/kan/statement/201107/13kaiken_e.html. Accessed 11 Apr 2016.

Kantei. 2012. The Prime Minister in Action: Energy and Environment Council. September 14. http://japan.kantei.go.jp/noda/actions/201209/14kaigi_e.html. Accessed 3 Mar 2016.

Katz, Richard. 2003. *Japanese Phoenix: The Long Road to Economic Revival.* Armonk, NY: M.E. Sharpe.

KEPCO (Kansai Denryoku) (n.d.). Sai ene ECO Puran (teiatsu) ni tsuite setsumei suru. https://kepco.jp/ryokin/menu/ecoplan/. Accessed 29 Feb 2020.

Koizumi, Jun'ichirō. 2018. *Genpatsu zero yareba dekiru*. Tokyo: Ōta Shuppan.

Kyodo. 2020. Kyushu Electric Halts Sendai Reactor Due to Delay in Terrorism Measures. *Japan Times*, March 16. https://www.japantimes.co.jp/news/2020/03/16/national/kyushu-electric-halts-sendai-reactor/#.XtZS3m5uJhg. Accessed 11 April 2020.

Kyodo News. 2020a. Costs for managing Japan's Nuclear Plants to Total 13 Trillion Yen. January 15. https://english.kyodonews.net/news/2020/01/8722fafaff9b-costs-for-managing-japans-nuclear-plants-to-total-13-trillion-yen.html. Accessed 29 Feb 2020.

Kyodo News. 2020b. Power, Water Outages Plague Typhoon Hit Chiba Pref. September 11. Accessed at: https://english.kyodonews.net/news/2019/09/309613laeede-450000-homes-still-without-power-in-typhoon-hit-chiba-pref.html.

LOOP. n.d. Loop denki kaitori sa-bisu. https://looop-denki.com/low-v/plan/. Accessed 1 Mar 2020.

Mainichi. 2019. Japan Gov't Reviewing Renewable Energy Support Policies Over Costs. June 14, 2019. https://mainichi.jp/english/articles/20190613/p2a/00m/0bu/018000c. Accessed 25 Nov 2019.

Matsubara, Hironao. 2018. *Renewable Energy Policies and the Energy Transition in Japan*. Tokyo: Fredrich Ebert Stiftung.

METI (Ministry of Economics, Trade and Industry). 2010. The Strategic Energy Plan of Japan: Meeting Global Challenges and Securing Energy Futures (Summary). June. http://warp.da.ndl.go.jp/info:ndljp/pid/3487098/www.meti.go.jp/english/press/data/pdf/20100618_08a.pdf. Accessed 29 Feb 2020.

METI. 2014. Strategic Energy Plan. April. http://www.enecho.meti.go.jp/en/category/others/basic_plan/pdf/4th_strategic_energy_plan.pdf.. Accessed 29 Feb 2020.

METI. 2015. Long Term Energy Supply and Demand Outlook. July. http://www.meti.go.jp/english/press/2015/0716_01.html. Accessed 29 Feb 2020.

METI. 2018. Strategic Energy Plan. July. http://www.enecho.meti.go.jp/en/category/others/basic_plan/5th/pdf/strategic_energy_plan.pdf. Accessed 29 Feb 2020.

Midford, Paul. 2013. Foreign Policy as an Election Issue. In *Japan Decides 2012*, ed. Robert Pekkanen, Steven Reed, and Ethan Scheiner, 179–194. New York: Palgrave.

Midford, Paul. 2014. The Impact of 3-11 on Japanese Public Opinion and Policy Toward Energy Security. In *The Political Economy of Renewable Energy and Energy Security: Challenges and National Responses in Japan, China and Europe*, ed. Espen Moe and Paul Midford, 57–96. Palgrave: London.

Movellan, Junko. 2016. Japan at the Electricity Crossroads: Nuclear Power to Lower Electricity Bills or Solar Power to Create Resiliency? *RenewableEnergy.com*, March 11. Accessed at: https://www.renewableenergyworld.com/2016/03/11/japan-at-the-electricity-crossroads-nuclear-power-to-lower-electricity-bills-or-solar-power-to-create-resiliency/#gref.

Nagata, Kazuaki. 2016. Japan's Electricity Shake-Up Gives Power to the People. *Japan Times*, March 31. https://www.japantimes.co.jp/news/2016/03/31/reference/japans-electricity-shake-gives-power-people/#.Xta-sG5uJhg. Accessed 2 Apr 2016.

OCCTO (Organization for Cross-regional Coordination of Transmission Operators). 2019. Role of the OCCTO in The Electricity System Reform. March 18. Accessed at https://www.occto.or.jp/en/about_occto/about_occto.html.

Oppenheim, Richard. 2013. Japan's Energy Crossroads: Nuclear, Renewables and the Quest for a New Energy Mix. In *After The Great East Japan Earthquake: Political and Policy Change in Post-Fukushima Japan*, ed. Dominic Al-Badri and Gijs Berends. Copenhagen: NIAS Press.

Pawa-shifuto. 2020. Denki wo erabe ha shakai ga kaeru. http://power-shift.org/. Accessed 1 Mar 2020.

Pollitt, Hector, Seung-Joon Park, Soocheol Lee, and Ueta Kazuhiro. 2014. An Economic and Environmental Assessment of Future Electricity Generation Mixes in Japan—An Assessment Using the E3MG Macro-Econometric Model. *Energy Policy* 67: 243–254.

Publicover, Brian. 2016. New Curtailment Rules Cast Pall Over Japanese Renewable Industry. *Recharge News* (September 28). https://www.rechargenews.com/solar/new-curtailment-rules-cast-pall-over-japanese-renewables-industry/1-1-870172. Accessed 29 Feb 2020.

PV Tech. 2020. Domestic Solar and Storage On The Rise As Japanese Market Bounces Back. https://www.pv-tech.org/editors-blog/domestic-solar-and-storage-on-the-rise-as-japanese-market-shifts-gear. Accessed 28 May 2020.

REI (Renewable Energy Institute). 2019a. Registered, Operational, and Nonoperative Capacity at the end of the Fiscal Year. October 8. Accessed at: https://www.renewable-ei.org/en/statistics/policies/.

REI. 2019b. Statistics: Electricity Market. October 8. https://www.renewable-ei.org/en/statistics/electricitymarket/. Accessed 29 Feb 2020.

REI. 2020. A Series of Disasters Shows How Renewable Energy Helps Us. June 2. https://www.renewable-ei.org/en/activities/reports/20200602.php. Accessed 3 June 2020.

Sakakibara, Ken. 2017. Taiyoko hatsuden utage no ato. *Nihon Keizai Shimbun*, May 13.

Sakurai, Yoshiko. 2018. Taiyoko hatsuden ashiki seido no wana. *Sankei Shimbun*, September 3. http://www.sankei.com/premium/news/180903/prm1809030007-n4.html. Accessed September 7, 2018.

Sakurai, Yoshiko. 2020. Saibankan no genpatsu tsubushi katsuryoku wo sogu. *Shūkan Shinchō*. January 30. https://yoshiko-sakurai.jp/2020/01/30/8496. Accessed 29 Feb 2020.

Samuels, Richard J. 1994. *"Rich Nation Strong Army": National Security and the Technological Transformation of Japan*. Ithaca, NY: Cornell University Press.

Seikatsu kurabu enaji-. n.d. Kyōkyū eria. https://scenergy.co.jp/about/supply. Accessed 29 Feb 2020.

Sentaku. 2014. A Losing Bet on Green Energy. *Japan Times*, November 25, p. 3.

Shibata, Tetsuji, and Hiroaki Tomokiyo. 2014. *Fukushima genpatsu jiko to kokumin yoron*. Tokyo: ERC shuppan.

SII (Sustainable open Innovation Initiative). 2017. Heisei 26 nendo hosei yosan saisei kanō enerugi- setsuzoku horyuu kinkyū taou hojyokin-jigyou hōkoku. March 31. https://sii.or.jp/re_energy26r/. Accessed 25 Nov 2019.

Shizen enerugi- kenkyūkai. 2012. *Kan Naoto no shizen enerugi-ron–kiwarareta sōri no oki miyage*. Tokyo: Mainabi.

Softbank. 2019. Saisei kanō enerugi. https://sii.or.jp/re_energy26r/. November 25, 2019. https://www.softbank.jp/energy/special/shizen-denki/fit/. Accessed 29 Feb 2020.

Sugiura, Nami and Kuroda Kenrō. 2019. Saisei ene yokusei 'hijō ni zannen' Kyūden shachō to kaidan chiji. *Asahi Shimbun*, January 18 (Saga-ken morning edition), p. 23.

Tanaka, Hisatoshi. 2018. Taiyōkō hatsuden yakusei, Kyūden ni shimin kōgi 'sekai, jidai ni gyakkō. *Asahi Shimbun*, October 31 (Kumamoto-ken morning edition), p. 31.

TEPCO. n.d. Gokatei muke puran-akua enaji- 100. https://www.tepco.co.jp/ep/eco/plan/private/detail.html. Accessed 29 Feb 2020.

Tsukimori, Osamu. 2018. Japan's Kyushu Electric May Restrict Renewable Energy Supplies After Nuclear Ramp-Up. *Reuters*, August 29. https://www.reuters.com/article/japan-nuclear-renewables-restrictions/update-1-japans-ky. Accessed 11 Oct 2018.

Watanabe, Chisaki. 2014. Clean Energy Boom Challenges Power Grid. *Japan Times*, October 2. https://www.japantimes.co.jp/news/2014/10/02/national/clean-energy-boom-challenges-power-grid/#.XtZOi25uJhg. Accessed 29 Feb 2020.

Yabe, Akira, et al. (2015) "Demand Response (DR) Strategy." Presentation for the Renewable Energy/Energy System and Hydrogen Unit, Technology Strategy Center, NEDO, November.

Yamashita, Hiroshi. 2018. Taiyōkō 1.3% Shutsuryoku yokusei, kyūden, 10gatsu no kei 4 kai de. *Asahi Shimbun*, December 13 (western morning edition), p. 9.

Yasuda, Yoh, et. al. 2015. International Comparison of Wind and Solar Curtailment Ratio. In *Proceedings of WIW 2015 Workshop Brussels*, October 20–22, pp. 1–6.

Yamazaki, Takuya. 2014. Electricity Market Reform in Japan. Agency for Natural Resources and Energy, METI. Presentation at the Konrad Adenauer Foundation Seminar, Capitol Hotel, Tokyo, September 30.

Yomiuri Shimbun. 2015. Higashi nihon daishinsai 4 nen yoron chōsa fukkō yosan no shito kimonshi. March 8, p. 30.

Yonden. n.d. Sai ene puremiamu puran. https://www.yonden.co.jp/customer/price/plan/saienepremium.html. Accessed 29 Feb 2020.

CHAPTER 6

Renewable Energy as a New Choice for Consumers: The Case of Minna Denryoku

Yuki Takebuta

Translated by Eivind Lande and Paul Midford

The Liberalized Electric Power Market and New Producers: "A Face That Can Be Seen on the Other Side of the Plug Socket"

It has been approximately five years since the full liberalization of the retail market for electric power in April 2016. While looking back at how the

Midford also added notes and several references to provide background and context regarding several points in this chapter.

Y. Takebuta (✉)
Minna Denryoku, Tokyo, Japan
e-mail: takebuta@minden.co.jp

E. L. P. Midford
Norwegian University of Science and Technology (NTNU), Trondheim, Norway

P. Midford and E. Moe (eds.), *New Challenges and Solutions for Renewable Energy*, International Political Economy Series,
https://doi.org/10.1007/978-3-030-54514-7_6

newly liberalized market has been operating, this chapter will show how new power producer and supplier enterprises can make use of renewable energy.[1]

The Globally Expanding RE100

On the world market, since the 2015 Paris Accord entered into force, the movement to promote sustainability has become more dynamic, centered on large European and American enterprises that actively use renewable energy. There also exists a network called RE100, and its affiliated companies make up a movement for increasing their value by declaring that they will use 100% renewable energy. RE100 is operated by the London-based global sustainable management support organization CDP (https://www.cdp.net/en), and 96 companies, including Apple, Starbucks and IKEA that have proclaimed that they aim to use 100% renewable energy, have been so registered by RE100.

Use of Renewable Energy by Corporations and Individuals in Japan

Within Japan the use of renewable energy by corporations has flourished since the second half of 2017. Several Japanese companies registered with the aforementioned RE100, including Marui Group, Aeon Corporation and Sony Corporation. So far government offices have selected bids for power supply according to criteria emphasizing price, but it is expected that the proportion of renewable energy generation will be added to bidding criteria for suppliers from now. For example, one item adopted in the Tokyo Metropolitan government's "Environmental Master Plan" of 2008 was to raise the renewable energy proportion of the energy consumed in Tokyo to approximately 20% by 2024, and the city is actively strengthening its effort to increase the use of renewable energy, centered on solar energy. The ambition in recent years has been to show Tokyo as a city where the introduction of renewable energy is progressing when it hosts the 2020 Olympics and Paralympics.

On the other hand, the use of renewable energy by private individuals has not spread so much. More than 600 new power producer and supplier enterprises have entered the market, but during the first nine months following energy market liberalization, the changeover ratio of consumers from the previous monopoly supplier (Tokyo Electric Power, TEPCO) had only reached approximately 20% in the Tokyo region, so the impact of new power producers and suppliers on the market can be said to remain small. I feel that one reason for this is that sufficient information is not being conveyed about what the changes resulting from energy liberalization mean for ordinary consumers nationwide, and what the merits and demerits of switching from the previous monopoly supplier to another energy company are.

To begin with, a major part of the reason is that almost all new power producers and suppliers try to compete with large power companies (especially the former monopolist EPCOs) on price. For electric power, which is a product with a high production cost, differentiation in a situation where price competition dominates is difficult and is not a strong enough incentive to motivate consumers to switch to new suppliers. Consequently, the current situation is that the liberalized electric power market has fallen into a price competition spiral (Fig. 6.1).

The New Electric Power Company (Minna Denryoku) Sells Renewable Energy

Even under these conditions, new power producers and suppliers that raise the topics of choice and use of renewable energy exist. Minna Denryoku (Everyone's Electricity) stock company, with its base in Setagaya Ward of Tokyo Metropolis, is one of them. The domestic top-class renewable energy-type power producer and supplier Minna Denryoku, which has achieved a renewable energy supply rate of above 75%, poses the following question to us: "Can you see the face on the other side of the socket?" The electricity that we use every day must necessarily come from a power plant, and it may be electricity generated by a coal-fired plant, or it may be electricity made using sustainable forms of energy such as sunlight or hydropower. Using the opportunity of electric power liberalization, we are offering the choice of renewable energy to the ordinary consumer.

The "Electric Power with a Face" System

Minna Denryoku's special feature is being organized as a "power company with a face," which it has trademarked. "Power company with

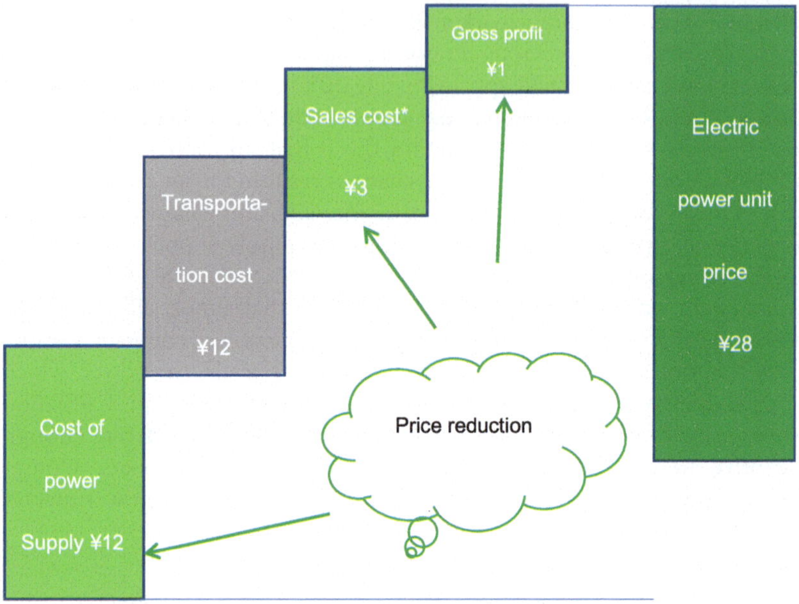

Fig. 6.1 Electricity price structure. *Sales cost includes operating costs, management of supply and demand, etc. Figure created by Yuki Takebuta, translated by Eivind Lande

a face" is the only domestic electric power retail business platform that makes clear who the "producer of electricity" is, and makes it possible for the buyer of electric power to choose the electricity generator that they wish to support. This arrangement operates according to the system "ENECTION," (combining the words "Energy" and "Connection") which Minna Denryoku, developed on its own. Instead of the conventional systems that power companies have needed to use up to now, this system uses a cloud service salesforce, which is flexible and makes it possible to operate at low cost with few people (Fig. 6.2).

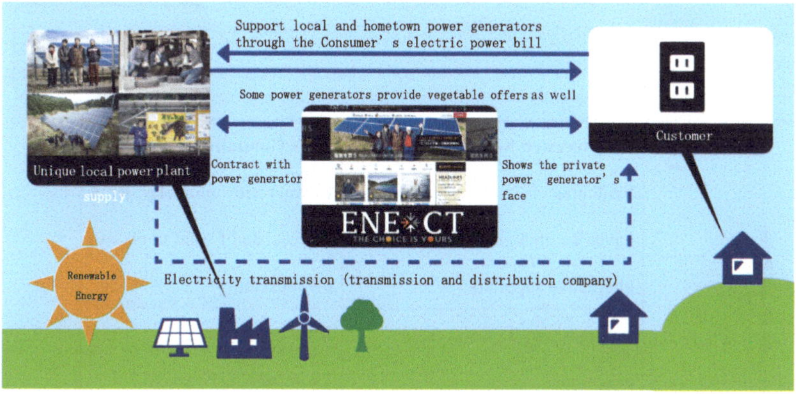

Fig. 6.2 Supporting local energy producers through the consumer's electricity bill (*Source* Figure created by author, translated by Eivind Lande)

On the Minna Denryoku website, the "faces" of those who produce electric power at generation plants are presented, in the same way as for "vegetables with a visible face,"[2] and each month one can "support" the power generation plant that one favors. When a customer "supports" a power plant, Minna Denryoku gives 100 Yen from the electricity charge that the customer has paid to the power plant.[3] Customers do not have to do anything in particular: through the act of consuming energy they are able to contribute financially to the place where power is generated, and to contribute to the spread of renewable energy.

Consumer Segments

At Minna Denryoku our research divides consumers into the segments depicted in Fig. 6.3. When the retail electric power market was liberalized the first to switch to new power companies was the "price conscious segment," while the main target of Minna Denryoku is the renewable energy "committed segment," which wants to use renewable energy even if it costs more. Other groups are the "renewable energy supporting but economizing segment," which would like to use renewable energy if it is cheaper, and the "disinterested segment," which still has not thought much about electric power.

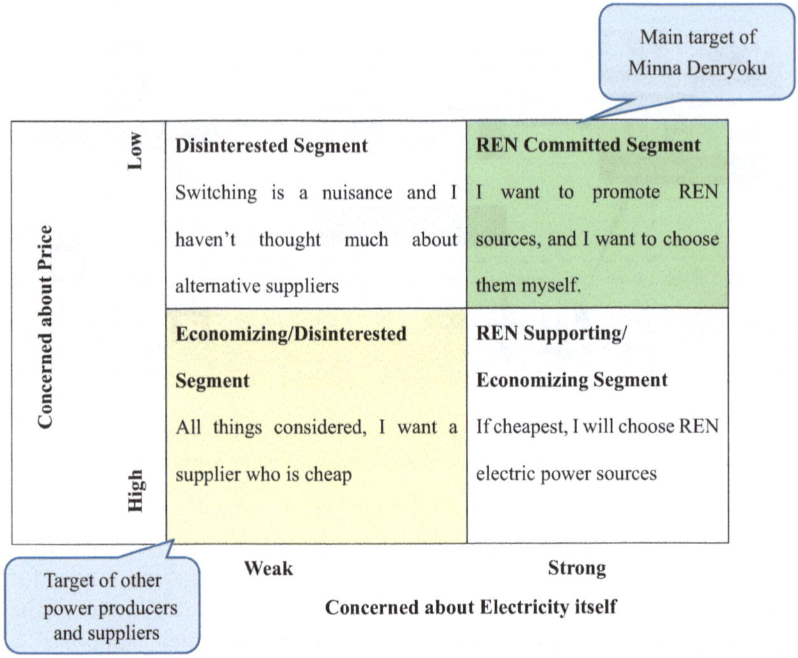

Fig. 6.3 Household consumer segments on switching power suppliers and interest in renewable energy (REN) (*Source* Figure created by author, translated by Eivind Lande)

Moreover, the proportions of the different segments were, according to a previously conducted questionnaire, approximately as listed in Fig. 6.4. Minna Denryoku thinks that, in addition to the "committed segment," it can also reach the "consistently economizing segment" through lowering costs by means of the "Enection" system and the "disinterested segment" by means of the enjoyable "electric power with a face" system.

Lineup and Examples of "Electric Power with a Face"

Currently, approximately one hundred electric power plants are listed under "electric power with a face." Depending on the electric power

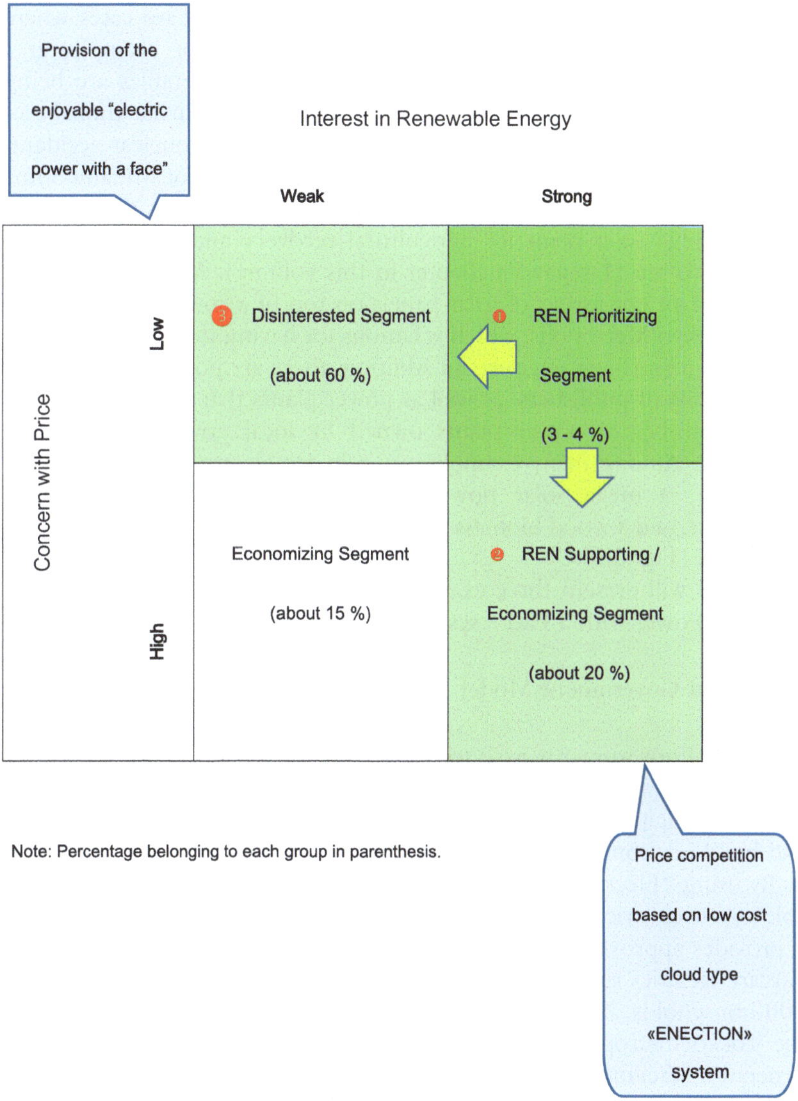

Fig. 6.4 Movement among targeted household consumer segments regarding interest in renewable energy (REN) (*Note* Percentage belonging to each group in parenthesis. *Source* Figure created by author, translated by Eivind Lande)

company, as with the "hometown tax program," there are cases where such power generating companies provide a gift.[4] For example, at a citizen's power plant in Chiba Prefecture, organic vegetables are being produced beside the plant. Second, the Great East Japan Earthquake of March 11, 2011, which led to the Fukushima Daiichi nuclear accident, resulted in an evacuation zone being created in the city of Minami Sōma in Fukushima Prefecture (Samuels 2013, pp. 15, 40) and this town is using a mega solar plant for agricultural recovery and regional rejuvenation (also see Hasegawa's chapter in this volume). A third example is a power plant consisting of solar panels on top of a cowshed in Hachioji City in Tokyo Metropolis, which is famous for having the world's smallest workshop for making yoghurt. In addition, there are power plants owned by musicians or publishers, as well as power plants that offer CDs.

The number of power plants owned by local governments is also increasing. Among many examples are a hydro-electric plant in Nagano Prefecture, a mega solar power plant in Setagaya Ward in Tokyo Metropolis, and a wood biomass power plant in Kawaba Village in Gunma Prefecture (Fig. 6.5).[5]

Here, I will present three examples of how Minna Denryoku uses the "Electric Power with a Face" system.

1. Local Government Model

Many local governments wish to make their power generation businesses and their locality better known, and by using "Electric Power with a Face" to sell electric power, they can gain greater visibility, provide information and build relationships with consumers (Fig. 6.6).

By using "Electric Power with a Face," Nagano Prefecture sells renewable hydro-electric power to Setagaya Ward in Tokyo. On an annual basis, it provides approximately 1.8 million kWh of electric power to about 40 daycare facilities run by the Ward, corresponding to the consumption of 500 households. Nagano Prefecture tries to attract local governments in the Tokyo metropolitan region that have few renewable energy power generation facilities to expand the market for renewable energy from the Shinshu region, with the aim of increasing income from sales. This is the first case of a local government in the countryside and one of Tokyo's 23 wards cooperating to provide renewable energy. Regarding cooperation between local governments in the countryside and those in

Fig. 6.5 Examples of Minna Denryoku electricity generation and co-use facilities. *Notes* Clockwise from top left: (1) Tatsuishi, Fujioka City, Gunma Prefecture, Tokyo Yuden Power Company (which also generates electricity from recycled cooking oil used in tempura), output capacity 145 kW, number of subscriber openings remaining: 142; (2) Eichi, Sodegaura City, Chiba Prefecture, Aigamo Power Plant, output capacity 49.5 kW, number of subscriber openings remaining: 30; (3) Chuo, Edogawa Ward, Tokyo Metropolis, Edo, Sora Plant no. 3 Parking Lot, output capacity 22 kW, number of subscriber openings remaining: 17; (4) Minamishitauramachi, Miura City, Kanagawa Prefecture, Setagaya Ward Miura Solar Panel Power Plant, output capacity 344 kW, number of subscriber positions remaining: 18; (5) Takato, Ina City, Nagano Prefecture, Takato Sakura Power Plant/Mizubasho Power Plant, output capacity 1160 kW, number of subscriber positions remaining: 106; (6) Kobiki, Hachioji City, Tokyo Metropolis, Plant no. 3, Mother Cow Yoghurt Workshop Power Plant, output capacity 19.8 kW, number of subscriber positions remaining: 151 (*Source* Figure created by author, translated by Eivind Lande)

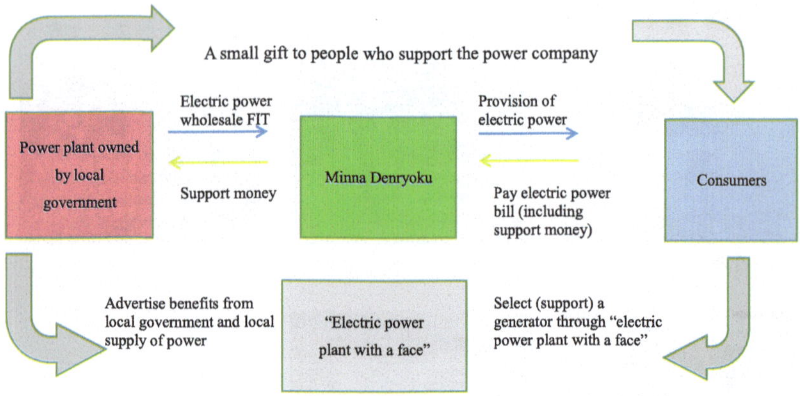

Fig. 6.6 "Electric power with a face" business model (*Source* Figure created by author, translated by Eivind Lande)

Tokyo Metropolis, another example is Kawaba Village in Gunma Prefecture, which is providing wood biomass-generated power to about 40 households in Setagaya Ward.[6]

2. Naming Rights Model

The sporting goods producer Adidas purchased naming rights, allowing it to designate an "Adidas Electricity Generating Plant," and was even able to acquire the naming rights for the electricity itself. The plant has a capacity of 72 kW and generates on average nearly 67,000 kWh per year. Adidas acquired the naming rights for electricity supplied to a private exhibition curated by the artist Chim ↑ Pom, held in November 2016, where artwork produced by the artist Hirai Arita was exhibited. The theme was "Everything is renewable / Sustainable," "Life is avant-garde." Energy, fashion, music and a variety of genres of art work were displayed (Yahoo nyu-su 2016). Adidas purchased the naming rights to the renewable electricity supplied to this exhibition from Minna

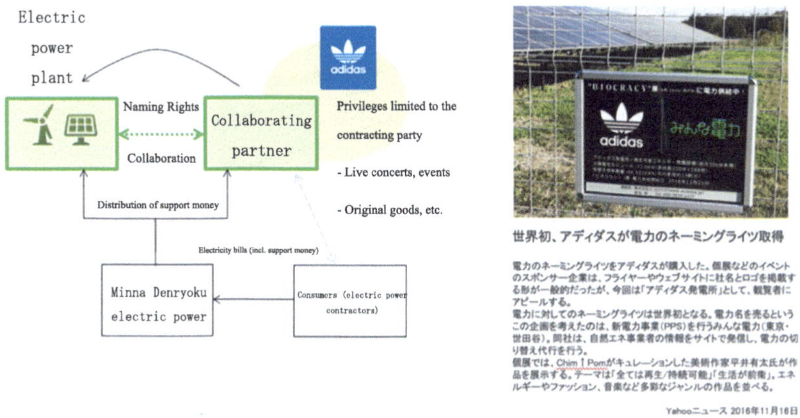

Fig. 6.7 How the naming rights model works. **Text under illustration with Adidas poster**: Adidas became the world's first company to acquire naming rights to electric power. It has been common that companies sponsoring individual exhibitions and events have been able to list their company name and logo on flyers and webpages, but in this case the sponsor has been able to increase its appeal to visitors by using the term Adidas Power Plant (*Source* Figure created by author, translated by Eivind Lande)

Denryoku. In general, naming rights are acquired for a facility, but this was the world's first case of providing naming rights for the electricity itself. In this way, businesses that cannot easily run their own electricity generation business, can, by sponsoring an existing electric power generating business "whose face cannot be seen," receive naming rights and be able to promote a renewable electricity generating business (Fig. 6.7).

3. Local Production and Consumption Model

Due to electricity market liberalization, and the "electric power with a face" system, "local production for local consumption" has become possible. For example, the electricity produced by a solar panel power generator owned by Minna Denryoku, which is installed on the roof of an apartment house in Kamisoshigaya, Setagaya Ward, Tokyo Metropolis, is supplied to "Setagaya school of Physical Creation" (also known as the Ikejiri Institute of Design) in the same Setagaya Ward, where Minna Denryoku has its headquarters. This can even be said to be not local

Fig. 6.8 Minna Denryoku PV solar facility and the roof of its customer (*Source* Pictures and figure created by author)

production for local consumption, but a "production in the Ward for consumption in the Ward" model (Fig. 6.8).

Conclusions

These examples show how, by applying "Electric Power with a Face," to electricity that we have often used without being conscious of it, we build relationships with a visible face, and in this way create added value through "the enjoyment of being able to choose," "the enjoyment of receiving premiums," "the enjoyment of becoming connected" and "the enjoyment of supporting." When one is not conscious of electricity, ultimately the money the consumer pays for it is used to cover the operating expenses of coal-fired electricity generating plants, which impose a great burden on the environment, and in some cases it is possible that payments for electricity may be contribute to nuclear power plant accidents.

Approximately half of the CO_2 emissions from the average household is said to come from its use of electricity. This means when one switches to renewable energy for electricity, one is "choosing" to use sustainable electricity, which leads to increased use of renewable energy, and makes possible the creation of a sustainable future. I would like everyone to look at the new electric power producers also from that viewpoint.

Notes

1. Power-shift.org provides a useful summary (in Japanese) of new entrants into Japan's liberalized retail electricity market divided into municipal government generators, regional companies and a few multi-regional companies, including Minna Denryoku. For more on the background and development of community-based renewable electricity generation in Japan see Hasegawa (2011, 2015), and Hasegawa's chapter in this volume.
2. For one example of this sort of webpage in Japan (in Japanese), see https://www.iyfoods.co.jp/dept_3.html (accessed February 29, 2020).
3. For more on this program (in Japanese), see Minna Denryoku (ud).
4. The hometown tax (furusato nōzei) program, launched in 2008, allows taxpayers to donate to their hometown or any municipality of their choice and receive a corresponding tax deduction. Municipalities have competed for donations by offering (increasingly lavish) gifts to contributors. See Kyodo (2018).
5. For more on this plant (in Japanese), see http://woodvillage-kawaba.com/ (accessed February 29, 2020).
6. For an overview of renewable energy tie-ups between Setagaya Ward and rural municipalities, see Setagaya ku (2019).

References

Hasegawa, Koichi. 2011. *Datsu genshiryoku shakai he: denryoku wo gurinka suru*. Tokyo: Iwanami Shoten.

Hasegawa, Koichi. 2015. *Beyond Fukushima: Towards a Post-Nuclear Society*. Melbourne: Trans-Pacific Press.

Kyodo. 2018. Japan to Curb Expensive Gift Incentives under 'Hometown Tax' Program. *Japan Times*, September 11. Accessed at https://www.japantimes.co.jp/news/2018/09/11/business/japan-curb-expensive-gift-incentives-hometown-tax-program/#.Xoz9t4j7TIU. 29 Feb 2020.

Minna Denryoku. (ud). Kao ga mieru hatsudensho. Accessed at https://minden.co.jp/powerdev/. 29 Feb 2020.

Samuels, Richard J. 2013. *3.11: Disaster and Change in Japan*. Ithaca: Cornell University Press.

Setagaya ku. 2019. Saisei kanō enerugi- wo katsuyō shita jijitai renkei (denryoku) wo susumeteimasu. Last updated November 7. Accessed at https://www.city.setagaya.lg.jp/mokuji/sumai/011/003/d00182578.html. 29 Feb 2020.

Yahoo nyu-su. 2016. Sekai sho, Adeedasu ga denryoku no ne-mingu raitsu shutoku. November 11. Accessed at http://alternas.jp/joy/date/66822. 28 Feb 2020.

CHAPTER 7

Betting on Hydrogen: Japan's Green Industrial Policy for Hydrogen and Fuel Cells

Robert M. Uriu

Introduction

Utilizing hydrogen gas in fuel cells to generate, use, and store electricity is an idea that has been around for more than a century. Despite recurring periods of high hopes, the dream of hydrogen and fuel cells becoming commercially viable on a large scale has remained elusive. This chapter assesses whether all of this is about to change. The past five years have seen a renewal of optimism as fuel cell products in the transportation, residential power, and other sectors have seen increasing adoption. As importantly, major national governments in Europe, Japan, Korea, and most recently China, have made long-term commitments to developing their national industries.

This chapter focuses on the industrial policies of Japan, which in 2014 became the first nation to commit to becoming a full-fledged "hydrogen society" by 2050. This was an unexpected policy shift in that many strong actors in the Japanese energy policymaking process, especially the nuclear

R. M. Uriu (✉)
University of California, Irvine, CA, USA
e-mail: rmuriu@uci.edu

P. Midford and E. Moe (eds.), *New Challenges and Solutions for Renewable Energy*, International Political Economy Series,
https://doi.org/10.1007/978-3-030-54514-7_7

industry, had a vested interest in opposing policy change. In the first section of this chapter, I argue that the decision to pursue hydrogen was spurred by a unique confluence of factors. First was the crisis caused by the 3/11 Fukushima nuclear disaster, which led to a search for a viable, clean alternative to nuclear power. At the same time, the Japanese hydrogen and fuel cell industry, which had been supported by industrial policy for three decades, was just at the cusp of becoming a commercially viable industry. This was a case of an industry being in the right place at the right time.

In the second part of the chapter, I step back to assess the impact of Japan's green industrial policies, which I argue have helped the industry achieve a degree of commercial viability in fuel cells for residential and automobile uses. If progress is to be made on the remaining parts of the hydrogen society—as a storage medium and creating a green hydrogen supply chain—industrial policy is likely to play a helpful and even necessary role. I also note examples from elsewhere in the world where more and more private companies are entering the industry and are now constantly rolling out new technologies and products. And with other major governments also making new national-level commitments, the future of the industry seems bright.

Hydrogen and Fuel Cells: Potential, Obstacles, and the Role of "Green Industrial Policy"

Most of the world seems to be finally coming to realize that preventing catastrophic global warming will require a drastic decarbonization of all aspects of our industrial economies. Yet while there has been strong growth in renewable sources of energy, the industrialized nations are not close to fulfilling even the modest commitments set forth in the Paris Agreements. Hydrogen proponents argue that if we are to achieve significant decarbonization, hydrogen and fuel cells will necessarily play a major role.[1]

Proponents see the industry as poised to contribute to three key areas. First is decarbonizing the residential power sector, where stationary fuel cells in homes and office buildings can be used to convert hydrogen into electricity to provide power and heating. Japan has led the way in this sector, through its ENE-FARM program that encourages the adoption of stationary fuel cells for use in homes, apartments, offices, and larger structures. Japan has also made efforts to deploy fuels cells on a city-wide basis.

Second, fuel cells can help decarbonize the transportation sector. Fuel cell electric vehicles (FCEVs) convert stored hydrogen with ambient oxygen to produce electricity to power a car's motor, with water being the only tailpipe emission. Major auto companies have introduced FCEVs, and are beginning to challenge their cousin, the battery electric vehicle. FCEVs show more promise in other transportation sectors that require greater range or hauling capacity, including material handling, trucking, ships, buses, and trains, as demonstrated by the recent introduction of products worldwide.

A third crucial role for hydrogen fuel cells is as a storage medium, which could help increase the use of renewable sources by the electrical utilities. Because of the problem of intermittency (the sun is not always shining), the utilities have been reluctant to rely too heavily on REN sources. A greater capacity to store energy that could later be converted to electricity should help alleviate these concerns. While there are many competing alternatives, hydrogen has the advantage of being a nearly indefinite storage medium that could be transported where needed.

Advocates argue that while there are alternatives to hydrogen in each of these roles, hydrogen is unique in that it stands to make a substantial contribution in all these realms. That is, hydrogen is described as being flexible across the energy spectrum, and thus will necessarily be a part of a renewable energy future. The only question for proponents is what form the hydrogen economy will take, and how long it will be before it is realized.

While the industry's prospects seem clear and exciting, even the most optimistic proponent recognizes that numerous technical and commercial obstacles still exist. Like any new technology, hydrogen fuel cells are in a race with other competing technologies. If not enough progress in overcoming obstacles is made, quickly enough, the industry's future is by no means guaranteed. That is, all realize that even the best technologies are not always adopted.

Each fuel cell sector faces significant obstacles. Stationary and residential fuel cells are still costly to produce, and demand has not been sustained in the absence of government subsidies. The main obstacles to the wider adoption of passenger FCEVs are the still high cost of vehicles and the scarcity of hydrogen refueling stations (HRS). These obstacles are linked in that unless there are enough HRS, drivers will be reluctant to buy an FCEV; but until there are enough FCEV's on the road, companies will not want to build costly HRS. Until this chicken and egg dilemma is solved, auto companies will not be able to produce at a scale large enough for them to lower vehicle production costs.

Another set of obstacles involves the difficulty of producing and transporting green hydrogen. While hydrogen is the most abundant element in nature, it must be extracted from other elements through processes that are costly and can be energy intensive. Most of today's hydrogen is derived from fossil fuels, especially methane, and so this is not yet a truly green resource. More and more companies are using electrolysis to produce green hydrogen, but the process is still expensive. Gaseous hydrogen is not dense, so transporting and storing it in this form is difficult and costly, and newer processes to make hydrogen denser or more easily transported are still complex and expensive (Romm 2004; Zubrin 2007).

Even if the industry can solve some of these obstacles, it will not succeed unless it can solve all of them simultaneously. If even one link in the chain fails, the entire approach is likely to fail. For instance, even if demand rises, firms might not be able to lower prices enough to be competitive. Even if production increases, the industry might not be able to produce enough hydrogen at a low enough cost, and so on. The list of potential bottlenecks is long.[2]

Even if these technical issues can be solved, the industry would then need to achieve *commercial viability*, which I define as a self-sustaining market where an industry is able to thrive without the need for government support or subsidies. To achieve this, the industry will need to produce and sell at a large enough scale so that production costs can be lowered. Commercial viability thus presents a chicken and egg dilemma: until products reach a level of demand where companies can mass produce, costs will remain high; but products will never be in high demand if they remain too costly. Further, firms will see the possibility of profits only after this dilemma is solved; until then, they will be reluctant to invest in a risky industry.

This industry is thus a classic case of a "market failure," in which there is a disconnect between private incentives and public needs. Faced with technological obstacles, high costs and the likelihood of failure, firms will not have an incentive to make the needed investments in R&D or product development. This may be especially true when it comes to technologies that can help alleviate climate change, since the potential public benefits are so large.

Overcoming these market failures often requires some form of government industrial policy support to change private incentives. In the case of the promotion of energy-related industries, this approach is known as "green industrial policy." Developing clean energy industries at the massive scale required and "with the speed needed cannot be done without industrial policies" (Hallegatte et al. 2013, p. 21).[3] In particular, governments can play a useful role in establishing a stable framework for public–private coordination, clear and consistent policy objectives and goals, and a long-term public commitment to the industry. Only when policy commitments are consistent and predictable will private sector actors make the needed investments.[4]

Industrial policy is exactly what Japanese public policymakers, working in conjunction with private sector companies, have been attempting to do throughout the postwar period, for a wide range of industries. And although industrial policy is not as overt as it was during its high-growth era, it has continued to support the promotion of future technologies. Given this long history of industrial policy, Japan may be in a unique position to pioneer green industrial policy. Indeed, Japan has been "repurposing" its industrial policies in this direction ever since the Oil Shocks. As Hughes argues, Japan's industrial policy has been "retained and redeployed" in this direction, in terms of spending, ministry personnel, and policy activity (Hughes 2012, pp. 109–111).

Japan's approach to industrial policy has traditionally been a collaborative one, bringing together Japanese government officials with the private sector and scientists. This approach is known as the *san-kan-gaku* (産官学) system, in which industry actors (*san*), government officials (*kan*) and the academic research world (*gaku*) cooperate in the formulation and implementation of industrial policy.[5] Typically, the Japanese government will convene some form of a deliberative policy council, made up of government officials, representatives of the private companies, related academic experts, and other notable figures. The role of the private sector, namely the firms and industries most affected by policy, has been particularly important. Private sector actors are always involved in these deliberations and are very often the driving force behind policy creation and implementation.

Past industrial policy efforts in Japan have focused on three stages of development. The first is the initial scientific research stage, typically bringing together the academic and business worlds. In the past, the Japan government has supported a wide range of research efforts and will then supply more concentrated support for those technologies that seem most promising. The second stage has involved public policy support for the development and testing of commercial products. While many governments have supported their industries in this manner, Japan is distinctive in providing concentrated support for the third stage: the full-scale commercialization of selected industries, often through the encouragement of exports, until the industry achieves true commercial viability.

The Origins of Japan's Industrial Policy for Hydrogen and Fuel Cells

Japanese industrial policy has often been most crucial in the earliest stage of an industry's development, as private firms have low incentives to invest in an industry that is so far away from commercial viability. It is thus important to look at the origins of Japan's policies for the hydrogen and fuel cell industry, which date back to the 1970s. Then, as now, Japanese energy policymakers have tried to strike a balance between energy security, economic development, and environmental concerns. A long-term dilemma for Japan has been how to access cheap (often imported) sources of energy without becoming overly dependent on external sources of supply. Moreover, since at least the 1990s, Japanese policy has reflected a concern for mitigating global climate change.

The story of Japan's hydrogen and fuel cell policy begins in 1974 when hydrogen was first mentioned in an official energy policy statement. That was of course the year of the First Oil Shock, an event that spurred Japan to search for ways to decrease its demand for imported oil. The initial policy focus was on reducing energy consumption, but the government also began to investigate new sources of energy. Japan's core policy was known as the "Sunshine Program," which aimed to research a range of "new energy" that included hydrogen along with solar, geothermal, coal, and nuclear power (Behling 2012; Hoffman 2012).[6]

After years of research and product development, most of which was taking place outside of the public view, Japan's interest in hydrogen and fuel cell technology took a major leap forward in the 1990s in the context of the negotiations on climate change that led to the Kyoto Protocol of 1997. By the early 1990s, public and private actors had begun to identify the hydrogen technologies that showed the most promise. In 1993, the Japanese government created an initiative as part of its "New Sunshine Program" that for the first time specifically targeted hydrogen energy. This New Energy and Industrial Technology Development Organization (NEDO) run program was named the "International Clean Energy Technology Utilizing Hydrogen," but was more popularly known as the World Energy Network project, or WE-NET. The program was labeled by the International Energy Agency (IEA) as the world's "first major, national (hydrogen fuel cell) R&D programme" (OECD/IEA 2004, p. 147). WE-NET set out comprehensive goals in all aspects of hydrogen technology, from generation, to transport, storage, and utilization. Budgetary support was US$3 billion but was spread out over a period of 28 years, through 2020. Perhaps as important as the amount of funding, however, was that a government had for the first time made a clear, long-term commitment to the technology (Okano 2016, p. 70; Hirai 1997, pp. 1–2).

By the late 1990s, with companies making progress on developing new products, Japanese policy shifted from basic research to the second stage, support for commercialization. In December 1999 the Director General of ANRE agreed to create the "Policy Study Group for Fuel Cell Commercialization," an advisory group composed of representatives of the major auto and electronics firms, the main utilities and petroleum companies, as well as academic advisors and governmental officials.[7] As its name implies, the group's purpose was to pursue more active efforts to bring fuel cell products to market.

As a result of these recommendations, the government in March 2001 created an important consortium, the Fuel Cell Commercialization Conference of Japan (FCCJ) to offer advice and policy guidance. The FCCJ was an expanded version of the Policy Study Group. It was chaired by the CEO of Toshiba, with vice chairs representing the main sectors involved in the sector, namely an auto firm (Toyota), a consumer electronics firm (Matsushita), a petroleum firm (Japan Petroleum), and a gas utility (Tokyo Gas). The group included 112 other firms or industry associations, representing a range of automobile firms, electronics makers, and private energy firms—in short, all private sector actors that had an interest in commercializing hydrogen or fuel cell technologies. It is important to note that this group has ever since been the driving force behind policy developments of Japan's hydrogen and fuel cell policy.[8]

The FCCJ was tasked with devising policies to overcome obstacles to the commercialization and widespread diffusion of fuel cells. The group's early recommendations were essentially adopted by METI and became the basis for Japan's first long-term plan for the industry, the "Fuel Cell and Hydrogen Policy Framework" of 2001 (Maeda 2003b, p. 14). This Framework, also known as Japan's "New Hydrogen Project," was an across-the-board attempt to foster commercialization of the full range of fuel cell and hydrogen technologies, from production to storage and transport. It also earmarked a total of US$540 million in funding through 2010.

According to the IEA, the 2001 Framework was a comprehensive effort to create "self-sustained growth" by integrating "the development of fuel cell, hydrogen production, and hydrogen transport and storage technologies, concurrently with demonstration programs, vehicle sales, construction of refueling infrastructure, establishment of codes and standards, and a general push to enlarge the consumer market for fuel cells and fuel cell vehicles." The IEA concluded that Japan's policy created "a world leading program focused on early commercialization of fuel cells in close partnership with the national industry" (OECD/IEA 2004, p. 147).

On the eve of 3/11, the Japanese industry had already established itself as a leader in fuel cell-related products and was beginning to show signs of actual commercial viability. In the residential fuel cell sector, the industry had become the first to market products for home use. In terms of transportation, the Japanese FCEV sector was deemed to be the most advanced and developed in the world, with Toyota on the verge of being the first to bring an FCEV to market.

Policy toward the industry up to the eve of 3/11 was thus a typical industrial policy story, with public policy encouraging investment, but with the actual work being done by the industry itself. The list of policies and projects was a long one, characterized by trial-and-error failures but also many successes. Although other countries were rated more highly in terms of research capabilities, Japan's private sector firms had become world leaders in terms of commercialization (Behling 2012).

In the context of Japan's overall energy policy, however, the industry was still very much on the back burner. Even though the DPJ government, which came to power in September 2009, supported renewable energy, Japanese policy was still heavily influenced by the nuclear and electric utility sectors. While the DPJ had mentioned hydrogen in various statements, other powerful actors in the Japanese energy economy still had a stranglehold on the policy process. Much of this was about to change.

The Great East Japan Earthquake: The Role of Crisis and Policy Change

The role of crisis is crucial in understanding Japan's decision to adopt its hydrogen society plan. As is often the case, an external crisis can be a helpful or even necessary impetus for policy change. Put differently, absent a crisis, institutions often find it difficult to initiate change, as existing political actors resist new policies that run counter to their interests, and new policy ideas seem unproven and risky. A crisis can be useful in upsetting the status quo, forcing actors to rethink objectives and interests, and focusing attention on new ideas and policy alternatives.

The Great East Japan Earthquake, tsunami, and the Fukushima Daiichi nuclear accident, known as the 3/11 disaster, represented just this sort of crisis. The most concrete change was that one crucial energy avenue, nuclear energy, was temporarily closed off. At the time, public opposition to nuclear power meant that policymakers could no longer rely on an energy source that had become a mainstay of Japan's energy economy. And with the political power of the nuclear industry weakened, a policy space opened in which new actors could be more active and new ideas more attractive.

The crisis led to a rethinking of Japan's overall energy policy direction. In October 2011, the DPJ had METI and ANRE appoint a 15-member "Basic Problem Committee" to carry out a comprehensive review of Japan's energy policy. The committee called for increasing Japan's reliance on renewable energy to above 20% by 2030 and recognized hydrogen as "an important secondary energy carrier" that would help achieve the goal of increasing reliance on renewable sources.

After Abe Shinzō recovered the prime minister's chair in 2012, it was assumed that his government would seek to revive the nuclear industry. It was thus a surprise when his administration announced its plan to elevate the hydrogen and fuel cell industry in terms of national energy policy. The Abe administration over the next two years gradually came to embrace the industry, and to eventually champion it. This support became official in April 2014 when the goal of creating a "Hydrogen Society" was first mentioned in an official document, METI's Strategic Energy Plan (METI 2014). The key policy pronouncement was the June 23, 2014 unveiling of the "Hydrogen and Fuel Cell Strategic Road Map."

However, while the 3/11 crisis elevated the visibility of the industry, it is important to note that actual industrial policies did not change very much. That is, the government and industry actors remained the same, and the problems facing the industry were also unchanged—how to lower costs and resolve the many technological bottlenecks that still existed. The only difference was the industry's new prominence.

As shown in Fig. 7.1, the 2014 Road Map was essentially a restatement of Japan's existing commercialization policies that the industry had been working on since 2001. The Road Map itself was put forward by the industry advisory group, which in December 2013 had changed its name from the FCCJ to the "Hydrogen and Fuel Cell Strategy Council." Like its predecessor, the group consisted of academic experts, representatives of Japan's main automobile firms, petroleum suppliers, electric and gas utilities and electronics makers, as well as representatives of the Japanese

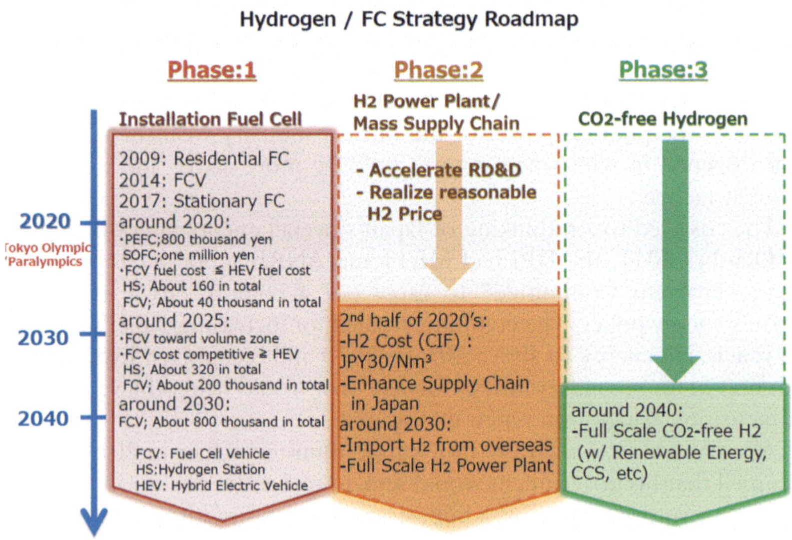

Fig. 7.1 Step by step approach to realizing a hydrogen society (*Source* METI 2017, p. 5)

government. In April 2014, the Council proposed the new policy of transitioning to a hydrogen society and proceeded to create a more concrete plan for the sector (Behling 2014; Behling et al. 2015, pp. 212–213).[9]

The new Road Map was divided into three phases. The first phase called for the promotion of two product lines, the FCEV and the residential ENE-FARMS, where Japan was already a world leader, and provided subsidies over a period of 10 years to help the industry develop products and infrastructure. The second phase of the Road Map was the more abstract and longer-term goal of establishing a worldwide hydrogen "supply chain." This stage focused on "hydrogen production, storage, and transport capabilities," and indicated that this phase would begin in earnest only in 2020. The third phase outlined the abstract goal of establishing a zero-carbon emission energy economy based on "green hydrogen." The roadmap indicated that this phase would take place around 2040, with the goal of creating a "completely CO_2-free, low-cost, safe, and stable hydrogen supply infrastructure" at some point thereafter.

The Road Map thus covered a period of over 25 years and provided a commitment of funding for the entire period. Of greatest impact was the immediate increase in spending, with METI requesting US$401 million for FY 2015, which represented a doubling of the prior year's budget. The plan envisioned the creation of a new market for hydrogen and fuel cells that would increase from US$10 billion in 2030 to US$80 billion by 2050. The Road Map did not mention any employment numbers, but the assumption is that this will be a major job creator (Behling 2015, p. 4).

The Road Map was announced with great fanfare, with numerous Japanese political leaders expressing optimism that the plan would work. Abe, for instance, argued that "the key to acting against climate change without sacrificing economic growth is the development of innovative technologies" such as the hydrogen society. Then Tokyo governor Masuzoe Yōichi added that "the 1964 Tokyo Olympics left the Shinkansen high-speed train system as a legacy. The upcoming Olympics will leave a hydrogen society as its legacy" (Daugherty 2016, p. 1).

Explaining Japan's Policy Change

This section discusses Japan's unexpected shift to support the hydrogen and fuel cell industry. Given the numerous technical obstacles and unclear commercial prospects, why did the Japanese government decide to gamble on the hydrogen society? What gave the Japanese government such faith in the future of the industry? Indeed, many questioned the wisdom of Japan's new approach, arguing that it had made a misguided bet that was doomed to fail. Second, this policy shift flies in the face of the orthodox view of Japanese energy policy, namely that policy is captured by strong private sector actors, especially the electric utility sector, represented by the politically powerful EPCOs, and the nuclear industry. These actors are seen as having outsized influence over policymakers, government regulators, and politicians and have long opposed the promotion of new energy sources, preferring to protect their existing (fossil fuel and nuclear) assets. Finally, why did this shift occur under the administration of Abe Shinzō, a politician who is routinely given low marks in terms of his commitment to fighting climate change?

One leading scholar of Japan's hydrogen and fuel cell policies, Noriko Behling, argues that the Japanese government was driven mostly by concerns for national energy security and was able to reach such a quick consensus because it simply saw no other alternative. That is, policymakers were above all looking for a way to decrease Japan's reliance on imported fossil fuels and realized that expanding nuclear energy was not then feasible. Behling argues that policymakers weighed different alternatives and concluded that "over the long term there was no other alternative but to rely on hydrogen energy to power Japan" (Behling 2015, p. 1). My sense is that while concern for reducing reliance on imported fossil fuels was certainly a factor in Japan's policy shift, this was only a partial motivation. Policymaking is rarely that easy, as there are always multiple actors involved in any decision, and there are almost always different alternatives that could have been chosen.

Another view holds that the Japanese government was responding to public opinion, not only in terms of opposition to nuclear energy, but also the desire to contribute to combatting global climate change. As the world's sixth largest emitter of greenhouse gases, Japan has felt considerable pressure to take corrective measures, and successive governments had made pledges to reduce carbon emissions. Again, this view may be partially correct in that policymakers and the industry did have high hopes for hydrogen as a way to fight climate change. Nevertheless, given that Japan's energy policy has often been far from green, some scholars of Japanese energy policy remain highly skeptical (Incerti and Lipscy 2018; Midford 2014).

An opposite, quite cynical view is that the Japanese government only wanted to *appear* to back a new, green technology, but that its real purpose was to gain enough good publicity so that it could avoid making painful changes in its energy economy. This process has been labeled "green washing," in this case an attempt to draw attention away from Japan's status as a laggard in fighting climate change and to achieve the LDP goal of reviving the nuclear or coal industry. If this view is correct, the promotion of renewables and the hydrogen industry was just a splashy sideshow, or a smokescreen.

A final explanation understands the hydrogen society idea as simply corporate welfare for certain Japanese industries. As Incerti and Lipscy argue, the hydrogen society idea was no more than an extension of Abenomics and the overriding goal of economic growth. As such the support for the hydrogen sector is described as "industrial policy under the guise of environmentalism" and in particular "in support of Japanese automakers" (Incerti and Lipscy 2018, p. 619).

My own view focuses on the unique situation Japan's hydrogen and fuel cell industry found itself in, in the aftermath of 3/11. First, the industry could argue that it was on the verge of commercial success, and that continued policy support would allow it to achieve true commercial viability. Many Japanese firms stood to benefit from the promotion of hydrogen—not just the automakers but also a wide range of manufacturers of fuel cells, large stationary fuel cell units, hydrogen filling stations, the gas utilities, and others. These firms represented some of Japan's largest manufacturers, and their voice was amplified by the industry association, the FCCJ. The industry could thus argue that its promotion would spur economic growth and employment, an important motivation for both political as well as bureaucratic leaders.[10] And from the government's point of view, given the resources that it had already put into the sector over the preceding decades, it was natural for it to want to double down on the commitment to hydrogen.

In addition, the hydrogen and fuel cell industry could paint a compelling picture of itself as a possible solution to climate change. As proponents argued, this new industry, if successful, could help to decarbonize important parts of the energy economy. This green potential no doubt had great appeal to multiple actors, although perhaps for different reasons. That is, environmental motivations could have been *both* sincere and a smokescreen, depending on the actor in question. Some were no doubt sincerely motivated by environmental concerns, but it is also easy to imagine that others saw the promotion of this industry as a way to simply show that Japan was doing something on climate change, while still others may have seen the new policy as a way to increase their corporate profits. In essence, the hydrogen vision appealed to many actors, even if for contradictory reasons.

In sum, in the aftermath of the 3/11 crisis, Japan's hydrogen and fuel cell industry basically found itself in the right place at the right time. The crisis simultaneously weakened the power of existing actors, most notably the nuclear industry, and forced Japanese policymakers to search for new energy pathways. At that time, the industry had developed just enough that it could claim to be on the verge of commercial viability and that it could help Japan fulfill its climate change obligations. Japanese leaders saw the promotion of the industry as a potential answer to Japan's energy dilemma, and the industry was able to take full advantage of its opportunities.

As I argue in the second half of this chapter, the exact motivations behind Japanese policy may be less important than the fact that Japan has made a national-level and very public commitment to the industry. By tying its national policy to the 2020 Tokyo Olympics, Japan gave itself a self-imposed deadline to make some real progress. In short, even if Japan was only motivated by the most selfish and least environmentally enlightened of reasons, its practical commitment to the industry may have an enormous impact. Japan's decision to commit to a hydrogen society, coming as it did at a time when similar programs in other countries were stagnating, no doubt helped to resuscitate global interest in the industry. In the end, this may prove to be the biggest long-term contribution of Japan's new policy direction.

Japan's Prospects for Success: Obstacles and the Continuing Role of Green Industrial Policy

For the remainder of this chapter, I assess Japan's progress in overcoming the still major cost and technical obstacles that stand in the way of its hydrogen society. I look at each sub-sector in terms of past policy and progress toward the goals of the 2014 Road Map. Industrial policy for two sectors, residential fuel cells and FCEVs, has reached the commercialization phase. Policy for the other sectors, including storage and the green hydrogen supply chain, is still at the research and product development state. I also note developments in selected foreign markets. This is a fast-moving industry, with developments appearing daily, and at times becoming outdated as soon as they are reported.

Decarbonizing Residential Heat and Power. One focal point of the 2014 Road Map is the ENE-FARM ("energy farm"), a stationary fuel cell unit that generates electricity to power the home and captures heat generated by the unit to provide heated air and water. These units are thus known as "combined heat and power (CHP)" stationary fuel cells.[11] In 2008, Japan became the first country to bring these units to market. Pioneers in the industry were Panasonic, Toshiba, and Aishin Seiki, working together with gas utilities Tokyo Gas and Osaka Gas, which installed the units and supplied natural gas.

Earlier industrial policy had focused on assessing consumer usage patterns and preferences, and to identify technical and commercial problems.[12] The government helped spur initial consumer demand by heavily subsidizing end users, with support totaling almost half of the cost of each unit (Hashimoto 2015, p. 9). As a result, the industry had achieved an installed base of some 115,000 units by 2014. The June 2014 Road Map continued to focus on creating a self-sustaining market, and set an ambitious goal of installing 1.4 million units by 2020 and 5.3 million by 2030. The industry's immediate goal was to lower unit costs to around US$8000 by 2015, which would represent a 50% cost reduction compared to 2008 (Behling 2015; Maruta 2016, p. 14).

The government and industry recognized that significantly lower unit costs would require a massive increase in consumer demand. To this end, the Japanese government worked with the private sector and municipalities to create entire cities based on stationary fuel cells. In April 2016, Panasonic cooperated with the town of Tsunashima, near Yokohama, to create a "Smart City" that would install CHP units to run its town energy center and other public facilities, with a goal of reducing the city's CO_2 emissions by 40% (Shayon 2016). In March 2017, the northeastern city of Kamisu announced that it would use hydrogen fuel cells in all city-owned facilities in the hopes of being known as "Advanced Hydrogen City Kamisu" (JFS 2017). The Tokyo government has announced it will use CHP units to power the 6000-unit Olympic Village being built for the 2020 Olympics.

While the Road Map's ambitious installation targets will not be met (METI reported installations of around 350,000 units in 2019), unit costs of some stationary fuel cells have already declined below the US$8000 target. As a result, in 2019 the Japanese government was able to end its subsidies for these units (E4tech 2019).[13]

The 2014 Road Map also called for the deployment of stationary fuel cell units for larger buildings such as stores, offices, hospitals, hotels, and data centers. Since then Japanese firms such as Denso, Fuji Electric, Kyocera, and Hitachi Zosen have introduced medium-sized units (around 5 kW), while Toshiba now offers a massive 100 kW system. The industry has secured some promising deals at home, such as the 2017 announcement that Seven-Eleven Japan will use solar-powered CHPs for all its stores. Demand for stationary fuel cells has also been growing internationally. Most notably, in October 2017 Microsoft announced that it would install experimental fuel cells in its data centers worldwide, a significant move because data centers worldwide "draw more than 1 percent of the world's electricity" (Day 2017). Some Japanese firms are moving to meet this demand. Fuji Electric, Panasonic, Toshiba, and Aisin have all begun to export, particularly to end users in Korea and Germany.

Decarbonizing the Transportation Sector. As the transportation sector accounts for some 20% of total worldwide CO_2 emissions, decarbonizing this sector is crucial (Hydrogen Council 2017, p. 30). There is a broad consensus that this will eventually require replacing all internal combustion engine (ICE) vehicles with some form of electric vehicle. A handful of countries, from Norway to China, have begun restricting the sale of ICE vehicles.

Of the two types of electric vehicles, the battery electric vehicle (BEV) is the most well-known, and today accounts for most non-ICE vehicles sold. Proponents of FCEVs, however, argue that BEVs suffer from two major disadvantages. First, batteries are better suited for lighter vehicles and for short-distance driving, whereas FCEVs enjoy a longer driving range that can be more easily extended. Second, recharging BEVs is still time-consuming compared to FCEVs, which can be refueled in a matter of minutes.

Japan's current industrial policy has continued to address the main obstacles to wider adoption of FCEVs: increasing the number of hydrogen refueling stations and providing incentives to consumers to buy or lease an FCEV. These efforts date back to the FCCJ's 2001 Hydrogen and Fuel Cell Framework, which was heavily influenced by Toyota and other Japanese auto firms (Maeda 2003b, pp. 2, 8, and 9).[14] In addition to building the first HRS between Tokyo and Fukuoka, industrial policy also sought to address public perceptions, including concerns for the safety of hydrogen cars and stations.

At the time the 2014 Road Map was published, Japan's industry was only just getting off the ground. Toyota introduced its first FCEV only in December 2014, and Honda began to sell its Clarity in March of 2016. The first policy task was to increase the number of HRS in operation, which at that point numbered only 17 nationwide. The government continued to subsidize the HRS companies, supplying some 50% of station installation costs. The Road Map sought to bring down the costs of building and operating an HRS, including reducing regulations in terms of restrictions on materials and locations.

In 2019, Japan was able to reach the 130-HRS mark, so the Roadmap's goal of 160 stations by 2020 is within range. These efforts were spurred by Japan H2 Mobility, an alliance of 11 Japanese companies that promised to build 80 new HRS by 2022 (Watanabe 2014, p. 17; Behling 2014, p. 30; METI 2017).[15] In terms of vehicle sales, Toyota's production of its Mirai finally reached the 10,000-unit mark in 2019, and the company has plans to increase annual production to 30,000 units.[16] However, total sales in Japan remained far below the Road Map's initial targets. Public subsidies to encourage consumers to buy or lease an FCEV will thus continue.

Japanese car makers have begun to see more growth in demand in overseas markets, the biggest of which so far is California. Exports of the Toyota Mirai to California began in late 2016, and by early 2018 the company had sold or leased more than 3000 units, then accounting for more than half of its worldwide sales. These inroads were in part spurred by the state's impressive public–private California Fuel Cell Partnership, which has created a statewide HRS infrastructure. Similar efforts are underway in the Northeastern US, as well as in Germany, the UK, Denmark, South Korea, and most recently China (IEA Hydrogen 2017).[17]

The industry consensus is that the FCEV enjoys even larger advantages in transportation sectors that involve longer driving ranges and heavier payloads. A 2018 DOE analysis indicates that while small BEVs will remain cheaper in terms of total cost of operations for the near future, FCEVs will be less expensive for nearly all of other transportation sectors (Green Car Congress 2018).[18] This helps to explain why a January 2018 survey of automotive executives found that 62% felt that BEVs will "fail due to infrastructure challenges," while 78% felt that FCEVs represent "the real breakthrough for electric mobility," in part because "range is everything" (KPMG 2017, pp. 14–15). Perhaps it is not the case that the

Japanese have misread the market; perhaps they have seen market trends that others are only now catching on to.

FCEV advantages in terms of longer range and larger payloads make them an attractive option for the trucking industry, which is currently looking for alternatives for its diesel fleets. In 2017, a Utah-based firm, Nikola Motors, introduced its Nikola One semi, which has a driving range of 750 miles and can carry up to 65,000 pounds, far outpacing similar BEV models. The company quickly received orders for 13,000 trucks, totaling $3 billion in deposits (E4tech 2019, p. 21).[19] Medium and heavy-duty trucks are also growing in popularity in Europe, spurred by an EU initiative, H2Haul. In 2018, Hyundai announced a deal to sell 1000 hydrogen trucks in Switzerland, and expects growing sales in markets in the US, Europe, Israel, and China (E4tech 2019).

Hydrogen fuel cells are also showing their competitiveness in the mundane area of material handling (basically moving goods in warehouses), with giant companies such as Amazon and Walmart deciding to switch to FCEV forklifts in their distribution centers. Because these companies operate around the clock and must pay extreme attention to efficiency, they have found that BEV forklifts are too often down for recharging. In contrast, FCEV forklifts can be refilled quickly by the operator, eliminating the need for additional personnel or equipment. For these reasons, there are already some 30,000 FCEV forklifts operating in the US, and this number is expected to grow (E4tech 2019; Ferris 2017).[20] Similar arguments can be made for the entire materials-handling sector, including operations at railroads, ports, and airports.

Another promising development is that some FCEV manufacturers are working to build in-house hydrogen refueling networks that may be made accessible to passenger vehicle owners. Nikola plans to build a nationwide network of 700 refueling stations for their trucks to use, and Amazon and Toyota have announced similar plans (Geuss 2017; Hsu 2016; Casey 2016).[21] The hope is that these HRS networks, built by companies for their own needs, will help ease the shortage of hydrogen refueling stations.

These recent developments are also notable because they are not dependent on government subsidies or inducements; rather, these were decisions that were driven by market concerns for cost and efficiency. Optimists are hopeful that the industry has reached the point where enough private actors perceive similar profit motives, perhaps putting the industry on the cusp of a true take off stage.

The industry has also shown advantages in the mass transit sector, as any vehicle that travels a closed-circuit loop could usefully adopt fuel cell technology. In Japan, Hino Motors and Toyota worked together to create a fuel-cell bus and carried out tests in Toyota City in January 2015. The 2020 Olympic Village will rely on a fleet of 100 such buses. Governments worldwide have encouraged the adoption of fuel-cell buses, especially in Europe and China, but also in South Korea, the UK, India, and California (IEA Hydrogen 2017, p. 11; EPCA 2018; METI/ANRE 2015, p. 7). As a result, worldwide sales of these buses are now approaching "commercial normality" (E4tech 2019).

Fuel cell trains are also growing in attractiveness, especially for countries that are looking to replace their diesel train fleets. In 2018, Germany put its first fuel cell passenger train into service, and the success of this unit quickly led to orders for 40 more. Operators in the UK, France, Canada, Korea, and China are also putting such trains into service (E4tech 2019). The maritime sector is another intriguing place where fuel cells can be utilized. Numerous projects have been reported in which companies have converted existing small ferries and cruise ships to run on fuel cells, including in Norway, Scotland, France, Denmark, Korea, and San Francisco (IEA Hydrogen 2017; E4tech 2019). In October 2017, Viking Cruises announced that it plans to build the world's first cruise ship powered by hydrogen fuel cells. The European Maritime Safety Agency also reports 12 different shipping projects that have investigated various fuel cell technologies (MAREX 2017).

A final intriguing development is recent interest on the part of the US military in using FCEVs for rear-area transportation and material handling. The Department of Defense (DoD) has been working with Chevrolet to create prototypes for convoy vehicles and unmanned aerial drones. As a huge user of energy, the DoD is evidently looking to save on fuel and energy costs (E4tech 2019).[22] The DoD has also discussed with General Motors the idea of building an unmanned fuel cell submarine. While these ideas are still at an early stage, demand from the military would be a big boost for the FCEV industry.[23]

In short, FCEVs are just now beginning to compete with BEVs in the small car market, but are likely to dominate other transportation sectors. If FCEVs are successful in these other sectors, this success may spill over back to the small car sector. If businesses build up their hydrogen refueling networks, the current lack of HRS may become less of a problem in the future.

Hydrogen's Role in Facilitating the Use of Renewable Energy. Industrial policy for the final pieces of the hydrogen puzzle, storage and the green hydrogen supply chain, is still at the stage of supporting basic research and product development. As a result, much of the work being done is not yet in the public eye; like other cases of industrial policy, results are often not visible until years or even decades later. The dilemma in this case is that we are running out of time, so there is a great deal of pressure on policymakers and the industry to show results now.

Perhaps the most compelling potential role for hydrogen is as an energy storage medium, a missing link that is holding back the more widespread adoption of renewable energy. As a stable energy carrier, hydrogen can help to ease the problem of "intermittency" that is often cited by electric utilities as a reason that they cannot shift entirely to renewable resources. Utilities have argued that relying on renewables makes it more difficult to maintain the delicate balance between energy put into the system and energy taken out by users. If the utilities rely too much on intermittent sources such as solar and wind, they worry about periods of plunging electricity supply as well as periods of excess production. Utilities can continue to use their natural gas "peaker plants" to cope with supply imbalances, but this "ramping up" is expensive. Hence, many utilities have resisted the widespread introduction of renewable sources.

During periods where there is an excess of electricity from renewable sources, utilities must figure out what to do with it. Some utilities have been able to transfer the excess to neighboring power utilities, but this can be expensive and inefficient. Utilities have also been experimenting with smart grids, smart meters, and other technologies that are able to make matching production and usage more efficient. While helpful, these solutions have not been enough. Because allowing too much electricity into the grid would be destabilizing, utilities often end up simply throwing their excess electricity away, a process known as curtailment.

The problem of intermittency is also a seasonal one. It is easy to imagine a country that has limited access to sunlight during the winter months. Even if this country had abundant solar capacity and can create excess energy during the summer months, without an effective way of storing that energy, it would still need to rely on non-renewable resources in other months. As one example, the IEA notes that "solar generation in Europe is about 60% lower in winter than in summer," while demand in winter is about 40% higher (IEA Hydrogen 2017).

The solution, all seem to agree, is energy storage. The ability to store energy would allow utilities to even out their daily, weekly, and seasonal fluctuations, and would remove a major objection to the wider use of renewables. The problem, however, is that all our current storage technologies have limitations in terms of cost, efficiency, or scale.

The storage approach most widely used today is "pumped hydro" in which water is pumped to an elevated position and then released when needed to run electricity-generating turbines. Especially when utilities create surplus electricity, this method can be very efficient. Pumped hydro today accounts for 95–99% of all electricity storage currently in use, and some countries with suitable geography have placed high hopes on the technology (Williams 2017).[24] Critics, however, point to some major limitations in this approach. At the moment, the amount of storage currently available is tiny compared to the massive amount of storage that will eventually be needed. Not all areas have access to the geography needed to make this work, and facilities are often far away from where the electricity will be used. Also, many of these projects may do considerable damage to the environment (Brouwer 2017).

Other technologies have been used on a smaller scale. One is compressed air energy storage (CAES) in which electricity is used to compress air, which is then stored, reheated, and used to power turbines. Only two large CAES plants are currently in operation, in part because of relative low efficiency rates (40–70%). Other solutions such as superconducting magnetic energy storage, super-capacitors, and flywheels are all capable of providing instant power, but are either expensive or are not able to store enough energy for long enough (Lazarou and Makridis 2017, p. 2; METI 2017).[25]

Currently, a lot of attention is being paid to batteries as a storage medium, especially to help utilities deal with short-term or daily fluctuations. Tesla has made the biggest splash, installing large-scale battery units in Australia and Southern California in 2017. However, critics argue that batteries are large and expensive, and most importantly cannot hold their charge for long enough. As one analyst puts it, "lithium ion batteries, to be quite honest, are probably going to be good for 4–6 hours of grid storage and up to a couple hundred megawatts in scale. When you get beyond that, the challenges of lithium ion scalability become apparent."[26]

As shown in Fig. 7.2, the main storage technologies have limitations in terms of scale or length of storage. Analysts thus realize that the full adoption of renewable sources increases the need for "weeks- and months-long

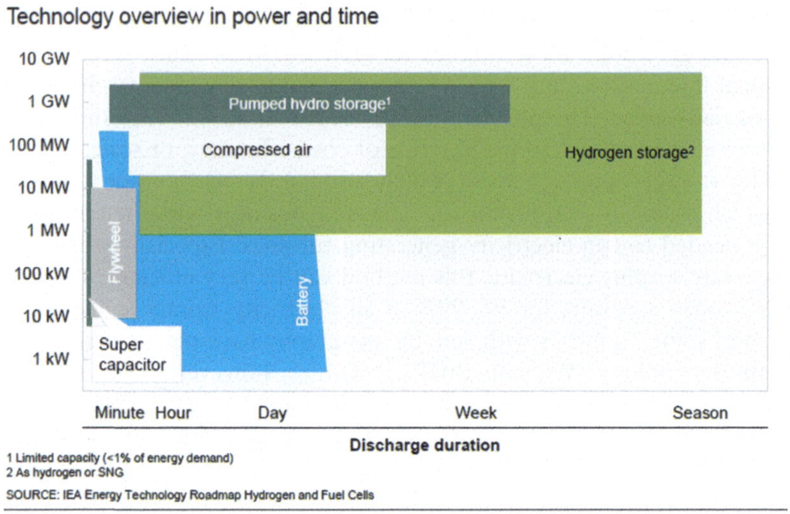

Fig. 7.2 Hydrogen can be stored for months without losing much of its power (*Source* Figure from Hydrogen Council 2017, p. 58. Used with permission)

energy storage," or as mentioned above, even on a seasonal basis (Spector 2018).

Using hydrogen as an energy storage medium is thus receiving growing support among analysts, as hydrogen is "exceptionally well suited to store large quantities of energy for long durations." That is, once created, hydrogen can be "time shifted" to meet later needs. Hydrogen created with excess summer solar electricity, for example, could be stored and then utilized in winter to run stationary fuel cell power plants.

Hydrogen as an energy carrier is also flexible in terms of location. Hydrogen storage may be very useful for areas that do not have the needed geography (for instance, it may be difficult to use pumped hydro in a flat desert area), or in remote areas that are not connected to a larger grid system. This flexibility allows for the distribution of energy across regions, including internationally. The Hydrogen Council hopes that transporting hydrogen across borders will be a major component of the hydrogen economy, aiming for a target of 55 million tons of hydrogen to be traded internationally by 2050, amounting to 10% of total annual hydrogen demand (Hydrogen Council 2017, p. 57).

Storage of the massive amounts of hydrogen is a final technological problem that must be overcome. The IEA reports that "underground storage of hydrogen in salt caverns and depleted oil wells is a well-established practice," while the German government states that storing hydrogen gas can utilize the country's existing "extensive gas grid" (IEA Hydrogen 2017, p. 58; Germany Trade & Invest 2017). If these obstacles can be resolved, the industry's Hydrogen Council, predicts that soon some "250 to 300 tWh of surplus renewable electricity could be stored in the form of hydrogen," with a 2050 goal of 500 TWh (Hydrogen Council 2017, p. 10).[27] If successful, hydrogen-based energy storage could allow for the flexible distribution of energy across sectors, regions, and even across time (Lazarou and Makridis 2017, p. 5).

Securing the Green Hydrogen Supply Chain. The final set of technological problems facing the hydrogen society is the creation of a stable supply chain for green hydrogen. Japan's decision to create its Hydrogen Society spurred Japanese companies to redouble their efforts, but all realize that the amount of hydrogen that will be needed is massive. Furthermore, many technical solutions, especially how best to transport hydrogen over long distances, are only now in the research and development phase. As a result, the 2014 Road Map simply calls for more sustained R&D efforts and relegates them to Phase 3, set to show fruit only around 2040.

Currently, most hydrogen produced worldwide is not considered green, with 95% using steam methane reforming. Steam reformers could use carbon capture and storage (CCS) technology, but while some research efforts in Japan aim to bring these costs down, the process is still expensive (U.S. Department of Energy 2016).[28] It should also be noted that the Road Map also calls for the use of "unutilized energy resources imported from overseas," which basically means "low-grade coals like lignite, crude petroleum, and associated gas in gas fields" (METI/ANRE 2015, p. 10). This far-from-green policy is in part a pragmatic recognition that the world simply cannot create enough green hydrogen currently.

How long Japan will rely on non-green hydrogen will depend on progress toward the final piece of the production puzzle: lowering the cost of creating hydrogen through the electrolysis of water, in which electricity is used to split water into hydrogen and oxygen. Analysts have discussed for decades creating hydrogen not only from solar and wind, but also hydro, nuclear, geothermal, and more recently from biomass, all processes known as "power to gas" or P2G technology. In any case, this key technology is the furthest from commercialization, and thus has been left to the third phase of the Road Map.

The industry must also improve its transport technology. Japan's efforts have been ongoing for more than 30 years, dating back to 1993, when the WE-NET program selected liquid hydrogen (LH2) as the best way to transport hydrogen (Nogrady 2017; Akaishi 2007; Koguchi 2010; Hirai 1997; Chiba et al. 1998).[29] Since then Japanese researchers and industry have been investigating a wide range of possible technologies and have begun work on building a hydrogen transportation infrastructure (Karagiannopoulos et al. 2017; Grandum 2016, pp. 29–30; Nishimura 2015, pp. 16, 22 and 27).[30] The jury is still out as to which of these technologies will emerge as the best for transporting hydrogen, so at the moment the Road Map only provides funding for a wide range of R&D and pilot projects.

Although the technical obstacles to Japan's hydrogen society still seem daunting, potential producers of hydrogen are beginning to eye the huge market opportunity it would represent. This is especially true for countries that have ample renewable energy resources. Norway, for example, may be able to turn energy from its hydroelectric resources and wind power into hydrogen (Karagiannopoulos et al. 2017). Further, hydrogen can be produced anywhere in the world, including even the most remote or underdeveloped of areas. All that would be needed would be ample sunshine or wind (Lazarou and Makridis 2017, p. 3).[31] As one example, a remote and undeveloped area like Australia's Pilbara "is considered the world's best solar resource in terms of the intensity of the sunlight and the fact that it's never cloudy. That means you can potentially set up a big solar farm there (to create hydrogen) … then you need to get that renewable energy to Japan" (Nogrady 2017). Other analysts are excited that exporting renewable energy could replace Australia's current reliance on exporting fossil fuels. "Here's an exciting thought: Australia has the potential to be one of the largest generators and exporters of renewable energy in the world. We could be a renewable energy superpower" (Rawson 2017). Such possibilities will be open to any country that has access to renewable energy resources.

Conclusion

Will Japan succeed in its effort to create its hydrogen society? Will the global industry finally fulfill the very high hopes that fuel cell proponents have discussed for decades? Skeptics can be forgiven for feeling that we have seen this picture before, as there have been many past periods where hopes have been raised, only to be dashed. Many of the problems the industry is discussing today—green production and the problems of transport and storage, for instance—are the same ones that have been discussed for decades.

Today, however, proponents believe that technologies have developed to the point of being commercially viable and are encouraged by the constant stream of announcements of new technical breakthroughs and products introduced to the market in all sectors of the energy economy. Proponents have also been heartened by Japan's political commitment to support the industry. The feeling is that Japan's past policy efforts have been crucial in the industry's initial progress, especially in the early R&D stages, when companies had few incentives to invest in expensive research on their own. Furthermore, going forward, the view is that continued support for commercialization is necessary if enough progress is going to be made quickly enough. As one industry participant puts it, "leaving (the development of the hydrogen society) solely to the ingenuity and technological development of the private sector wouldn't give us the speed and sense of urgency needed. With the government taking the lead in an industry-academy-government program, we should be able to speed development by a significant margin" (Muraki 2015).[32]

The final reason for optimism is that two other regional governments have recently made similar industrial policy commitments to increase the use of hydrogen fuel cells. The Korean government in 2018 announced a Hydrogen Roadmap of its own, which focused on FCEVs and large-scale stationary fuel cells. And China in 2019 made a huge international splash when its government announced that it would promote the adoption of fuel cells in cars, buses, and trains. These announcements raised the possibility that the dynamics of regional competition among these historic rivals may also spur further interest, investment, and progress in the industry.

The stakes involved are high, not only in terms of desperately needed progress on combatting climate change, but also in terms of the economic stakes that are involved. The Hydrogen Council, for instance, claims that a full “scaling up” of the industry would reduce annual CO_2 emissions by 6 Gt yearly, and could lead to “revenues of more than $2.5 trillion per year, and jobs for more than 30 million people globally” (Hydrogen Council 2017, p. 9). Although analysts are concerned about which company or country ends up being more successful, what may ultimately be more important to all of us is whether the hydrogen and fuel cell industry can be developed quickly enough so that it can become part of the solution to our current environmental crisis.

Notes

1. Hydrogen and fuel cells are tied together in terms of usage but are separate technologies. Fuel cells can utilize many gasses, including methane, natural gas, and ammonia, but today the most promising focus is on hydrogen. Hydrogen can also be used as a clean source for power generation, as a feedstock for industrial uses, and as a storage medium. Hydrogen could replace fossil fuels used in the production of certain chemicals, petrochemicals and steel (Hydrogen Council 2017; Brouwer 2017; IEA Hydrogen 2017).
2. Steven Chu, President Obama’s Secretary of Energy, for instance, argued that we will need “four miracles” to occur before fuel cells will be viable (Mench 2015).
3. As Rodrik (2014), argues that governments today often use policy to favor the fossil fuel sector. Green industrial policy advocates argue that policy should be used to support cleaner, future technologies.
4. The Hydrogen Council, a leading international policy group that represents most companies in the hydrogen sector, makes this argument. While self-serving, the Hydrogen Council’s call to action is exactly what Green industrial policy scholars agree is necessary.
5. In the case of energy policy, METI has long been the most influential government agency. Other energy-related government agencies include the ANRE, which is a part of METI, and the New Energy and Industrial Technology Development Organization (NEDO), a semi-governmental organization affiliated with METI. Another major actor is the National Institute of Advanced Industrial Science and Technology (AIST), which oversees Japan’s research facilities.
6. In this first basic research stage, the Japanese government provided some US$3.6 billion for hydrogen research and “modest funding” for R&D on a range of basic fuel cell technologies. Further support for fuel cell technologies were a part of METI’s 1981 “Moonlight Plan” (Maeda 2003a, p. 10).

7. This group was created through an "administrative decision" of ANRE's Director General, rather than by statute. In addition to one official from NEDO and AIST, the advisory group consisted of nine academic experts and representatives of Japan's four main automobile firms, three petroleum suppliers, five electric and gas utilities, and four electronics makers.
8. The group was renamed in 2013, but there has been great continuity in terms of leadership, membership, and policy roles.
9. Although there were some changes in the makeup of the group, the important point is the continued role that this Advisory Council has played in hydrogen and fuel cell policy, dating back to 1999.
10. This is a corollary of the Incerti and Lipscy (2018), view, stressing more the influence of the industry on policy makers. However, the authors are correct that Japanese energy policy under Abe has been mixed together with policy to stimulate economic growth, which they label "Abenergy-nomics." Japanese industrial policy has always incorporated a concern for economic growth, and this policy was no different.
11. These are also known as "co-generation" units. Because CHP units recapture heat that would otherwise be lost, their energy efficiency rates are relatively high, from 80% and even up to 95%, according to (Hashimoto 2015).
12. One such project began in 2005, which NEDO described as the "Demonstration of Residential PEFC System for Market Creation" (Koguchi 2010, p. 8). Another started in 2008, when NEDO helped Nippon Oil and Seibu Gas install 150 prototype ENE-FARM units in the Kyushu city of Maebara (Pasternack 2009).
13. While PEM stationary units had reached the target price, SOFC fuel cells are still above it, so small subsidies are still available.
14. Toyota began its efforts to develop fuel cells for automobile use in 1992, in part as a response to the "toughening of environmental regulations in North America," including the Zero-Emission Vehicle (ZEV) mandate created by California in 1990.
15. The alliance is led by Toyota and JXTG Nippon Oil and includes Honda, Nissan, Idemitsu Kosan, Iwatani, Tokyo Gas, Toho Gas, Air Liquide Japan, and Toyota Tsusho (Tajitsu et al. 2018).
16. Japanese auto firms have recognized the need to reduce vehicle costs in the coming years. Toyota, for instance, has set very ambitious targets, aiming for production costs at 1/100 of its 2008 costs (Behling 2014, p. 16). This would bring costs well below the costs of most hybrids and BEVs.

17. As of December 2019, California had 44 stations in operation, with 20 more pending. The German effort, H2 Mobility, now has put 78 hours into operation. Norway plans 200 hours to fuel 50,000 FCEVs, with 20 scheduled for completion by 2020. Korea eventually plans to have 200 hours nationwide, and China is building 100 hours by 2020.
18. It is important to note that the TCO for small FCEVs will be lower than BEVs by 2040. One sector to watch is the SUV and pickup market, as these are two very popular vehicles where fuel cells may prove very useful. Major auto firms have been hinting that products in this segment are on their way.
19. A significant customer is Anheuser-Busch, which ordered 800 of the semis.
20. FCEV forklifts have also been in operation in France, Japan, Norway, and elsewhere, but the largest demand is in the US (IEA Hydrogen 2017).
21. Nikola aims to build this network by 2028.
22. For the moment, the DoD is not looking to use fuel cells in combat vehicles, but some have argued that their quiet operation and power may be seen as possible advantages.
23. Analysts have mentioned other uses for fuel cells, including in self-driving or even flying cars, but also in sectors of the aviation industry, but these are not yet close to marketability (Ferris 2017; Klippenstein 2017).
24. In pumped hydro "energy is stored as gravitational potential energy." Operators can pump the water back to the elevated position when electricity is abundant, then generate electricity when needed.
25. Williams (2017), provides a useful synopsis.
26. The problem is that Tesla's battery unit is gigantic, costly, and cannot hold its charge over long period. New types of batteries, including flow batteries, may be able to hold energy longer, but not for the long periods needed Mammoser (2018).
27. In addition, another 200 TWh could be generated in large power plants using hydrogen.
28. Although this process is not completely clean, when used with a FCEV, the total greenhouse gases emitted is less than a natural gas vehicle. Furthermore, it should be noted that a great deal of electricity used to charge BEVs also relies on less than green resources.
29. The drawbacks of liquid hydrogen are that the process, which requires very low temperatures and very high pressures, is costly and also energy intensive.
30. Some possibilities that are prominently mentioned are ammonium, methanol, and other liquid organic hydrogen carriers.
31. If successful, hydrogen could open some areas that are currently not even being considered. An example would be a deep-sea facility or ship that converts wind or tidal movements into electricity, and then into hydrogen.
32. The speaker is Muraki Shigeru of Tokyo Gas.

References

Akaishi, Koishi. 2007. Governments Role for Innovation. PowerPoint presentation presented at the OECD Seminar, September 20, Berlin.

Behling, Noriko Hikosaka. 2012. Strengths and Weaknesses of Major Government Fuel Cell R&D Programs: Europe, Japan, and the United States. *In Fuel Cells: Current Technology Challenges and Future Research Needs*, 601–619. Amsterdam, the Netherlands: Elsevier Science.

Behling, Noriko Hikosaka. 2014. Japan's Fuel Cell and Hydrogen Commercialization Strategy. PowerPoint presentation presented at the Fuel Cell Seminar, November 10.

Behling, Noriko Hikosaka. 2015. Japan's New Plan to Become a Hydrogen Society: Will Japan Succeed? *ECS Transactions* 65 (1): 1–23.

Behling, Noriko Hikosaka, Mark C. Williams, and Shunsuke Managi. 2015. Fuel Cells and the Hydrogen Revolution: Analysis of a Strategic Plan in Japan. *Economic Analysis & Policy* 48: 204–221.

Brouwer, Jack. 2017. Interview with Jack Brouwer, National Fuel Cell Research Center (NFCRC), UC Irvine.

Casey, Tina. 2016. *One Step Closer to the Hydrogen Economy Dream*. Clean Technica, June 30.

Chiba, Mitsugi, Harumi Arai, and Kenzo Fukuda. 1998. WE-NET: Japanese Hydrogen Program. *International Journal of Hydrogen Energy* 23 (3): 159–165.

Daugherty, Cyrus. 2016. Hydrogen Is the Energy of the Future for Japan. *International Association for Hydrogen Energy* 8 (3): 1.

Day, Matt. 2017. Microsoft Makes a 'Crazy' Bet on Fuel Cells to Feed Power-Hungry Data Centers. *Seattle Times*, September 23.

EPCA, India. 2018. Hydrogen Fuel-Cell Buses. 88. New Delhi, India. http://www.epca.org.in/EPCA-Reports1999-1917/report88-hydrogen-buses.pdf. Accessed 29 Feb 2020.

E4tech. 2019. *The Fuel Cell Industry Review 2019*. London, England: E4tech.

Ferris, Robert. 2017. *Elon Musk Hates Hydrogen, but Automakers Are Still Investing in It—And for Good Reason*. CNBC, May 26.

Germany Trade & Invest. 2017. *The Energy Storage Market in Germany*. Berlin: Germany Trade & Invest.

Geuss, Megan. 2017. *Toyota Not Giving Up on Fuel Cell, but Now Banking on Electric, Too*. Ars Technica, December 18.

Grandum, Svein. 2016. Japan's Developments to Secure a Renewable Energy Future. PowerPoint presentation presented at the NTNU-Experts in Team, January 18.

Green Car Congress. 2018. DOE Analysis Suggests Rapid Convergence of FCEV and BEV TCOs; FCEVs Less Expensive for Majority of LDV Fleet by 2040; Mass Compounding. Green Car Congress, January 14.

Hallegatte, Stéphane, Marianne Fay, and Adrien Vogt-Schilb. 2013. Green Industrial Policies: When and How. Policy Research Working Paper 6677. The World Bank.

Hashimoto, Michio. 2015. Japan's Hydrogen Policy and Fuel Cells Development in NEDO. PowerPoint presentation, April 23.

Hirai, Toshihiro. 1997. *WE-NET: The National Hydrogen Project of Japan, Its Vision and Status*. Tokyo, Japan: Engineering Advancement Association of Japan.

Hoffman, Peter. 2012. *Tomorrow's Energy: Hydrogen, Fuel Cells, and the Prospects for a Cleaner Planet*. Cambridge, MA: The MIT Press.

Hsu, Tiffany. 2016. *Nikola's Hydrogen Fuel Stations Could Propel Fuel Cell Cars*. Trucks.Com, December 20.

Hughes, Llewelyn. 2012. Climate Converts: Institutional Redeployment, Industrial Policy, and Public Investment in Energy in Japan. *Journal of East Asian Studies* 12 (1): 89–117. https://doi.org/10.1017/S1598240800007633.

Hydrogen Council. 2017. *Hydrogen—Scaling Up: A Sustainable Pathway for the Global Energy Transition*. Belgium: The Hydrogen Council.

IEA Hydrogen. 2017. *Global Trends and Outlook for Hydrogen*. Paris, France.

Incerti, Trevor, and Phillip Y. Lipscy. 2018. The Politics of Energy and Climate Change in Japan Under Abe: Abenergynomics. *Asian Survey* 58 (4): 607–634. https://doi.org/10.1525/as.2018.58.4.607.

JFS. 2017. Japanese City Formulates Comprehensive Strategy to Promote Safe, Sustainable Hydrogen Use. *Japan for Sustainability*, June 11.

Karagiannopoulos, Lefteris, Sonali Paul, and Aaron Shildrick. 2017. *Norway Races Australia to Fulfill Japan's Hydrogen Society Dream*. Reuters, April 28.

Klippenstein, Matthew. 2017. *Fuel Cell Fall Update: Drones, Cruise Ships and Carbon Capture*. Greentech Media, November 2.

Koguchi, Haruhisa. 2010. Research and Development of Fuel Cells and Hydrogen in Japan. PowerPoint presentation presented at the 3rd FCH JU Stakeholders General Assembly, November 9, Brussels, Belgium.

KPMG. 2017. *Global Automotive Executive Survey 2017*. Amstelveen, Netherlands.

Lazarou, Stavros, and Sofoklis Makridis. 2017. Hydrogen Storage Technologies for Smart Grid Applications. *Challenges* 8 (13): 1–11.

Maeda, Akira. 2003a. Fuel Cell Technologies in the Japanese National Innovation System. PowerPoint presentation presented at the International Conference on Innovation in Energy Technologies, September 29, Washington, DC.

Maeda, Akira. 2003b. *Innovation in Fuel Cell Technologies in Japan: Development and Commericalization of Polymer Electrolyte Fuel Cells*. Tokyo, Japan: OECD/CSTP/TIP Energy Focus Group.

Mammoser, Alan. 2018. *What's Holding Renewable Energy Back? An Interview with Dr. Steve Griffiths, Part 1*. OilPrice.Com, August 7.

MAREX. 2017. World's First Hydrogen-Powered Cruise Ship Scheduled. *The Maritime Executive*, October 3.

Maruta, Aki. 2016. Japan's ENE-FARM Programme. PowerPoint presentation presented at the Fuel Cells: Why Is Austria Not Taking Off? Austria, October 10.

Mench, Matthew M. 2015. High Hopes for Hydrogen: Fuel Cells and the Future of Energy. *Foreign Affairs* 94 (November): 117–123.

METI. 2017. New Era of a Hydrogen Energy Society. PowerPoint presentation, Tokyo, Japan, February 28.

METI/Agency for Natural Resources and Energy (ANRE). 2015. New Era of a Hydrogen Energy Society. PowerPoint presentation, Tokyo, Japan, October 19.

Midford, Paul. 2014. Conclusions. In *The Political Economy of Renewable Energy and Energy Security: Common Challenges and National Responses in Japan, China, and Northern Europe*, ed. Espen Moe and Paul Midford, 318–324. New York, NY: Palgrave Macmillan.

Ministry of Economics, Trade and Industry (METI). 2014. *Strategic Energy Plan*. Tokyo, Japan: METI.

Muraki, Shigeru. 2015. *Leading the World in Hydrogen Energy Usage; Creating a Low-Carbon, Hydrogen-Based Society*. SIP.

Nishimura, Motohiko. 2015. Hydrogen Energy Supply Chain from Overseas. PowerPoint presentation presented at the Japan-Norway Energy Science Week, May 27.

Nogrady, Bianca. 2017. How Australia Can Use Hydrogen to Export Its Solar Power Around the World. *The Guardian*, May 18.

OECD/IEA. 2004. *Hydrogen & Fuel Cells: Review of National R&D Programs*. Paris, France: OECD/IEA.

Okano, Kazukiyo. 2016. Development Histories: Hydrogen Technologies. In *Hydrogen Energy Engineering: A Japanese Perspective*, ed. Kazunari Sasaki, Hai-Wen Li, and Akari Hayashi, 53–92. New York, NY: Springer Japan.

Pasternack, Alex. 2009. *World's Largest Hydrogen-Powered Town Starts in Japan*. TreeHugger, February 5.

Rawson, Jane. 2017. Could Renewable Energy Be Our Next Great Export?—Australian Renewable Energy Agency. *ArenaWire*, July 23.

Rodrik, Dani. 2014. Green Industrial Policy. *Oxford Review of Economic Policy* 30 (3): 469–491.

Romm, Joseph T. 2004. The Hype about Hydrogen. *Issues in Science and Technology* 20 (3): 74–81.

Shayon, Sheila. 2016. *Smart City: Panasonic to Co-Develop Sustainable Town in Japan*. Brandchannel (Blog), April 1.

Spector, Julian. 2018. *Inside Form Energy, the Star-Studded Startup Tackling the Toughest Problem in Energy Storage*. Greentech Media, June 18.

Tajitsu, Naomi, Osamu Tsukimori, and Richard Pullin. 2018. *Japan Venture Aims to Build 80 Hydrogen Fueling Stations by 2022*. Reuters, March 5.
U.S. Department of Energy. 2016. Carbon Capture, Utilization, and Storage—Climate Change, Economic Competitiveness, and Energy Security.
Watanabe, Shigenobu. 2014. Hydrogen Infrastructure in Japan. PowerPoint presentation presented at the 2014 AMR, June 19, Washington, DC.
Williams, Hayley. 2017. Beyond Batteries: How Energy Storage Can Make Australia's Renewables Reliable. *Gizmodo Australia*, June 29.
Zubrin, Robert. 2007. The Hydrogen Hoax. *The New Atlantis* 15 (Winter): 9–20.

PART II

New Challenges and Opportunities in East Asia

CHAPTER 8

Between the Rhetoric and the Reality: Renewable Energy Promotion vs. Adoption in South Korea

So Young Kim and Inkyoung Sun

Introduction

With a population of fifty-one million, South Korea is the 12th largest economy when measured in nominal gross domestic product (GDP) as of 2015. It ranks higher on energy consumption. South Korea is the 9th largest energy consumer with per capita energy consumption being the 7th largest. Because it lacks domestic energy reserves despite relatively high demand for energy, it has become one of the top energy importers in the world. About 95% of energy sources come from abroad. South Korea relies on imports for about 98% of its fossil fuel consumption. It is

S. Y. Kim (✉)
Korea Advanced Institute of Science and Technology (KAIST), Daejeon, South Korea
e-mail: soyoungkim@kaist.ac.kr

I. Sun
Science and Technology Policy Institute (STEPI), Sejong, South Korea

P. Midford and E. Moe (eds.), *New Challenges and Solutions for Renewable Energy*, International Political Economy Series,
https://doi.org/10.1007/978-3-030-54514-7_8

among the world's top five importers of liquefied natural gas (2nd), coal (4th), and crude oil (5th).

Nonetheless, South Korea is a laggard in the adoption of renewable energy with the lowest rate of renewable energy use among member countries of the International Energy Agency (IEA) as seen in Fig. 8.1. It derives only 1.1% of its energy from renewable sources, though this figure is a substantial improvement over the previous rate of renewable energy use that has stalled for 25 years since 1990. In addition, South Korea ranks the lowest in electricity generation from renewable sources as a percentage of the total electricity generation.

Yet, South Korea has spent a generous amount in the research and development (R&D) of renewable energy. According to the IEA energy statistics, the size of South Korea's investment in research, development, and demonstration (RDD) for renewable energy is the 7th largest as of 2011, after only the US, Japan, Germany, France, the UK, and Canada. That is, other than these major advanced countries, South Korea spends the largest amount on renewable energy R&D. In particular, its investment in renewable energy R&D has accelerated in the 2000s as illustrated in Fig. 8.2, reaching the peak level of $172 million in 2010.

South Korea has not only invested heavily in renewable energy R&D but put great emphasis on renewable technologies as a way to deal with climate change. Notably during the Lee Myung bak administration (2008~2012), the so-called Green Growth was the mantra of government policymaking on virtually every level. Renewable energy and related

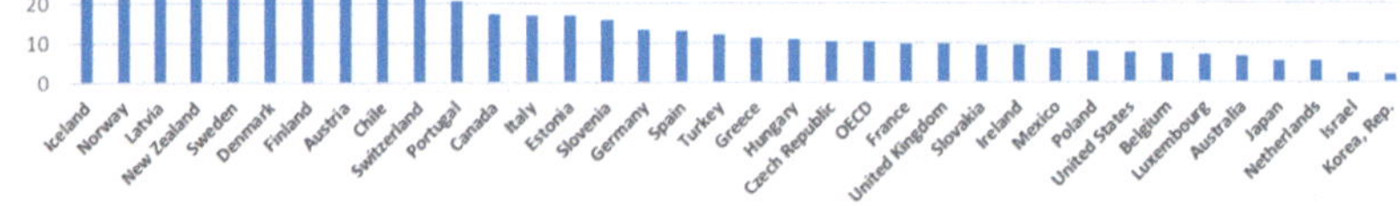

Fig. 8.1 Renewables as % of total primary energy supply (TPES) in OECD countries (2017) (*Source* Author created based on data extracted on September 10, 2019 from OECD iLibrary DB)

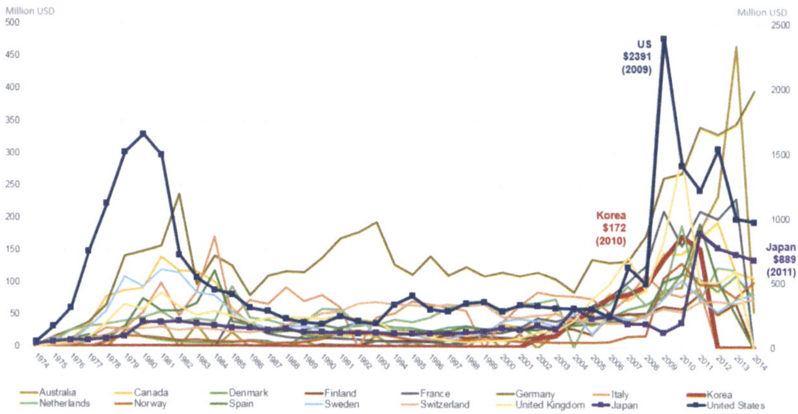

Fig. 8.2 Renewable energy RDD (1974~2014) (*Source* Author created based on data extracted on February 16, 2017 from OECD iLibrary DB)

industries were viewed as the new growth engine, forming one of the central components of the governmental initiative to promote green growth.

It is then quite puzzling that despite the aggressive promotion of renewable energy in the face of climate change as well as generous investment in renewable technologies South Korea still remains a country with the lowest rate of renewable energy adoption. Why does the country investing heavily on renewable technologies draw so little on renewable sources? This study addresses the apparent gap between the promotion of renewable energy and related technologies and industries and the penetration of renewable energy in energy consumption in South Korea by exploring political, economic, and technological dimensions of renewable energy adoption.

Renewable Energy Promotion in South Korea

Legal and Policy Measures. As one of the few countries that successfully transformed an agricultural economy into a modern industrializing economy in just three decades, South Korea has always been keen on the stability of energy supply for industrial production. It is no wonder that the central goal of energy policymaking in postwar South Korea was to secure cheap energy. Indeed, the most important consideration in the choice of energy was the economic criteria such as cost efficiency and price competitiveness. Given the low economic feasibility and unstable supply, renewable energy did not receive much attention in the early years of South Korea's postwar industrialization drive. The first systematic initiative of the South Korean government to promote renewable energy did not come until almost two decades after the launch of the government-driven economic growth plans.[1] Largely driven by the need to diversify energy sources in the aftermath of two oil shocks, the Korean government enacted the Act on the Promotion of the Development of Alternative Energy in 1987.

With the adoption of the Kyoto Protocol, the first international agreement to set mandatory goals for greenhouse gas reduction, the Korean government amended the aforesaid law in 1997 with the emphasis on environment-friendly energy. The name of the law itself was revised to the Act on the Promotion of the Development, Use and Diffusion of Alternative Energy. Several new clauses were added to this revised law. In particular, the law mandated the government to create a long-term plan for the systemic development of alternative energy technologies and the widespread utilization of alternative energy (Article 4). Another notable change was the introduction of the feed-in-tariff (Clause 6 of Article 11), according to which the difference between the base price and the transaction price of electricity generated by alternative energy would be supported by the Government Electricity Fund.

This revised law was extensively amended in 2004 as the Act on the Promotion of the Development, Use and Diffusion of New and Renewable Energy. The number of articles was almost doubled from 19 to 35, making the amended law almost a novel one. Most notably, the vague term of alternative energy was replaced by "new and renewable" energy. This was more than just the change of the name, however. As shown in Table 8.1, the types of alternative energy were rather simple in the first two laws, but were substantially expanded to cover almost all energy sources other than petroleum, coal, nuclear power, and natural gas.

In addition, the scope of the special fund for technology development and diffusion of new and renewable energy stipulated in the law was substantially enlarged. Previously such a fund was devoted only to purely technological development. In the 2004 law, the fund was allowed to support a wide variety of programs and activities to promote new and renewable energy such as special training programs for the skilled workforce, international standardization, support for the implementation of

Table 8.1 Definition of alternative/new/renewable energy in each promotion act

Act on the Promotion of the Development of Alternative Energy (1987)	*Act on the Promotion of the Development, Use and Diffusion of Alternative Energy (1997)*	*Act on the Promotion of the Development, Use and Diffusion of New and Renewable Energy (2004)*
a. Solar energy b. Bio energy c. Wind power d. Water power e. Fuel cells f. Gasified coal g. Marine power h. Energy from waste materials i. Other energy prescribed by Presidential Decree	a. Solar energy b. Bio energy c. Wind power d. Water power e. Fuel cells f. Energy from liquefied or gasified coal, and energy from gasified heavy residual oil g. Marine power h. Energy from waste materials i. Hydrogen energy j. Other energy prescribed by Presidential Decree	New energy a. Hydrogen energy b. Fuel cells c. Energy from liquefied or gasified coal, and energy from gasified heavy residual oil prescribed by Presidential Decree d. Other energy prescribed by Presidential Decree, other than petroleum, coal, nuclear power, or natural gas Renewable energy a. Solar energy b. Wind power c. Water power d. Marine power e. Geothermal energy f. Bio energy converted from biological resources prescribed by Presidential Decree; g. Energy from waste materials prescribed by Presidential Decree h. Other energy prescribed by Presidential Decree, other than petroleum, coal, nuclear power, or natural gas

Source Author created table

mandatory production and consumption targets, and subsidies for firms specialized in the installation of new and renewable facilities.

The latest government plan for new and renewable energy promotion as stipulated by the aforesaid law is the Fourth Basic Plan for New and Renewable Energy prepared in 2014. With the target period of 2014~2035,[2] the plan set energy supply targets for each type of renewable energy source with the overall target of 11% of primary energy supply produced by new and renewable energy by 2035. As displayed in Fig. 8.3, the most distinctive change is found in the reduction of energy from waste materials (67% in 2014 to 29.2% in 2035) and the sizeable expansion of solar and wind energy in total primary energy supply (from 8% in 2014 to 40.2% in 2035). This strategic focus on solar and wind energy is largely out of consideration for the industrial impacts on the related industries, for the share of photovoltaics (PV) and wind energy reached 91% of the total investment in renewable energy and 97% of exports of renewable energy during the previous planning period.

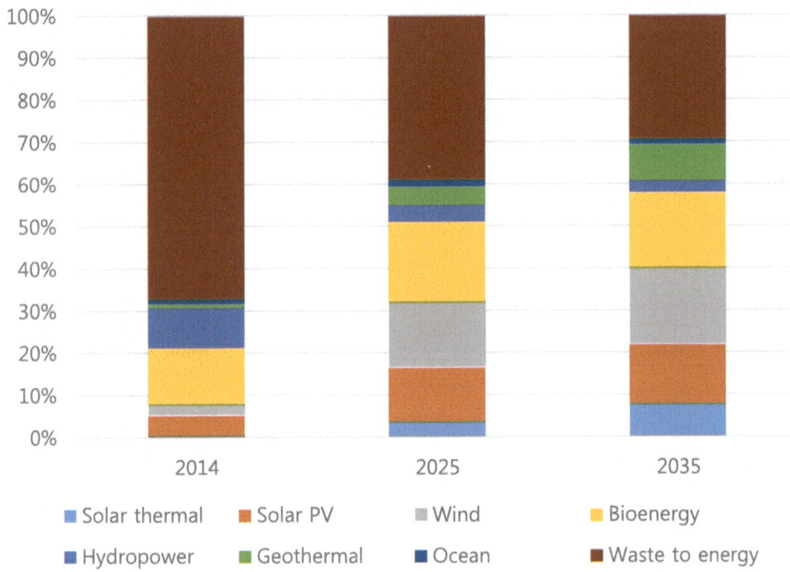

Fig. 8.3 Renewable energy targets as a percentage of total primary energy supply (*Source* Author created figure based on data from MOTIE 2014)

Technological Development. According to the 2013 Delphi survey of 500 experts by the Korea Energy Economics Institute, Korean levels of renewable energy technologies are assessed to be 86% of the best technologies available around the world. This is roughly a 10% gap with the technology levels of the US (96.7%) and Japan (96.5%). This level of technology can be considered a substantial improvement over the past goals, as the cross-ministry goal of technology development goal for new and renewable energy was merely 76.7% (MOEST 2010). Another technology level assessment published as part of the annual National Technology Level Assessment evaluated the level of South Korea's solar energy technology to be 80.7% for basic research, 90.9% for applied research, and 80.8% for technological deployment (Korea Institute of Science and Technology Evaluation and Planning 2013). South Korea also turns out to rank second on the number of patent applications for solar PV and seventh for wind energy for the period of 2000~2013, according to our own analysis using OECD Patents Statistics.[3]

Renewable Energy Adoption in South Korea

Despite various proactive policy measures and the aggressive promotion of renewable energy technology over the past three decades, the level of renewable adoption is quite low in South Korea. Measured in terms of renewable energy as a share of total primary energy or the share of renewable sources in electricity generation, South Korea ranks the lowest among OECD members as shown in the two figures of the introduction. This gap between the government's promotion of renewable energy and the actual diffusion of renewable energy is particularly disturbing given that South Korea is one of the largest energy consumers. As of 2016, South Korea ranks 7th in electricity consumption after China, the US, India, Japan, Russia, and Germany. Why is renewable energy adoption so low despite strong effort to facilitate the use and diffusion of renewable energy? This section tackles the puzzle in two aspects—political and economic dimensions of renewable energy adoption.

Political Dimension. There are three political aspects underlying the slow diffusion of renewable energy in South Korea. One can be traced to the scope of renewable energy, which has been politically ambiguous from the start in the history of renewable energy policy in South Korea. While the three acts explained in the previous section created legal grounds for the government to intervene in the private sector for renewable energy development, the definition of renewable energy was quite problematic. Initially, the law set out to promote alternative energy where "alternative" was meant to be energy sources other than oil, coal, nuclear, and natural gas. When the law was substantially revised in 2004, the types of energy to be promoted by the government for the cleaner environment were expanded to include "new" energy—namely, hydrogen, fuel cell, and residual oil gasification which are all converted from fossil fuels. Although renewable energy was defined to cover eight areas—solar thermal, solar PV, wind, bioenergy, hydropower, geothermal, ocean, and waste to energy—in MOTIE's *Fourth Basic Plan for New and Renewable Energy*, the terminology essentially lumping new energy and renewable energy generated much confusion in renewable energy statistics. It is also responsible for the illusion of the country's renewable energy production rate appearing higher than the actual level of penetration. Furthermore, it is criticized to lead to unclear and inconsistent policymaking as well as improperly defined roles for relevant government agencies (Yoon and Sim 2015).[4]

At a deeper level, the slow adoption of renewable energy is closely related to the very process of South Korea's industrialization in the 1970s and 1980s. For fast-track industrialization, it relied heavily on nuclear energy as a source of cheap electricity. Even the latest national energy blueprint, the Second National Plan for Energy, aims to increase the share of nuclear power in total energy generation to 29% by 2035 despite the general tone of renewable energy promotion. In comparison, the target for renewable in total energy penetration is only 6% by 2020 and 11% by 2035. In fact, nuclear energy promotion in South Korea was much more than a choice of necessity for cheap energy. One of the widely read studies in the science and technology literature on the dominance of nuclear power in South Korea explains it as driven by the sociotechnical imaginary of "atoms for development" in which the state not only imported nuclear technology from abroad, but also incorporated it into its scientific, technological, and political practices (Jasanoff and Kim 2009).

The study by Yoon and Sim (2015) traces the slow adoption of renewable energy in South Korea to the policy shift from a feed-in-tariff (FIT) to renewable portfolio standard (RPS). The FIT was introduced in 2002 and ran about ten years. The FIT was discontinued due to concerns with the cost. Total government expenditure amounting to KRW 1.59 trillion between 2002 and 2012, but the FIT budget supported out of the Electricity Industry Fund totaled only KRW 395 billion. Out of concern with skyrocketing costs, the government replaced the FIT with an RPS in 2012. According to Yoon and Sim, this shift not only benefitted large-scale producers disproportionately, but also made it harder to promote renewable energy diffusion. The FIT was very costly, yet it enabled small-scale energy producers to get finance easily because profitability was guaranteed for 20 years. Under RPS, the minimum quantity of total energy generation is allocated from renewable sources. In such a system, large-scale producers are in a better position to take advantage of the wholesale electricity price and renewable energy certificate sales, whereas it is difficult for small-scale producers to even secure the sales of renewable energy certificates (REC).

Economic Dimension. In addition to the political and historical conditions discussed previously, one would not be able to explain the low level of renewable energy adoption in South Korea without understanding the preferences of South Korean consumers. As is well-known, one of the biggest drawbacks is the high cost, which emanates from unstable supply as well as relatively immature markets and technologies. How willing consumers are to pay for various forms of renewable energy has been much studied, as summarized in Table 8.2. Adapted from Lee and Heo (2016), the table shows the results of various studies measuring the willingness to pay (WTP) for renewables, with some studies examining WTP for renewable energy, others WTP for renewable technologies or WTP for different types of renewable sources.[5] Consumers of EU countries show relatively high WTP with the WTP for small-scale wind turbines reaching 176 euros per month. In the UK and the US, consumers turn out to

Table 8.2 Willingness to pay for renewables (non-Asian countries)

Country	*Estimated amount*	*Willing to pay for (or if)*	*Source*
EU	€8424 (176) €6208 (129) €3839 (80)	Small-scale wind turbines Solar battery panels Solar water heaters	Claudy et al. (2011)
UK	£2381 (50) £2903 (60) £1288 (27)	Solar electricity Solar hot water Wind turbines	Scarpa and Willis (2010)

Source Author created table

Table 8.3 Willingness to pay for renewables (East Asian countries)

Country	*Estimated amount*	*Willing to pay for*	*Source*
Japan	JPY2000 ($17 per month)	Renewable energy sources	Nomura and Akai (2004)
China	CNY7.91~10.30 ($1.15~$1.51 per month)	Green electricity generated by renewable energy sources	Zhang and Wu (2012)
S. Korea	KRW1681 ($1.8 per month)	Green electricity	Yoo and Kwak (2009)
S. Korea	$1.35 per month	Renewable energy through the implementation of the renewable portfolio standard	Kim, et al. (2012)
S. Korea	KRW1345 ($1.25 per month)	Premium-priced green electricity	Kim et al. (2013)

Source Author created table

be willing to pay about 8~10 dollars more for electricity generated by renewable energy.

The levels of WTP of East Asian countries as shown in Table 8.3 are in marked contrast to those of the advanced countries in the previous table, however. This table, based on Huh et al. (2015) and Lee and Heo (2016), shows the results for the WTP for renewable or green energy and technologies. Except Japan, whose consumers show a similar level of WTP ($17 per month), the WTP of South Korean and Chinese consumers for renewables is roughly one-tenth of the level of advanced countries.[6]

In light of South Korea's level of development, with its per capita GDP more than twice that of China, the finding that South Korean consumers' WTP for renewables is just the same as that of Chinese consumers' WTP is particularly troublesome for those advocating renewable energy.

Development During the Current Administration

At the time of this writing, South Korea underwent the presidential election with the politically more liberal party candidate, Mr. Moon Jae-in, winning the election and ending the conservative government rule that lasted more than a decade. Since the inauguration of the new administration in May 2017, several major initiatives were launched, including the phaseout of nuclear energy. The nuclear phaseout policy was officially announced by the new president on June 19, 2017, the day of shutdown of South Korea's first nuclear power plant, Gori Number One.

As the Moon administration actively promoted energy transition in 2017, the South Korean government has set more aggressive goals for renewable energy adoption. As to the distribution of renewable energy sources, the most distinctive features of the new energy policy are twofold. First, a more ambitious target was set for electricity generation from renewables. The so-called Renewable Energy 3020 announced by the Ministry of Trade, Industry and Energy in December 2017, included the plan to increase renewable energy to 20% by the year 2030 by investing 18 trillion won from the national budget and 92 trillion won from facility investment (which combines funds from public institutions, private sectors, financial institutions, cooperatives, and individual investors). According to the 3020 Plan, the amount of renewable energy capacity would increase from 13.3 GW (2016) to 27.5 GW (2022), and eventually to 63.8 GW (2030). Moreover, over 95% of the newly built energy facilities would consist of solar (63%) or wind (34%) energy if the plan is successful.

Second, the government has promised to increase its support for renewable energy generation by increasing the RPS goal as well as financial support for solar energy. The RPS goal was revised from originally 10% by 2023 to 28% by 2030 to promote large-scale renewable energy projects such as on-shore and off-shore wind power generations, and floating photovoltaic panels. The 3020 Plan made some changes to the REC system, by providing incentives for energy produced by participatory citizen funds or social corporations. The plan also proposed to boost financial support on solar energy panels in an urban landscape, by temporally (5-year plan) introducing the FIT to support individual or small-scale production of renewable energy.

In addition to these two notable proposals, the 3020 Plan added the "New Energy Industry Promotions," which entails infrastructural plans such as energy storage systems (ESS), fuel cells, and Internet of Things (IoT), with the expectation that IoT and ESS technologies would open up new coordination of energy sources, including decentralized power sources, energy-broker markets, and real-time energy demand management systems. In particular, the development of a smart grid to link diverse sources of renewable energy is at the center of renewable energy infrastructure. According to the Second Smart Grid Power Plan

(2018~2022), the government is planning to launch a power intermediary market in which intermediary agencies can trade electrical power on behalf of individuals generating renewable energy in various forms. The same plan also contains a proposal to build an integrated renewable energy control system by 2020 (Ministry of Trade, Industry and Energy 2018).

What is notable in the energy policy of the new administration is the effort to shift the national energy policy from the centralized energy system (notably based on nuclear energy and coal) to more decentralized, democratic, and participatory mode of energy production and consumption. Despite the new policy attempts to increase renewable energy, South Korea's low percentages of renewable energy production are likely to linger on longer than the written plans. In addition to the already low level of the WTP of South Korea, a low level of the willingness to accept (WTA) for renewables has recently become a significant challenge for the diffusion of renewable energy in South Korea.

In fact, 37.5% of renewable electricity generation projects were rejected or postponed in 2016 due to resistance from nearby residents to new construction of solar or wind power plants (Jung 2017; Lee 2017). Given the lack of incentives for residents from renewable power plants, their resistance often leads to a higher level of regulation by local governments, which is more likely to limit new construction of renewable energy facilities. Therefore, there are several calls for creating more incentives for local communities by allowing them to participate in local renewable business (Rogers et al. 2008; Warren and McFadyen 2010; Ejdemo and Söderholm 2015; Howard 2015).

According to a recent survey conducted by Korea Energy Economics Institute, local acceptance for renewables power generation turned out to be significantly lower than that of nationwide public acceptance (Jung 2017). The expected rates of return from participating in renewable business differ noticeably between the general public and local residents living close to renewable power plants. For example, local residents are likely to accept a renewable power project when they annually expect a 12.3% rate of return, whereas the general public accepts the same at a rate of return of 3.1%. Also, the expected rate of return varies depending on the renewable energy source—photovoltaics are the most costly for local WTA, followed by biomass and wind energy (see chapters on Japan, China, and Norway in this book regarding Not-in-my-backyard [NIMBY] movements).

Citizen participation in renewable energy policies has also been both a difficult and ambivalent task for the new administration. Although President Moon made a public pledge to terminate all new constructions of nuclear plants, the government instead decided to let the people make the decisions on the construction of the Shin Gori Units 5 and 6, eventually leading to two more nuclear plants in Korea. Moreover, there remains controversy about the land uses of renewable energy generators, including solar panels and wind turbines. Although the government seems to promote a more participatory environment constructing renewable energy facilities, recent cases in local areas in Korea suggest that site selection processes for large-scale renewable energy is not always democratic or transparent.

Conclusion

Examining why South Korea has been so slow in adopting renewable energy despite the proactive measures and funding to develop renewable energy technologies, this study looked into political and economic factors that have made it hard to spread renewable energy. Politically, the terminology of renewable energy has been very vague and confusing in government policymaking documents and practices for renewable energy, often yielding the illusion of the better-than-real performances of renewables. In addition, South Korea's reliance on nuclear power as cheap energy is historically deep-rooted in its so-called sociotechnical imaginary of "atom for development." What is worse is that South Koreans' WTP for renewables is significantly low in light of its level of development. South Koreans' WTP turns out to be only one-tenth of the WTP levels of advanced countries.

How can this gap between the governmental promotion of renewable energy and the actual level of renewable energy adoption be filled? With the new energy policy shift, the fundamental transition away from nuclear energy to renewable energy seems to have just begun. Yet the process of this transition appears to be quite troubled given the generally low levels of WTP, or even acceptance renewable energy coupled with insufficient incentives for local residents. One thing that is very clear from this study is that the journey to renewable energy for South Korea would require more than just rhetorical appeals about the necessity or environmental value of renewable energy, especially given its political and economic hurdles as discussed here.

Notes

1. The first such long-term plan was the Five-Year Economic Development Plan (1962–1966), which has since been repeated every five years. Five-year plans of various sorts are the hallmark of the developmental state that characterizes Korea's political economy (Woo-Cumings 1999). For instance, in areas related to science and technology, more than 100 five-year plans were identified as being in place across 19 ministries (Korea Institute of Science and Technology Evaluation and Planning 2011).
2. While the long-term governmental plan for renewable energy was mandated by the earlier law, the specific length of planning frequency was set to 5 years in the 2014 amendment of the Act on the Promotion of the Development, Use and Diffusion of New and Renewable Energy. Note that this planning frequency is different from the target period, which is set to at least 10 years or more.
3. This analysis utilized the International Patent Classification (IPC) Green Inventory published by the World Intellectual Property Organization (WIPO). The patent categories used are as follows: F03D for wind energy, and C01B, C23C, C30B, F21L, F21S, G05F, H01G, H01L, H01M, H02J, H02N for solar energy.
4. Related to this perplexing terminology for renewable energy, another confusion arises from the meaning of "sustainable energy" in government policymaking documents. A recent study of the origins of sustainable energy policy in South Korea tracking a large number of official documents produced in relation to the promotion of sustainable energy and green growth finds that the adjective "sustainable" in these documents meant continuous development or growth rather than environmental cleanness or safety (Baek 2013).
5. Lee and Heo (2016) divides WTP estimation for renewable energy sources (RES) into two types: (1) using a conjoint analysis that usually examines WTP for each separate renewable source, and (2) those using the contingent valuation method that focuses on WTP for the additional cost required to expand RES.
6. Notably the studies of the South Korean case shown in the table are produced during the Lee administration (2008~2012) that made a big push for green growth. Lee and Heo (2016) point out from their review of the WTP studies that South Korean consumers appear to perceive green electricity as a good differentiated from fossil fuel, but they do not perceive each individual renewable-energy technology as a differentiated good.

References

Baek, Kwang Hoon. 2013. *Old Policy and New Rhetoric: The Origins of Sustainable Energy Policy in Korea*. Master's Thesis. Korea Advanced Institute of Science and Technology.

Claudy, Maurius C., Cauls Michelsen, and Aidan O'Driscoll. 2011. The Diffusion of Microgeneration Technologies—Assessing the Influence of Perceived Product Characteristics on Homeowners' Willingness to Pay. *Energy Policy* 39 (3): 1459–1469.

Ejdemo, Thomas, and Patrik Söderholm. 2015. Wind Power, Regional Development and Benefit-sharing: The Case of Northern Sweden. *Renewable and Sustainable Energy Reviews* 47: 476–485.

Howard, Tanya. 2015. Olive Branches and Idiot's Guides: Frameworks for Community Engagement in Australian Wind Farm Development. *Energy Policy* 78: 137–147.

Huh, Sung-Yoon, Jongsu Lee, and Jungwoo Shin. 2015. The Economic Value of South Korea's Renewable Energy Policies (RPS, RFS, and RHO): A Contingent Valuation Study. *Renewable and Sustainable Energy Review* 20: 64–72.

Jasanoff, Sheila, and Sang-Hyun Kim. 2009. Containing the Atom: Sociotechnical Imaginaries and Nuclear Power in the United States and South Korea. *Minerva* 47: 119–146.

Jung, Sungsam. 2017. Korean Citizens' Willingness to Accept Renewable Energy. Korea Energy Economics Institute Report [In Korean].

Kim, Jihyo, Jooyoung Park, Haeyeon Kim, and Eunnyeong Heo. 2012. Assessment of Korean Customers' Willingness to Pay with RPS. *Renewable and Sustainable Energy Review* 16: 695–703.

Kim, Jihyo, Jooyoung Park, Haeyeon Kim, and Eunnyeong Heo. 2013. Renewable Electricity as a Differentiated Good? The Case of the Republic of Korea. *Energy Policy* 54: 327–333.

Korea Institute of Science and Technology Evaluation and Planning. 2011. *Long-Term Government Plans in Science and Technology* [In Korean].

Korea Institute of Science and Technology Evaluation and Planning. 2013. *Annual Technology Level Assessment* [In Korean].

Lee, Chul-Yong, and Hyejin Heo. 2016. Estimating Willingness to Pay for Renewable Energy in South Korea Using the Contingent Valuation Method. *Energy Policy* 94: 150–156.

Lee, Namsuk. 2017. Increase Renewable Energy? Need to Communicate with Citizens First. *Asia Times*, June 15. Available at: http://www.asiatime.co.kr/eView.html?idxno=151196. [In Korean], Accessed on 11 Apr 2019.

Ministry of Education, Science and Technology. 2010. *Cross-Ministry New and Renewable Energy R&D Strategy* [In Korean].

Ministry of Trade, Industry and Energy. 2014. *Fourth Basic Plan for New and Renewable Energy*, September 2014 [In Korean].
Ministry of Trade, Industry and Energy. 2018. *The Second Smart Grid Plan: 2018~22* [In Korean].
Nomura, Noboru, and Makoto Akai. 2004. Willingness to Pay for Green Electricity in Japan as Estimated Through Contingent Valuation Method. *Applied Energy* 78 (4): 453–463.
Rogers, Jennifer C., Eunice A. Simmons, Ian Convery, and Andrew Weatherall. 2008. Public Perceptions of Opportunities for Community-Based Renewable Energy Projects. *Energy Policy* 36 (11): 4217–4226.
Scarpa, Riccardo, and Ken Willis. 2010. Willingness-to-Pay for Renewable Energy: Primary and Discretionary Choice of British Households' for Micro-Generation Technologies. *Energy Economics* 32 (1): 129–136.
Warren, Charles R., and Malcolm McFadyen. 2010. Does Community Ownership Affect Public Attitudes to Wind Energy? A Case Study from South-West Scotland. *Land Use Policy* 27 (2): 204–213.
Woo-Cumings, Meredith (ed.). 1999. *The Developmental State*. Ithaca, New York: Cornell University Press.
Yoo, Seung-Hoon, and So-Yoon Kwak. 2009. Willingness to Pay for Green Electricity in Korea: A Contingent Valuation Study. *Energy Policy* 37 (12): 5408–5416.
Yoon, Jong-Han, and Kwang-ho Sim. 2015. Why Is South Korea's Renewable Energy Policy Failing? A Qualitative Evaluation. *Energy Policy* 86: 369–379.
Zhang, Lei, and Wu Yang. 2012. Market Segmentation and Willingness to Pay for Green Electricity Among Urban Residents in China: The Case of Jiangsu Province. *Energy Policy* 51: 514–523.

CHAPTER 9

China's Promotion of Wind and Solar Power: Supportive Policies, Geographical Challenges and Market Competition

Gang Chen

Introduction

China's economic miracle over the past three decades has made the country the world's largest energy consumer in the world. Despite China's constant effort to expand and diversify energy supply, its coal-dominant energy structure has put it in an awkward position in the context of global climate change, and its dependence upon fossil fuels is furthered by its increasing imports of petroleum and natural gas. In the long run, energy scarcity could become a real bottleneck for China's sustainable development due to its voracious demand for fossil fuel energy resources and the fact that it has per capita energy resources that are lower than the global average.

G. Chen (✉)
East Asian Institute, National University of Singapore, Singapore, Singapore
e-mail: eaicg@nus.edu.sg

P. Midford and E. Moe (eds.), *New Challenges and Solutions for Renewable Energy*, International Political Economy Series,
https://doi.org/10.1007/978-3-030-54514-7_9

Though fossil fuels (coal, oil and natural gas) still make up about 90% of China's total energy production and consumption, the country has been formulating one of the world's most aggressive strategies to promote low-carbon energy production from renewable sources. As part of its national effort for energy self-sufficiency and against climate change, China's low-carbon electricity strategy is undergoing transformation, which is sensitive to the dynamics of energy policy prioritization, domestic supportive policies, geographical characteristics and industrial barriers. With a carbon-constrained new normal that emissions of greenhouse gases can no longer be assumed to be costless, renewable energy sources like wind and solar power have witnessed fast growth in both production and consumption in China. In promoting the production and sales of solar and wind power in the context of global climate change, the Chinese government has been adjusting its policy prioritization from time to time in targeted growth areas, subsidy amount, on-grid tariffs and other financial incentives for various low-carbon power sectors. Other variables, like inherent geographical and meteorological advantages and disadvantages, domestic supporting industries, technological barriers related to power storage and transmission, have also helped to reshape the country's trajectory of renewable energy development.

Considering the exorbitant costs and risks involved in developing low-carbon energy, China's progress in promoting non-fossil fuels has been closely related to the government's robust support for, and heavy subsidies to, power generation from non-fossil fuel resources like wind power and solar PV. It's also a natural extension of the country's preferred "no-regret" strategy that emphasizes mitigation actions providing fringe benefits like profitability and employment to the country, regardless of whether the threat of climate change is real. "No-regret" options are steps to reduce greenhouse gases that would pay for themselves even without a climate change concern. Bringing down energy/carbon intensity and improving energy efficiency are the main elements of China's long-term "no regret" strategy, which implies actions providing fringe benefits to the country regardless of whether the threat of global warming is real (Chen 2012, p. xi). For China, such an approach would take as a given that economic growth remains the top priority (Hatch 2003, p. 55). With the energy policy long time being positioned at the center of its development strategy, China today has formulated low-carbon policies and targets to address the dual challenges from ecological problems and national energy security. When such renewable energy policies were designed, policy makers may have considered more about economic factors such as energy supply and jobs rather than the environment, but in practice, these no-regret policies have been functioning as efficient means to help the country slow down its greenhouse gas emission growth and gain competitive positions in the clean energy sector.

China institutionalized its efforts to develop clean energy capacity through passing the groundbreaking Renewable Energy Law in 2005, which defined renewable energy as non-fossil energy from wind energy, solar energy, water energy, biomass energy, geothermal energy, and ocean energy, etc. (PRC Renewable Energy Law 2005, Article 2). The Renewable Energy Law, together with its amendment, which listed the development of renewable energy as "the preferential area for energy development," introduced new schemes like "cost-sharing," a Feed-in Tariff (FIT), "mandatory grid-connection" and "renewable portfolio target" systems that had been successful in advancing renewables in some European nations and US states. The law prompted the government to utilize financial subsidies and tax incentives for the development of renewable energy. Considering the comparable disadvantage of generating electricity from renewable sources, the Renewable Energy Law addressed the core issues of pricing and fee-sharing for on-grid renewable energy through the government-set or government-guided pricing system.

In terms of promoting the production of low-carbon alternatives, the Chinese government has been taking differentiated supportive policies with variation in targeted growth areas in mid- and long-term plans, which may, in the long run, profoundly change the existing structure of the electricity generation market and the country's energy mix which is overwhelmed by coal burning. Non-fossil fuel sources, including hydro, nuclear, wind and solar power, witnessed a fast growth in both production and consumption over more than two decades, playing an important role in alleviating China's energy shortage and achieving low-carbon targets. From 1991 to 2016, the proportion of hydro, nuclear, wind and solar power in China's total energy consumption rose from 4.7 to 13.3% (National Bureau of Statistics of China 2017, p. 72). During this period, hydropower contributed a lot more than what wind, solar and nuclear power combined contributed to the nation's electricity market. By the end of 2017, China's installed hydropower capacity was 341.2 gigawatts (GW), accounting for 19% of its total installed capacity in electricity generation that stood at 1777 GW. See Fig. 9.1 (China Electricity Council 2018). Meanwhile, on-grid wind power made up 9% of the total

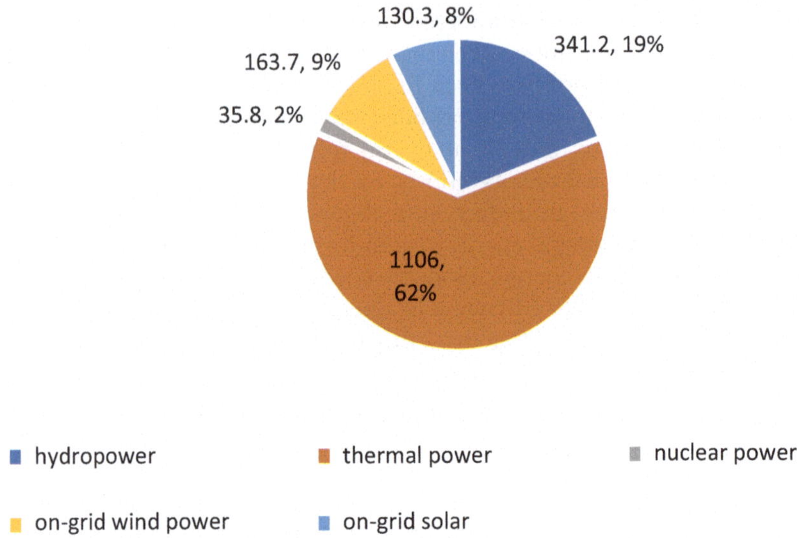

Fig. 9.1 Existing installed capacity of various electricity sources in China at the end of 2017 (*Note* Figures in GWs and percentage share. *Source* Author created figures based on data from China Electricity Council 2018)

and solar about 8%. If not for the strong opposition from the environmentally conscious civil society against massive dams, the share of hydropower in total installed capacity should have been much larger.

Prior to 2010, as compared to solar (mainly solar PV) power generation, wind power had been positioned at the center of renewable energy development, which had resulted in differentiated growth patterns in favor of wind. During this period, the installed capacity of wind power grew at 89.8% per annum, while solar's installed capacity, until 2010, remained 0.86 GW, a paltry proportion of 0.09% in the country's total installed capacity (China's 12th Five-Year Plan on Energy Development 2013). Since the beginning of the 12th Five-Year Plan (FYP) period (2011–2015), the Chinese government, in response to new industrial conditions of leapfrogging PV manufacturing capacity, has been adjusting its renewable energy strategy with focus being shifted from wind to solar. A reverse trend has occurred during the 12th FYP, when solar grew 122%

and wind only 33% annually (China's 13th Five-Year Plan on Energy Development 2016). By the end of 2017, China's installed capacity of wind power and PV power generation reached 163.7 and 130.3 GW, both ranking the first in the world. See Fig. 9.1 (China Electricity Council 2018).

When China has been experiencing a great leap forward in developing wind and solar PV power generation, the policy makers and market participants have been encountering industrial and technological barriers as well as geographic challenges that prevent full utilization of the quickly expanding installed capacity from these renewable resources. Moreover, when competing with conventional coal-fired thermal power that is much more polluting, wind and PV power production usually gets less support and subsidies from local governments, which exacerbates the market distortion in favor of thermal power plants, one of the major emission sources of China's urban smog and PM 2.5 (particulate matter with diameter less than or equal to 2.5 micrometers). Each year a huge proportion of wind and PV capacity remains unused. The integration of wind and PV generation especially in the western provinces of China has lagged in capacity expansion, resulting in appalling waste. According to China's Renewable Energy Industries Association (CREIA), the country's average wind curtailment rate stood at a record high of 15% in 2015 (Li 2016).

The Chinese government has realized the urgency of addressing these barriers and problems, and has been ready to adjust supportive policies, speed up grid construction and encourage more renewable capacity to be installed in economically vibrant coastal provinces rather than the hinterland. However, such efforts may take years to have effect upon existing renewable power production, transmission, consumption and geographic redistribution. Wind and solar power generation in China's coastal areas is largely restrained not only by local meteorological factors such as the intensity of wind and sunshine, but also by the NIMBY (Not in My Back Yard) movements against large-scale energy projects among the Chinese middle class in the populous eastern areas. Some southeast provinces along the coastlines, including Fujian, Guangdong, Hainan and Guangxi, actually have enormous wind energy potentials where wind energy intensity can go as high as 600 W/m^2, but such strong winds are difficult to harness as they are often the outcomes of frequent typhoon and tropical depression phenomena in summer (Chen 2019, p. 47).

Supportive Policies

Targets in the Five-Year Plans. As a typical model of state capitalism, China's FYP system, together with various forms of subsidies and other policy supports offered by the state to a wide range of energy industries, has been vital in fostering fierce competition among various energy sectors, most of which are highly regulated by the state. As evident in the specifications of targeted growth rates and supporting policies in various FYPs for different non-fossil fuels, as well as policy changes in tariffs and subsidies of on-grid electricity, China's active low-carbon strategy, being sensitive to dynamics of domestic supporting industries, technological competitiveness and environmental protection, has reshaped China's power market.

The government set the target of cutting energy intensity, which refers to energy consumption per unit of GDP, by 20, 16, and 15%, respectively, in its 11th, 12th, and 13th FYP. At the Copenhagen Climate Summit in 2009, China added carbon intensity, another efficiency index measuring carbon emissions per unit of GDP, into its energy planning system. Accordingly, China's 12th and 13th FYP set the five-year targets of cutting carbon intensity by 17 and 18%, respectively (Chen 2019, p. 111). In order to implement this, the government used economic and taxation tools such as preferential purchase of renewable power, preferential taxation policies toward energy efficiency projects, favorable financing policies, tax rebate cuts on high-energy intensity exports and compulsory government procurement of energy-saving products to achieve the goal. Detailed energy efficiency actions included petroleum substitution, surplus heat utilization, construction of energy efficiency buildings, public transport development, energy-saving auto development, innovation of coal-fueled industrial boilers and green lighting project.

Since 2007, the government started to use energy intensity index together with overall economic and investment figures to measure performances of local officials. Some regional energy efficiency and low-carbon projects like the state-level pilot zone in Hubei and Hunan Provinces and the Singapore-China Eco-City in Tianjin were launched. Finally, the country achieved a 19.1% decrease in energy intensity in the five years to 2010, 0.9 percentage point lower than the original target. In order to improve its image in global climate politics before the Copenhagen Climate Summit in 2009, China, for the first time declared that it was targeting a hefty 40–45% cut in carbon intensity (the amount of carbon dioxide emitted per dollar of GDP) by 2020. The main limitation of an intensity-based target is that, although it can lower an energy and emission growth trajectory below the projected business-as-usual level, it is unlikely to result in an absolute decrease in energy consumption and carbon emissions.

In the fight against climate change, China has proposed the targets of lifting the proportion of non-fossil fuels in total energy consumption to 15% by 2020, and 20% by 2030. Large parts of the 12th FYP (2011–2015) and 13th Five-Year Plan (2016–2020) have been set to be devoted to the building up of a low-carbon and energy efficiency economy. Nevertheless, this kind of symbiotic relationship between China's energy sufficiency and efficiency strategies could, in the long run, lead to overdependence upon newly-added capacities of low-carbon energy, including hydropower, nuclear, wind, solar and geothermal power, and overlook of real efficiency issues addressed in documents like the PRC Law on Conserving Energy (1997) or FYPs.

For the 12th FYP period, the Chinese government, in response to new industrial conditions of leapfrogging PV manufacturing capacity and nuclear power technology, as well as environmental concerns, was adjusting its low-carbon electricity strategy with focus being shifted from wind to solar. The 12th FYP anticipated the solar would grow 89.5% annually, whereas wind's growth rate would drop to 26.4%. Emerging trends, driven by more sophisticated energy markets, volatile energy production costs, reassessment of innate geographical conditions and environmental impact, and increased attention to supporting domestic industries, were changing the value of these non-fossil fuels in the eyes of Chinese policy makers.

As compared to the goals set by the 12 FYP, China's National Development and Reform Commission (NDRC), the country's energy regulator, has proposed the targets of increasing installed capacity of wind power by 63% to 210 GW, and that of PV power by 144% to 105 GW in the 13th FYP (2016–2020) period (China's 13th Five-Year Plan on Energy Development 2016). In terms of low-carbon electricity generation, hydropower by comparison has contributed much greater capacity to China's electricity market than wind, nuclear and solar PV power combined. However, if the above targets are achieved by 2020, the total installed capacity of wind and PV power would be 315 GW, close to the expected hydropower capacity of 340 GW by 2020 (China's 13th Five-Year Plan on Energy Development 2016).

More importantly, the 13th FYP paid more attention to the efficiency issues regarding the utilization of installed wind and PV capacity, intermittence problem caused by renewable power fluctuations, transmission difficulty and geographic uneven distributions of wind and PV power generation projects. To address the long-existing problem of high curtailment rates in wind and PV, the 13th FYP proposed for the first time the targets of annual power outputs for the wind and PV power generation as compared to the 12th FYP, which only set targets for the installed capacities of the two renewables. To promote sales of electricity from renewable sources, the 13th FYP set the goal of getting wind power to compete with local coal-fired power generation on the same price platform and PV power price to be equal to local selling price from grid companies. The 13th FYP emphasized the urgency of addressing the high curtailment rates of renewable energy projects and called for locating more wind and PV projects in China's eastern and central provinces instead of resourceful but sparsely populated western provinces.

Special Programs

China's abundant inland and offshore wind energy resources provide enormous potential for large-capacity wind farms. As a latecomer in the utilization of wind power, China's wind power installed capacity increased at a much slower rate than the world's average level before 2004 but, even during that early stage, strong policy support played a vital role in promoting wind power generation and turbine manufacture. To import technology from foreign companies and to establish a high-quality Chinese wind turbine generator sector, the former State Development and Planning Commission (SDPC), predecessor of NDRC, initiated the "Ride the Wind Program" (*chengfeng jihua*) in 1996, leading to the establishment of two joint ventures, NORDEX (Germany) and MADE (Spain), which effectively introduced 600 kilowatts wind turbine generator manufacturing technology into China.

The landmark policy came in 2003 when the NDRC launched the "Wind Power Concession Program" (*fengdian texuquan xiangmu*) to build large-capacity wind farms and achieve economies of scale through reducing the on-grid wind power tariff. Under the program, the power grid company signed a long-term power-purchase agreement with the wind power project investor and agreed to purchase the prescribed amount of electricity generated by the project, whose capacity must reach

100 megawatts. Investors and developers of wind farms were selected through a competitive bidding process that determined the on-grid tariff. As incentives, the government waved the import customs tariff and value-added tax on the equipment and accessories. The "Wind Power Concession Program" that minimized the risks to investors through government-guaranteed power-purchase agreements had pushed China's wind power industry into a fast-growing stage between 2005 and 2009 when the country's installed wind power capacity more than doubled annually in the four years.

Effective since February 2005, China's Renewable Energy Law stipulates "mandatory grid-connection" in Article 4, Item 14, which local grids should sign contracts with renewables generators to buy up all of the power that can be supplied, within the grid's coverage. The law sounded like good news but lacked the punitive powers needed to implement obligatory purchases.

In certain localities, feed-in tariff and other subsidization programs benefitting regional solar power projects had been experienced before the central government finally made up its mind to grant nation-level supports. A bidding process for a 10 MW program in the Dunhuang region of western China's Gansu Province resulted in an on-grid price of 1.09 yuan per kWh in 2009. Such a price level at the moment was applicable to local projects in surrounding regions with similar solar resources and equipment. One year later, in April 2010, the NDRC announced that four solar power stations in western Ningxia Autonomous Region would adopt a new price level of 1.15 yuan per kWh, higher than that applied to the Dunhuang program, the first large-scale solar power generation program commercialized in China. Despite all these sporadic examples of regional feed-in tariffs, the central government as well as most local governments had been reluctant to subsidize solar power generation in the same manner as they subsidized wind power production until recently, when the government, in the face of weak external demand and serious industrial overcapacity at home, had to speed up its policy shift onto the stimulation of domestic demand.

Diverse On-Grid Tariffs

Despite progress in reforming electricity price mechanisms, including trials of setting tariffs through competitive bidding processes, China's on-grid electricity tariffs are still strictly regulated by the government.

To foster the development of renewable energy resources, differentiated on-grid electricity tariffs have been adopted to provide incentives for renewable power producers including wind and solar.

Considering the comparable disadvantage of generating electricity from renewable sources, China's Renewable Energy Law (2005) addressed the core issues of pricing and fee-sharing for on-grid renewable energy through a government-set or government-guided scheme of feed-in tariffs. With the stipulation that "grid power price of renewable energy power generation projects shall be determined by the price authorities of the State Council" (PRC Renewable Energy Law 2005, Article 19), the feed-in tariff imperative requires that on top of benchmark on-grid electricity tariffs of coal-fired power plants, a fixed amount that varies according to renewable energy types and geographic locations be added to the selling prices from renewable electricity generators to the grid companies.

As a financial incentive, the feed-in tariff allows the renewable generator to achieve a positive return on its investment, despite the higher production cost compared with conventional fuels. When the policy makers had found that compared to solar energy production, wind power was more mature at the moment to be widely commercialized with modest government subsidies needed, they decided to promote the wind power vigorously as the main renewable source while deemed subsidization of solar generation as too costly and inefficient. In 2006 the cost of electricity generated from solar power was some 3 yuan per kWh, while that from a typical coal-fired power plant was only around 0.22 yuan per kWh, and that from a wind power plant was averaged at about 0.6 yuan per kWh (Li 2007).

Under China's differentiated on-grid pricing system, which essentially maintains the cost-accounting approach, nuclear power plants are rewarded with higher on-grid tariffs than coal-burning stations, most of which follow the benchmark prices that are higher than hydropower tariffs, while wind power prices are even higher than nuclear and solar power prices the highest. In 2009, the on-grid thermal tariff averaged about 0.35 yuan per kWh in China, higher than the hydro tariff averaged at 0.26 yuan/kWh but lower than nuclear price ranging between 0.39 and 0.49 yuan, and wind price from 0.51 to 0.61 yuan. There was no unified on-grid tariff for solar PV until 2011, when the government formally announced a feed-in tariff of 1 yuan/kWh for solar PV projects to be completed after 2011 and another tariff of 1.15 yuan/kWh for projects completed before the end of 2011 (NDRC 2011). Currently, on-grid tariffs for wind, solar PV and even nuclear are in a downward trajectory due to the sharp decline in related power production costs, while the government is considering the possibility of raising on-grid rates for hydropower or thermal projects to rebalance the competition in favor of conventional power sectors.

Since 2008, the Chinese government has adopted a set of preferential pricing schemes to encourage wind power generation. The tariff for wind power was fixed by the NDRC in 2009 and classified into four levels to give differentiated support to projects located in different regions that vary in local wind resources.

Solar PV. Prior to 2009, when the solar power market was small, the approved feed-in tariff rate ranged between four to nine yuan per kWh based on the different characteristics of individual projects, ten to twenty times that of coal power. The exorbitant cost of producing on-grid solar power, together with enormous cost disparities among projects in various localities, meant that the government had refused to adopt the fixed feed-in tariff, which policy makers thought would lead to over-capacity and costly production in some regions. In 2010, grid companies received government subsidies of 5.3 billion yuan for purchasing power from wind power generators, while they only got 63 million yuan and 83 million yuan in subsidies when buying electricity from solar and hydropower plants (China's State Electricity Regulatory Commission 2011, p. 10). In certain localities, feed-in tariff and other subsidization programs benefitting regional solar power projects had been experienced before the central government finally made up its mind to grant nation-level supports. A bidding process for a 10 MW program in the Dunhuang region of western China's Gansu Province resulted in an on-grid price of 1.09 yuan per kWh in 2009. Such a price level was applicable then to local projects in surrounding regions with similar solar resources and equipment. One year later, in April 2010, the NDRC announced that four solar power stations in western Ningxia Autonomous Region would adopt a new price level of 1.15 yuan per kWh, higher than that applied to the Dunhuang program.

In 2011, the Chinese government introduced a nationwide feed-in tariff of 1 yuan per kWh, which offered above-market-price contracts for the generation of solar electricity. Against the backdrop of exacerbated solar PV overcapacity and the launch of antidumping and countervailing duty investigations by the United States and EU over China-produced PV products, the Chinese central government planned to give additional nationwide subsidy of 0.45 yuan for each kWh of electricity generated by solar PV equipment while many provincial governments decided to offer additional subsidy of 0.25–0.3 yuan on top of that, with the unified term of 20 years.

The NDRC institutionalized the nationwide subsidy to all solar PV power plants through announcing a subsidy level of 0.42 yuan per kWh in a circular released in August 2013 (NDRC 2013), a 20% increase from the planned level of 0.35 yuan per kWh which had been proposed in its draft version half a year ago. With such a higher-than-expected government subsidy granted to local solar power plants, China, now the world's fourth-largest solar PV power producer, could witness its PV installed capacity grow exponentially in the next few years, just as its wind power market had performed from 2005 to 2010.

Other Incentives for Renewables. Besides feed-in tariffs, China's Renewable Energy Law also introduced other schemes like Renewable Portfolio Standard (RPS) and "mandatory grid-connection" that had been successful in advancing the cause of renewables in Europe and North America. From the Chinese version of RPS, which had originally drawn up plans to increase the proportion of renewable energy (including large hydropower) in the primary energy consumption from 7.5% in 2005 to 15% in 2020, it was evident that the government has prioritized the development of wind power over the solar power generation in the mid-term. Excluding hydropower, the aforementioned RPS target was further broken down by the NDRC as 30 GW from wind, 30 GW from biomass and 1.8 GW from solar photovoltaic (Zhang 2005, p. 1), in which solar PV would be projected to produce only 6% of the total wind power generation. The Chinese government subsequently revised up the RPS goal that aimed at 20% of its energy from renewable sources by 2020, with a widened gap between elevated wind capacity of 100 GW and solar capacity still at 1.8 GW.

The Renewable Energy Law's "mandatory grid-connection" part required grid companies to "buy the grid-connected power produced with renewable energy within the coverage of their power grid," and to "provide grid-connection service for the generation of power with renewable energy" (PRC Renewable Energy Law 2005, Article 14). To help grid companies to share such costs with end users, the law allowed grid companies to include grid-connection expenses paid by them for the purchase of renewable power and other reasonable expenses in the power transmission cost and recover them from the selling price (PRC Renewable Energy Law 2005, Article 21). After four years of implementing the Renewable Energy Law, the government found that although the country had rapidly increased its installed capacity of renewable energy, much of this capacity was not promptly connected to the grid and that not all power being generated was being purchased as required by the law.

When the Renewable Energy Law was amended in 2009, the top legislature required electricity grid companies to buy all the power produced by renewable energy generators and authorized the State Council energy department, in conjunction with the state power regulatory agency and the State Council finance departments, to determine the percentage of the quantity of electricity generated from renewable energy in the total quantity of electricity generated during the planned period (PRC Amended Law on Renewable Energy 2009, Article 14), a regulation system similar to the Renewable Portfolio Standards (RPS) adopted by some US states. The amendment made it clear that power enterprises refusing to buy power produced by renewable energy generators would be fined up to an amount double that of the economic loss of the renewable energy company (PRC Amended Law on Renewable Energy 2009, Article 29). To encourage grid companies to accept more power generated from renewable sources, the amendment waived partially the value-added and income taxes levied on grid companies on their revenue (the tax rate is about one-third of the revenue) generated from the surcharge on the retail power tariff for supporting renewable energy. China, a latecomer to modernization, has been swiftly emerging as the world's clean energy powerhouse, with strong national policies aimed at incentivizing the use of new renewables including solar and wind.

Hydrogen Energy Development in China

The Chinese government pays intensive attention to the research and development of hydrogen fuel technology as means of transporting, carrying and storing clean energy. China has shown its ambition in manufacturing and deploying hydrogen fuel cell technologies, which was included in the 13th FYP and "Made in China 2025" initiative issued by the State Council in 2015. According to China's guideline on strategic emerging sectors in the 13th FYP, the country tried to promote research and development of fuel cells, step up building hydrogen stations and achieve the mass production of fuel cell electric vehicles (FCEVs) by 2020

(*China Daily* 2018). The "Fuel Cell Technology Roadmap" included in the 13th FYP called for over 1000 hydrogen refueling stations to be in operation by 2030, with at least 50% of all hydrogen production coming from renewable resources. In addition, the Roadmap set a target for over 1 million FCEVs in service by 2030 (Jackson 2018). To overcome the gaps in the country's fuel cell infrastructure, China began to invest heavily in foreign fuel cell technology in 2018. The application of hydrogen fuel technology could in the long run make better use of China's enormous yet inefficient renewable energy capacity.

Hydrogen technology can potentially provide effective solutions to the intermittency problems associated with power generation from renewable energy sources like wind and solar. Usually, there are two ways of using hydrogen to store energy, both of which involve electrolysis to convert excess electricity into oxygen and hydrogen. This hydrogen is then stored and can either be converted back to electricity, using it in FCEVs, or the hydrogen can be burned directly, for example by pumping it into the natural gas infrastructure (A Medium Corporation 2018).

Geographical Challenges

China's prioritization of certain low-carbon electricity technologies is by and large related to the country's congenital advantages in geography and meteorology in tapping these energy resources. Having a sloping topography that descends in height from west to east, in which humid air currents above the sea can penetrate deep into China's interior areas and thus bring abundant runoffs to big rivers that flow invariably eastward, China leads the world in hydropower potential, with most hydroelectric potential remaining untapped in the less developed and populated southwest and western region. China's major river basins are well within the monsoon zone of the Pacific, with over 50% of annual precipitation in most areas concentrated in the four months of June to September. Sometimes more than 70% of regional rainfalls are concentrated in the two months of July and August. While two-thirds of China's land is threatened by floods and typhoons ravage coastal areas seven times a year on average, hinterland droughts occur almost every year in the dry seasons. Frequent devastating floods have justified the Chinese government's establishment of gigantic dams.

While most of China's bountiful hydroelectric resources are located in the southwest, the country's wind and solar resources are concentrated in the vast but sparsely populated northwestern and northern area, exacerbating the imbalanced distribution of energy resources that constantly afflicts the economically vibrant eastern (coastal) area with energy shortage. China abounds in wind resources, with potential wind energy resources on land, measured at 50 m above ground level, estimated about 2380 GW (China Meteorological Administration 2010). The most important wind-rich area in China is the long inland belt including northwestern Gansu, Xinjiang and Ningxia Province, and northern Inner Mongolia and Hebei Province. Inner Mongolia alone accounts for about one-third of China's total installed wind power capacity. For solar resources, strongest solar radiations are found in northwestern Tibet, Xinjiang, Qinghai and Gansu Province, while the southeast provinces, China's economic powerhouse, have much lower solar radiation amount per square meter.

The overconcentration of conventional coal, hydroelectricity and emerging wind and solar resources in the western part of Chinese territory has imposed enormous pressure upon the country's transmission networks, spurring China to launch the immense West-East Electricity Transmission Project that aims to transmit power from electricity-rich west regions to the economically prosperous eastern areas. Nevertheless, inefficiency in northwest provinces is still astronomically high and the NDRC has realized the urgency of building more ultra-high-voltage (UHV) transmission lines to carry electricity long distances, and to position new turbines close to major metropolises (*The New York Times* 2017, p. A8). The inability of grid system to incorporate intermittent renewable electricity in China has become a more severe bottleneck due to the lack of competitiveness of state-owned grid companies, the overinvestment in renewable power projects, the overcapacity of China's power industry as a whole and remote locations of many wind and solar power plants. The backward and aging grid system has resulted in insufficient inter-grid exchange capacity and under-investment in smart grid, an ideal solution to the intermittency problem associated with renewables. The Chinese government put its strategic focus upon UHV power transmission rather than smart grids, an ideal intra-regional solution to the intermittency problem associated with renewables. China has decided to upgrade transmission lines and build the world's most capable UHV transmission network that can transmit 1000 kilovolts or more over long distances.

Wind and solar only generate electricity when the wind is blowing or the sun is shining, and the energy produced from these renewables fluctuates daily according to the varying meteorological conditions. Intermittent wind and solar cannot be stockpiled in the absence of energy storage systems and must also be used when available, as otherwise they lose energy potentials (Evans et al. 2012, p. 4141). Such load balancing can be addressed with pumped-hydroelectric storage, where water is pumped to a higher elevation during low demand, and then released through turbines to produce electricity when required. Nevertheless, specific geographic conditions, including a suitable terrain that can accommodate an elevated reservoir, are required for the application of such storage technology. China was targeting 40 GW of pumped-hydroelectric storage by 2020 as part of ongoing efforts to cut high rates of wind and solar power curtailment (PVTECH 2019). Besides, the abovementioned hydrogen technology has been discussed in China as a potential storage solution for intermittent wind and solar electricity generation, which can overcome the geographic constraint linked to the construction of pumped-hydroelectric storage projects.

Due to the inconvenient locations of most hinterland wind and solar power plants, China has been making a big bet on offshore wind power projects that are at the doorsteps of its coastal industrial bases. The 12th five-year special plan of wind power technology development formulated by the Ministry of Science and Technology included the key technologies research and development of high power wind turbines, such as "10 MW wind turbine overall design technology," "3–5 MW permanent magnet direct drive (PMDD) wind turbine industrialization technology" and "7 MW-class wind turbine development and industrialization technology" (Wu et al. 2014, pp. 454–455). China has already had the ability to design and manufacture large-scale offshore wind turbines, with hoisting and trial operation for 6 MW offshore wind turbines having been completed. Nevertheless, as compared to onshore wind power, the development of offshore wind power is still facing difficulties in building power transmission, harnessing offshore harsh natural environments and fulfilling multi-sectoral coordination and management.

Industrial Barriers: Grid Restrictions, Unused Capacity and Local Politics

Unlike its persistent interest in hydro and nuclear for the sake of energy security, China's attention to wind and solar power generation has come much later largely in response to a new context of climate change and excess manufacturing capacity of renewable energy equipment. Since the global financial crisis in 2008, when China's export-oriented manufacturing was hit hard by sluggish external demand, the government has deemed renewable energy equipment sectors as emerging industries of strategic importance (*zhanlue xinxing chanye*) that need strong policy supports. Owing to massive producer subsidies, research grants, tax rebate, low-interest loans and cheap land under the model of state capitalism, China's manufacturers of wind turbines and solar panels, after occupying the largest share of the world market, have found themselves caught in a plight of overcapacity, selling price slump and exacerbated trade rows.

China's wind turbine industry, a new growth point immune from global economic downturn, has witnessed galloping expansion since 2008, bringing down turbine prices as much as 20–25% in western markets and more than 35% in China between 2008 and 2012 (REN21 2013, p. 54). Among the world's top 10 turbine manufacturers that captured 77% of the global market as in 2011, four hailed from China. The costs of operating and maintaining wind farms dropped significantly due to increased competition among contractors and improved turbine performance. The domestic manufacture boom has justified the government's approach to promote inland wind-generated power, which became gradually cost-competitive vis-à-vis conventional power and thus needed fewer subsidies on a per kilowatt-hour basis.

China's solar panel manufacture has been facing an even worse problem of overcapacity as a result of excessive government subsidies and overinvestment. One of the major chronic challenges facing China's solar PV industry is the asymmetrical supply-demand problem, in which China for long time had to export more than 90% of its finished products to other parts of the world due to an underdeveloped domestic market. This problem attracted attention from a number of researchers, who attributed China's solar overcapacity and over-dependence on foreign markets to the lack of appropriate interactions between renewable energy policy and renewable energy industrial (manufacturing) policy (Zhang et al. 2013, pp. 342–353).

Over the past decade, the gravity of solar PV production has shifted from the United States, to Japan, to Europe and to Asia, especially China, which quadrupled its poly-silicon solar panel manufacturing capacity between 2009 and 2011. China's existing production capacity of solar panels is about 150% of the global real demand, and in just four years between 2009 and 2012, the international price of solar panels had been cut by more than 75% largely due to an aggressive capacity built-up in China. To alleviate this overcapacity problem, the Chinese government thus has been redirecting energy subsidies in favor of PV power generation in order to breed a domestic renewable power market of sufficient scale to absorb manufacture oversupply that has encountered overseas boycott (Interviews with policy makers and researchers in NDRC's Energy Research Institute, Beijing, March 2012).

The country's average wind curtailment rate stood at a record high of 15% in 2015, with rates soared to over 30% in wind-rich northern provinces such as Gansu, Xinjiang or Jilin. See Table 9.1. Up to 33.9 billion kWh of wind electricity failed to connect to the grid, leading to losses of US$2.8 billion (Li 2016).

Table 9.1 Wind power curtailment rates in wind-rich provinces in 2015

Province	*Curtailment rates of wind power (%)*	*Wind capacity as a percentage of all installed capacity integrated into the grid in 2015 (%)*
Gansu	39	26.97
Xinjiang	32	26.07
Jilin	32	17
Heilongjiang	21	19
Inner Mongolia	18	23.31
Ningxia	13	26.03
Hebei	10	17.69
Liaoning	10	14.78
Shanxi	2	9.6
Qinghai	0	2.27
Tibet	0	0.51

Source China Dialogue (2016)

Government departments and grid dispatchers in favor of vested coal interests have been accused of exerting administrative power to disrupt wind production. Like many other countries, China's response to energy challenges is intertwined with its energy policy process, in which even the powerful NDRC lacks the authority, resources and tools to prevail over other state energy oligarchies. State-owned enterprises (SOEs) that dominate China's fossil fuel supply as well as thermal power generation and transmission sectors are major forces blocking local renewable energy production. In the coal production industry, large-sized SOEs, most of which are supervised by provincial governments, account for about half of China's domestic coal output. The remaining half of country's coal output comes from small-scale SOEs supervised by local governments and numerous privately- or collectively-owned coal mines in villages and townships. China's coal production industry has been partially liberalized to allow the participation of privately operated mines, thus creating many multi-millionaires, whose interests are often threatened by the booming local renewable energy projects.

When designing renewable energy policies, the Chinese policy makers expected local governments to play a very cooperative role in enforcing those policies, laws and regulations. In practice, however, this high degree of administrative cohesion does not exist in places where local government interests diverge sharply from those of the central planners and create substantial obstacles to strict enforcement of both national and local energy policies. As stakeholders of local fossil fuel companies, local authorities in many cases tend to help these companies bypass strict energy directives and curb competition from local renewable players. China's weakness in enforcing national energy policies is subject to the level of coordination within three kinds of relationships, namely the relationship among central government agencies, the relationship between central and local governments and the relationship between SOEs and government agencies.

First, responsibilities are not clearly defined among central government agencies and long bargaining process emerges when conflict of interests occurs. Second, the multilayer local regimes have great scope for distortion or non-implementation of national energy policy (Andrews-Speed 2003, p. 53). Third, the interest groups like gigantic energy SOEs in China's energy sectors have gained more influence in policy implementation. In China's state-centric society, the forces of environmental non-governmental organizations (NGOs) are still very weak and media's supervision is quite limited, because the government imposes strict regulations on them. Although many of these NGOs have also become increasingly visible players in China's energy and environment-related politics, their roles and functions are still quite restraint under China's one-party political systems, and therefore, they are not as strong and influential as their international peers. The void of effective civic supervision and lobby activities impedes the enforcement of China's low-carbon energy ambition and other environmental initiatives, as political and business elites, who benefit from patronage and income from natural resource rents and favorable policies, often have little incentive to engage with citizens and to build effective public authority over energy and environmental issues.

Besides, technical difficulties are also responsible for wind curtailments. To transfer electricity generated in northwest provinces to energy-hungry coastal areas in the east, long-distance transmission capacity is needed. China has decided to upgrade transmission lines and build the world's most capable UHV transmission network that can transmit 1000 kilovolts or more over long distances. A total of 12 such UHV transmission lines are being built to connect the northwestern provinces and coastal areas. China is now building the world's longest UHV line running from the Changji Prefecture in Xinjiang to Xuancheng city in east China's Anhui province. The 3324 km line is designed to accommodate currents of 1100 kilovolts. However, it normally takes at least three years to build a UHV line, which is also very costly. Intra-province integration within local grids is a more feasible solution. Local distribution grids can be converted into smart grids by implementing the latest technology, which helps integrate more renewable energy while avoiding costly transmission investments.

Conclusions

In the global climate change context, mitigation imperatives, enhanced by public health concerns over urban smog and PM 2.5 (particulate matter with diameter $\leq$ 2.5 micrometers), have pushed the Chinese government to cut reliance upon high-polluting coal-fired power plants and foster the development of low-carbon electricity generation. In 2015, China's solar energy capacity increased by a massive 74% over the year before, with a modest gain of 34% in wind. At the same time, China is importing 30% less coal and consuming 3.7% less overall (National Bureau of Statistics of China 2016). To fulfill the target of meeting 15% of primary energy demand with renewables by 2020, the Chinese government has proposed the targets of increasing installed capacity of wind power by 63% to 210 GW, and that of solar PV by 144% to 105 GW by 2020. In terms of promoting the production of solar and wind power, the government has been adjusting its policy prioritization from time to time, with variations in targeted growth areas, subsidy amount, on-grid tariffs and other financial incentives for the two types of renewables, which will profoundly change the current structure of the electricity market in the long run.

Since an economic slowdown in China has been exacerbating the issue of excess industrial capacity, more domestic equipment suppliers in the wind and solar sectors are expected to be pushed to the edge of collapse without robust governmental subsidization of domestic solar PV and wind power generation. As indicated in the case of interactions between China's wind energy industrial policy and wind power generation policy (Zhang et al. 2013, pp. 342–353), there should also be a natural affinity between the country's solar PV manufacturing policy and solar power generation policy, in which the improved competitiveness and capabilities of the manufacturers of solar PV equipment, components and parts result in lower costs for the installation and generation of solar power, and meanwhile, the development, generation and consumption of solar power are conducive to enhancing solar panel manufacturing competitiveness by providing a sustainable and stale domestic market demand. Thanks to massive government subsidies flowing to renewable energy sectors in recent years, China's wind turbine and solar PV manufacturers have gained competitive advantages over their foreign rivals and thus grown dramatically to leading positions in the world.

However, the huge investment of about 2.5 trillion yuan ($361 billion) plowed into renewable power generation between 2016 and 2020 (Reuters 2017) may bring about huge waste and inefficiency if the renewable energy developers fail to take into account industrial bottlenecks caused by uneven geographic distribution of renewable sources, grid constraint and policy disruptions from other local interest groups. Weak external renewable equipment demand and severe internal overcapacity have forced China's energy planners to make more efforts to develop domestic low-carbon power sectors. Therefore, it's likely that the central government in the next few years will take concrete actions to gradually absorb redundant local power capacity by building more UHV transmission networks, incentivizing local grid companies to purchase power from renewable power plants and installing more solar panels and wind turbines in coastal provinces.

With regard to the implementation of renewable energy policies at local levels, it is subject to the interests of local governments and the level of coordination between local governments and central government. The central government is responsible for policy regulation while the local governments are in charge of policy implementation. Unlike in the past when local levels used to be subject to a high degree of control under central government, after the economic reform and transition from a planned economy to a market economy, such control has been diminishing and local government has become more politically and economically independent. As they have their own interests to pursue, they need tactically to strike a balance between developing the local economy and serving as political agents for the central government. Sometimes, they prioritize local economic development and chose non-compliance or partial-compliance with orders from central government. Sometimes the coordination between central government and local governments has also been poor and it limits the efficiency of policy implementation at local levels (Zhao 2001, p. 6). As local governments have different priorities, sometimes they may justifiably reinterpret the national policies and bundle them with more pressing issues that are of greater importance to local development; sometimes they may actively or passively obstruct the implementation of central government policy when that policy is not in their interest of development; or sometimes it is simply because the local government lacks the resources to implement national policies (Meidan et al. 2009, p. 616).

China's power industry, despite being increasingly competitive and fragmented as a result of government-initiated structural reforms, is still highly regulated by the state, with on-grid electricity tariffs, in most cases still set by governmental regulators, being vital to profit margins and even survival of various power generators. As compared to major hydropower and nuclear power corporations that are state-owned behemoths subject to state plans and government edicts, wind and solar power plants, mostly small and medium-sized generators with diverse ownership, are more sensitive to fluctuations in power pricing, production cost and demand conditions. Government plans, which are more precise in predicting future development trajectories of hydro and nuclear power, often lag behind unexpected proliferation rates of wind and solar power plants that face higher supply and demand elasticity under conditions of government subsidies, feed-in tariffs and production cost slump.

China put in place a four-level wind power tariff system suitable for regions with different meteorological conditions in 2009, and in terms of solar PV power to which China has applied a unified nationwide feed-in tariff, the government has been inspired by its own approach to wind power to formulate three categories of on-grid tariffs targeted at regions with varied solar conditions since August 2013. When promoting low-carbon power generation, the Chinese government needs to respond to new industrial and market conditions as well as environmental concerns faster. Emerging trends, driven by more sophisticated energy markets, volatile energy production costs, reassessment of innate geographical conditions and environmental impact, and increased attention to supporting domestic industries, should be better reflected in the making of non-fossil fuel policies.

References

A Medium Corporation. 2018. China Homes in on Hydrogen. Available at: https://medium.com/@cH2ange/china-homes-in-on-hydrogen-977b37ddcca9. Accessed 19 Mar 2019.

Andrews-Speed, Philip. 2003. *Energy Policy and Regulation in the People's Republic of China*. The Hague: Kluwer Law International.

Chen, Gang. 2012. *China's Climate Policy*. London: Routledge.

Chen, Gang. 2019. *Politics of Renewable Energy in China*. Cheltenham: Edward Elgar.

China Daily. 2018. Project Launched to Develop Hydrogen Fuel Cells in E China, August 30. Available at: http://www.chinadaily.com.cn/a/201808/30/WS5b873f25a310add14f3888b1.html. Accessed 6 Mar 2019.

China Electricity Council. 2018. *2017 quanguo fadian zhuangji rongliang zengzhang 7.6%* [China's Installed Power Capacity Increases 7.6 Per cent in 2017]. Available at: http://www.cec.org.cn/nengyuanyudianlitongji/hangyetongji/dianlixingyeshuju/2018-03-02/178238.html. Accessed 10 Apr 2019.

China Meteorological Administration. 2010. *New Progress Made in Surveying China's Wind Power Resources* (in Chinese) [online]. Available at: www.cma.gov.cn/mtjj/201001/t20100104_55673.html. Accessed 9 Jan 2012.

China State Electricity Regulatory Commission. 2011. *Power Industry Monitoring and Supervision Report 2010* (in Chinese). Beijing: State Electricity Regulatory Commission.

China's 12th Five-Year Plan on Energy Development (in Chinese). 2013. Available at: http://www.gov.cn/zwgk/2013-01/23/content_2318554.htm. Accessed 6 Dec 2018.

China's 13th Five-Year Plan on Energy Development (in Chinese). 2016. Available at: http://www.ndrc.gov.cn/zcfb/zcfbtz/201612/W020161216659579206185.pdf. Accessed 19 Dec 2018.

Evans, A., V. Strezov, and T.J. Evans. 2012. Assessment of Utility Energy Storage Options for Increased Renewable Energy Penetration. *Renewable and Sustainable Energy Reviews* 16: 4141–4147.

Hatch, M. 2003. Chinese Politics, Energy Policy, and the International Climate Change Negotiations. In *Global Warming and East Asia: The Domestic and International Politics of Climate Change*, ed. P.G. Harris, 43–65. London: Routledge.

Jackson, C. 2018. Chinese Fuel Cell Industry Developments. *Fuel Cell and Hydrogen Energy Association*, December 9. Available at: http://www.fchea.org/in-transition/2019/2/4/chinese-fuel-cell-industry-developments.

Li, L. 2007. China Urges Electricity Suppliers to Buy 'Green' Power. Available at: http://www.renewableenergyworld.com/rea/news/article/2007/09/china-urges-electricity-suppliers-to-buy-green-power-49879. Accessed 6 Jan 2012.

Li, Y. 2016. Blowing in the Wind. *China Dialogue*, May 31 [online]. Available at: https://www.chinadialogue.net/article/show/single/en/8965-Blowing-in-the-wind. Accessed 10 May 2018.

Meidan, M., Philip Andrews-Speed, and M. Xin. 2009. Shaping China's Energy Policy: Actors and Progresses. *Journal of Contemporary China* 18 (61): 591–616.

National Bureau of Statistics of China. 2016. *Statistical Communiqué of the People's Republic of China on the 2015 National Economic and Social Development* (in Chinese). Available at: http://www.stats.gov.cn/english/PressRelease/201602/t20160229_1324019.html. Accessed 20 Jan 2017.

National Bureau of Statistics of China. 2017. *China Statistical Yearbook 2017*. Beijing: China Statistics Press.

NDRC (National Development and Reform Commission of China). 2011. *The NDRC Circular on Improvement of Solar PV On-Grid Tariff Policy* (in Chinese). Available at: http://www.gov.cn/zwgk/2011-08/01/content_1917358.htm. Accessed 8 Jan 2014.

NDRC (National Development and Reform Commission of China). 2013. *NDRC's Circular on the Promotion of Solar PV Sector Through Price Leverage* (in Chinese). Available at: http://www.sdpc.gov.cn/zfdj/jggg/dian/t20130830_556127.htm. Accessed 29 Nov 2013.

New York Times. 2017. Windmills Stand Idle in China as Even More Are Being Constructed, January 16, p. A8.

PRC (People's Republic of China) Amended Law on Renewable Energy (in Chinese). 2009. [online]. Available at: www.npc.gov.cn/npc/xinwen/2009-12/26/content_1538199.htm. Accessed 8 Jan 2018.

PRC (People's Republic of China) Renewable Energy Law. 2005. Available at: www.renewableenergyworld.com/assets/download/China_RE_Law_05.doc. Accessed 1 Jan 2019.

PVTECH. 2019. China to Build 6 GW of Pumped Hydro Storage in Bid to Slash Solar and Wind Power Curtailment, January 9. Available at: https://www.pv-tech.org/news/china-to-build-6gw-of-pumped-hydro-storage-in-bid-to-slash-solar-and-wind-p. Accessed 9 Apr 2019.

REN21 (Renewable Energy Policy Network for the 21st Century. 2013. *Renewables 2013—Global Status Report* [online]. Available at: https://www.ren21.net/wp-content/uploads/2019/05/GSR2013_Full-Report_English.pdf. Accessed 9 Nov 2019.

Reuters. 2017. China to Plow $361 Billion into Renewable Fuel by 2020, January 5. Available at: https://www.reuters.com/article/us-china-energy-renewables/china-to-plow-361-billion-into-renewable-fuel-by-2020-idUSKBN14P06P. Accessed 8 Jan 2019.

Wu, J., Z. Wang, and G. Wang. 2014. The Key Technologies and Development of Offshore Wind Farm in China. *Renewable and Sustainable Energy Reviews* 34: 453–455.

Zhang, G. 2005. How Can Energy Shortage Be Blamed on China? (in Chinese). *People's Daily Overseas Edition*, September 21, p. 1.

Zhang, S., Philip Andrews-Speed, X. Zhao, and Y. He. 2013. Interactions Between Renewable Energy Policy and Renewable Energy Industrial Policy: A Critical Analysis of China's Policy Approach to Renewable Energies. *Energy Policy* 62: 342–353.

Zhao, J. 2001. Reform of China's Energy Institutions and Policies: Historical Evolution and Current Challenges. BCSIA Discussion Paper 2001-20, Energy Technology Innovation Project, Kennedy School of Government, Harvard University.

CHAPTER 10

Solar PV in Singapore in the Absence of Subsidies

Gautam Jindal, Jacqueline Tao, and Anton Finenko

The Republic of Singapore is a city state, located on one main island surrounded by many small islets with a total land area that has grown from 581.5 sq. km. in the 1960s to 719.1 sq. km. in 2015. Singapore has a small albeit developed economy. With a 2017 GDP of 447.3 billion dollars (Ministry of Trade and Industry 2018), Singapore's population of 5.6 million has one of the highest Gross National Income per capita in the world. Singapore does not have any traditional energy resources domestically, and its energy sector almost entirely relies on natural gas imports. Traditionally, most of Singapore's supply of natural gas has come from pipelines that connect from Indonesia and Malaysia. However, in a bid to diversify and improve the security of fuel supply, Singapore has moved toward Liquefied Natural Gas, with an LNG terminal opening in 2013. As of early 2018, almost a quarter of Singapore's natural gas supply is now in the form of LNG (Ng and Heng 2018).

G. Jindal (✉) · J. Tao · A. Finenko
Energy Studies Institute, National University of Singapore, Singapore, Singapore

P. Midford and E. Moe (eds.), *New Challenges and Solutions for Renewable Energy*, International Political Economy Series,
https://doi.org/10.1007/978-3-030-54514-7_10

Singapore is also considered to be "alternate energy disadvantaged." Its average wind speeds of 2–3 m/s are not enough to install wind power which generally requires winds faster than 6 m/s. Furthermore, the seas are calm and tidal range is relatively narrow; thus, there is no potential for tidal energy. In fact, the low wind speeds and calm seas are factors that have made Singapore a historically significant trading port. Much of Singapore's terrain lies within 15 meters of the sea level, and there is no fast-flowing river, thus hydropower is also not feasible. Finally, Singapore's high population density and land scarcity render nuclear power and locally sourced biomass unviable.

Thus, currently the most viable domestic renewable energy option for Singapore is solar PV. Consequentially, the development of solar PV is a crucial mitigation strategy for Singapore to achieve its Nationally Determined Contribution target, namely its pledge to the 2015 Paris climate agreement to achieve a 36% reduction in emissions intensity from 2005 levels by 2030, and to stabilize its emissions with the aim of peaking around the same time. Toward achieving this target, Singapore has set in place, a target to install 350 MW of solar PV capacity by 2020, and to reach 1 GW of installed capacity "beyond 2020." Before discussing the strategy and challenges to large-scale growth of solar PV in Singapore, the following section provides a brief introduction to Singapore's electricity sector, its liberalized electricity market, and the electricity pricing framework.

Singapore's Electricity Sector

In 2017, Singapore's annual electricity consumption was approximately 49.5 GWh, with a peak demand of about 7.2 GW. Electricity demand has risen with a compound annual growth rate of 2.8% since 2005 (35.5 GWh), largely driven by consumption from industry, which comprises of about 40–42% of total electricity consumption, followed by the commerce & services sector, and households, which constitute the remaining 37 and 15%, respectively (Energy Market Authority 2018a). A study by the electricity market regular estimates that over the next ten years, electricity demand will grow at a compound annual growth rate of 1.3–1.9%, depending on factors such as population growth, temperature changes, and GDP growth rates among others. By 2028, Singapore's peak demand is expected to range between 8400 and 8980 MW and annual electricity demand could grow to 64.6 GWh (Energy Market Authority 2017a).

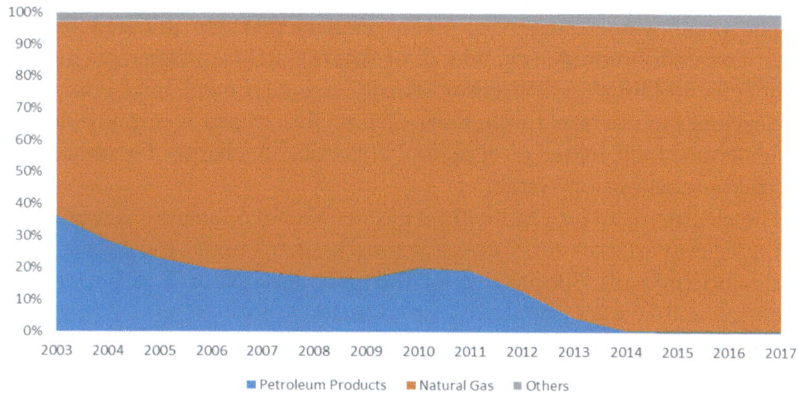

Fig. 10.1 Singapore electricity fuel mix (2003–2007) (*Note* Author created figure based on data from Energy Market Authority [2018b])

As of June 2018, Singapore's total installed generation capacity was approximately 13.5 GW (Energy Market Company 2018a). This is almost double the 2017 peak demand and well above the required minimum reserve margin of 30%.[1] This capacity almost entirely consists of natural gas combined-cycle turbines. Figure 10.1 shows that the share of natural gas in the total fuel mix for electricity generation has grown from about 36.5% in 2003 to approximately 95.3% in 2017. Waste incineration and a coal-biomass project contribute the remaining 4.1%, and fuel oil, used occasionally, contributes the remaining 0.6% (Energy Market Authority 2018b).

Deregulation

Singapore's electricity sector was deregulated starting in 1995, with generation and retail privatized in subsequent years and the transmission & distribution remaining under government control. Today, Singapore's generation sector consists of seven private companies with three large players occupying about 58% of the market share, as compared to 83% market share in 2005 (Energy Market Authority 2018a). The transmission and distribution system is effectively owned and managed by government-held companies, SP PowerAssets Ltd and SP PowerGrid Ltd, respectively.[2]

The retail sector consists of seven retailers that are affiliated to the seven-generation companies, and as of August 2018, another twenty-six companies holding retail licenses. Finally, a government-held company, SP Services Ltd., acts as the Retailer of Last Resort and is responsible for settlement of bills, meter reading, and collection of charges for use of the electricity transmission system.

The electricity sector (and natural gas market) in Singapore is regulated by the Energy Market Authority, a statutory board under the Ministry of Trade and Industry. Specifically, for the electricity sector, EMA approves technical and market standards, issues market licenses, and is responsible for protecting consumers' interests. It also houses the Power System Operator, which is responsible for ensuring the reliability of electricity supply. The overall energy policy is set by the Energy Division of MTI with the aim of supporting economic growth, addressing energy security, economic competitiveness, and the environmental sustainability (Energy Market Authority 2016).

Singapore's Electricity Market

Singapore is home to Asia's first liberalized electricity market. With the deregulation process in 1995, the incumbent's electricity and gas operations were split into six different companies—three generation companies, a transmission & distribution, and a retail company. Three years later, Singapore launched the first wholesale market in Southeast Asia, a simple day—ahead market called the Singapore Power Pool (SPP). In 2003, electricity trading moved to the National Electricity Market of Singapore (NEMS), which is a pool-type wholesale market that uses bid-based, security-constrained economic dispatch with locational marginal pricing. The NEMS is a real-time only market and does not operate a day-ahead market as is the practice in many electricity markets in the United States. Rather, it bears close resemblance to the Australia National Electricity Market and the New Zealand market.

NEMS co-optimizes the procurement of energy and two other ancillary services: Reserves for maintaining frequency in times of generator/demand outage, and Regulation for maintaining real-time balance with generation and demand. For every half-hour dispatch interval, generators are required to submit ten Price–Quantity bid pairs for energy, five Price-Quantity bid pairs each for of the two categories of Reserve, and five Price-Quantity bid pairs for Regulation. The Power System Operator provides demand forecasts and network data. These datasets are received by the Energy Market Company which runs the market-clearing engine and produces the optimal dispatch solution. The market is cleared for a system of more than 700 nodes across the island, although the price differential between different nodes is generally very small.

On the retail side, since 2001, Singapore has progressively lowered the threshold for consumers who are allowed to purchase electricity from retailers, known as contestable consumers. Currently, commercial and industrial consumers with an average monthly electricity consumption of 2000 kWh can be contestable. These consumers account for about 80% of Singapore's total electricity demand. Early in 2018, a soft launch of the open electricity market was conducted in one township where all consumers are now allowed to change their retailer. Full retail liberalization was completed in the second half of 2018.

In addition to the physical market, an electricity futures market is run by the Singapore Exchange since 2015. The futures market was introduced to allow new independent retailers, i.e., those who were not retail arms of generators, to secure electricity supply at competitive prices and hedge their risk. This was a particularly important development given that 20 out of the 23 licensed retailers only entered the market after full retail contestability (i.e., the ability to compete for a consumer's business) was announced in October 2015.

Electricity Pricing

Singapore does not provide any subsidies for any form of energy, and by extension, electricity. The regulated tariff, i.e., the tariff payable by non-contestable consumers and contestable consumers who chose to stay with the retailer of last resort, is set by the EMA quarterly based on the average forward fuel oil price over the past three months. The reason for linking the electricity tariff to fuel oil price is the prevalence of fuel oil-indexed pricing for natural gas in Asia as a standard practice. The non-energy component of the tariff consists of other charges such as a grid charge, fees for power system operation and market administration, and the cost of meter reading and billing.

On the other hand, retailers procure electricity from the wholesale market at the average price of energy at all market nodes, known as the Uniform Singapore Electricity Price, plus other cost components such as the overall cost of procuring reserves. They generally offer plans such as a fixed tariff, a discount on the regulated tariff, or time-of-use tariffs to consumers. Large consumers are also allowed to directly participate in the wholesale market.

In addition, Singapore implemented a carbon tax starting in 2019. The tax is applicable to all power generation companies, among other large emitters, and will be priced at S$5 per tonne of CO2e emissions. For power generation, this is expected to add an additional tax liability of S$2.10–2.15 per MWh,[3] and is expected to be passed on to end consumers by the power generators, in order to incentivize efficient electricity use.

It is important to note that while Singapore does not provide any form of direct energy subsidies, it provides vouchers to low-income households that can be used to offset utility bills. Compared to subsidizing the cost of electricity directly, this approach presents all consumers with the true cost of electricity, thus incentivizing them to save and be efficient.

Solar PV in Singapore

As mentioned previously, Singapore is "alternate energy disadvantaged," and thus, solar PV is the most promising source of zero-carbon electricity and a key pillar of Singapore's strategy for achieving its 2030 climate pledge. At the same time, electricity from PV installations within Singapore can reduce Singapore's dependence on imported natural gas, albeit to a small extent. Finally, Singapore's load profile peaks mostly in the afternoon, which coincides with the maximum PV output. Thus, PV can play an important role in demand "peak shaving" and may possibly reduce the need for peaking conventional generation plants in the future.

As a result, solar PV has been experiencing rapid growth in Singapore. As shown in Fig. 10.2, the total installed capacity grew from 0.3 MWac in 2009 to about 115 MWac by the first quarter of 2018 (Energy Market Authority 2018c). Due to scarcity of land, there are no utility-scale solar PV plants in Singapore. PV systems mostly comprise of small-scale rooftop PV projects with system capacity ranging from a few kW to a maximum of slightly more than 4 MW. The installed capacity is dominated by the private sector and by town councils, which account for slightly more than 90% of the total share.

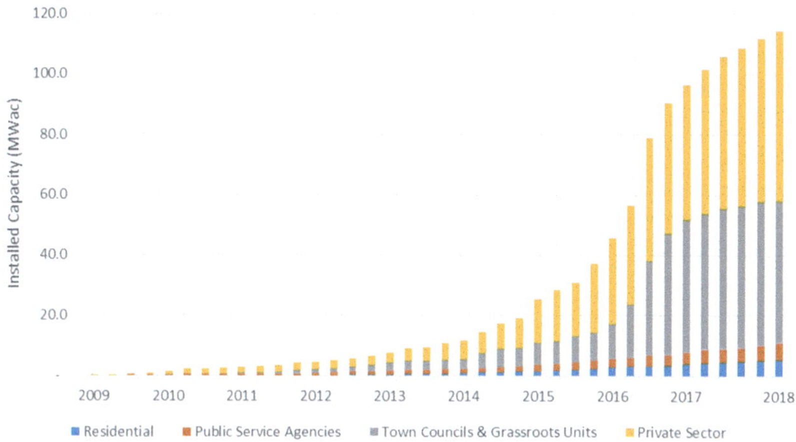

Fig. 10.2 Growth of installed solar PV capacity in Singapore (*Note* Author created figure based on data from Energy Market Authority [2018c])

Solar PV Potential

The Solar PV Roadmap for Singapore (Luther and Reindl 2014) estimates that out of Singapore's total land area of 719 sq. km, approximately 45 sq. km. can be used for solar PV development. Roughly two-thirds of this area comprise buildings' rooftops and facades, while the remaining share is taken up by floating solar panels on island waters and developments on offshore islets. The installed capacity potential over time in that area depends on the area efficiency of the solar modules and the expected technological level at the time of development. The roadmap gives an installed capacity range of 650 MW–950 MW by 2020 and 5000 MW–10,000 MW by 2050. However, these numbers should be interpreted with caution, as they merely give a theoretical maximal potential, and do not account for commercial and policy consideration, fossil fuel prices, most notably of natural gas, and development of technologies such as grid storage.

Business Models

Development of solar PV systems in Singapore is undertaken by private developers who enter contracts with either the government, private sector, or residential customers willing to deploy PV solutions. Due to limited space potential, private developers mainly focus on deployment and maintenance of small-scale rooftop PV systems, including technical, financial, and legal expertise. Most contractual agreements are either a direct purchase of the solar system or a solar leasing model. The direct purchase of solar systems requires the customer to pay the whole system upfront and only contribute maintenance costs over the 25-year lifetime of the system. The advantage for customers is that all electricity generated by the solar system is now virtually free, thus offsetting their monthly electricity bill.

In comparison, the most prevalent business model for PV in Singapore is the "leasing model" which requires the customer to pay only for the electricity generated by solar systems. The electricity is priced in such a way that it includes the investment cost of the PV system and also accounts for maintenance expenses, while providing the PV project rate of return high enough to compensate the initial investment and the risk. Theoretically, even if solar leasing may not lead to a reduction in the customer's electricity bill, it relieves the customer from the burden of heavy initial investment and transfers it to the project developer instead. Additionally, the entire project lifecycle risk can be borne by the project developer.

A variation of solar leasing that has recently gained popularity in Singapore is off-site leasing. This model allows customers who do not own rooftops to enter contracts and purchase solar power from panels installed at rooftops owned by a third party. As mentioned previously, Singapore does not provide subsidies to any form of energy, including no Feed-in Tariffs (FITs) for solar PV. Thus, project developers offer electricity at a discount to the customer, generally with respect to the EMA's regulated tariff or with respect to the customer's current tariff arrangement with its retailer.

Promoting Solar PV

SolarNova. As mentioned previously, one of the main limiting factors for installation of PV in Singapore is the limited availability of land or roof space. The Solar PV Roadmap for Singapore (Luther and Reindl 2014) estimates that 8% of Singapore's total land area is covered with buildings (approximately 56 km^2), out of which 25% are residential buildings developed and managed by Singapore's Housing and Development Board

(HDB). The HDB is the public housing agency and manages housing for more than 80% of Singapore's population.

Singapore's Economic Development Board, in partnership with the HDB, has developed a program called SolarNova that aggregates demand for solar PV from HDB and other government agencies; and calls for tenders from private solar PV developers to install the required capacity under a solar leasing model. The idea is to allow project developers to bring in economies of scale, tap low-cost sources of finance, and consequently offer more discounts with respect to prevalent electricity tariffs.

The first tender called under the SolarNova program was for a 40 MW facility in June 2015. The eventual award was given for installing a larger capacity plant of 76 MW instead, covering rooftops of 831 residential HDB blocks and eight other government facilities. In October 2016, the next tender for a capacity of 40 MW was called and was awarded in June 2017. This second phase will cover 636 HDB blocks and 31 installations from eight other government agencies, such as the Ministry of Home Affairs, Ministry of Education, and Ministry of Finance. The third tender, for an installed capacity of 50 MW, was called in November 2017 and was awarded in June 2018. Under this phase, panels will be installed over 848 HDB block rooftops and 27 other government sites (Housing and Development Board 2018).

The eventual aim of the SolarNova program is to have 220 MW of installed PV capacity over HDB and other public sector buildings by 2020, which will account for the bulk of the national targeted 350 MW capacity. In addition, the program will provide the much-needed "scale" to develop all aspects of the PV industry within Singapore, including manufacturing, system integration, and financing.

Tariff settlement. To promote the uptake of distributed renewable energy systems, it is generally important that such systems are allowed to sell excess electricity to the grid and receive appropriate compensation. In 2014, Singapore's energy regulator proposed a comprehensive framework that provides distributed PV systems the option of net metering, based on the nameplate capacity of the PV system and the contestability status of the consumer.

For non-contestable consumers that install PV systems of less than 1 MWac[4] capacity, any excess electricity that is sent to the grid receives payment at the prevailing electricity tariff minus the grid charge. This payment is in the form of a credit adjustment to their monthly electricity bill. Contestable consumers with PV systems of less than 1 MWac capacity are treated the same as conventional generators with the respect to the electricity market. Nonetheless, they are accorded some benefits such as the option to not have to register as a market participant in the wholesale market but receive or make payments through a central intermediary.

For each half-hour wholesale market trading interval, electricity withdrawal from the grid and PV electricity exported to the grid are netted off for these consumers. In case of net withdrawal, a consumer is charged either the USEP or retail price, whereas in case of net export, the consumer is paid at either the nodal price or the weighted average nodal price. At the same time, these consumers are required to pay reserve charges for frequency balancing services (e.g., regulation)[5] and market charges such as market operators' fees and grid charges, which are applicable to conventional generators. Reserves charges are calculated based on gross import from grid and gross PV generation, whereas all other market charges are calculated based on net import or export.

Alternatively, these contestable consumers with PV systems of capacity less than 1 MWac can choose not to receive any compensation for net export to the grid by registering themselves as an "IGS Non-Exporting" system. The advantage being that they are also not subjected to reserve and market charges in case of net export to the grid. These are only applicable in case of net withdrawal (Energy Market Authority 2015).

Consumers with PV systems larger than 1 MWac have no choice but to register as a participant in the wholesale market and are charged or paid at the nodal price in case of net import or export, respectively. They are subject to reserve charges based on gross import from grid and gross PV generation, whereas all other market charges are calculated based on net import or export. In 2017, the threshold for this scheme was modified for PV systems with capacity of up to 10 MWac (Energy Market Authority 2017b), thus making it easier for consumers to install larger PV systems and allowing them to decide the merits and demerits of having their PV system treated as a generator or an IGS Non-Exporting system. For consumers who utilize most of their PV system's output for self-consumption, the option to register as an IGS Non-Exporting system may be more attractive so as to avoid being subjected to volatile market charges.

Challenges for PV Adoption

Market and Financial Challenges. Despite accelerating growth in recent years, achieving a business case for solar PV in Singapore remains challenging. The difficulty of building a financial case for solar developers is one of the most cited reasons. To provide an illustration of the level of profitability of solar projects, a discounted cash flow model developed by the Solar Energy Research Institute of Singapore has found that even with a levelized cost of electricity of 10.2 Singapore cents per kWh for a "self-owned" 1 MW industrial roof system, a project installed in 2017 is most likely to have an internal rate of return of approximately 12% and a payback period of about eight years (National Solar Repository 2018).

Solar developers typically sell their projects by selling electricity based on a discount (typically ranging from 5 to 10%) on the prevailing electricity tariff offered to the consumer. This creates a twofold problem. First, most solar electricity consumers are industrial and commercial consumers. Under the current wholesale market conditions, these large consumers are already exposed to tariffs that are lower than the regulated electricity tariff, thus making payback periods longer than they would be if the discount was on the regulated tariff. Second, this exposes solar developers to fluctuations in profit margins. While the costs associated with implementing a solar project is fixed, the returns are likely to fluctuate based on the volatility of electricity tariffs. The EMA regulated tariff and the average retail price offered to contestable consumers are almost entirely dependent on the price of natural gas. In turn, these are indexed to oil prices, as is the prevalent practice in Asia. This exposes the electricity tariffs to fluctuations in the global oil price. Between 2014 and 2017, electricity prices dropped significantly due to the collapse of oil prices, thus creating increased pressure on solar project developers' profit margins.

One potential solution adopted in other countries would be the introduction of FITs or other support mechanisms. However, as mentioned previously, such a policy would run counter to established energy policy in Singapore wherein energy market and prices are to be operated under a free market environment with minimal government intervention. An alternative is to introduce fixed tariffs for solar electricity sales. One may argue that a fixed price mechanism may offer benefits to both developers and consumers by providing stable returns and costs. However, this pricing strategy, though offered by most solar developers, has somehow not gained much acceptance from consumers.

This lack of appeal of fixed price solar electricity comes from a bias inherent in consumers' decision-making process. As mentioned earlier, most solar electricity consumers are commercial or industrial users, which typically employ corporate finance metrics to make decisions on investments in projects. A percentage reduction model generally has a stronger appeal, since it ensures a cost saving whereas a fixed price may not. Over the years, as Singapore's solar market has evolved and matured, variations of the fixed and floating pricing structures have emerged, including mechanisms such as price ceilings and price floors. However, the fixed price regime has not garnered much interest.

Domestic market dynamics adds another challenge to solar development in Singapore. The geographical potential for PV adoption in Singapore is very small compared to its neighboring countries such as Indonesia, Malaysia, and Thailand. Moreover, the types of projects common in Singapore are almost entirely distributed rooftop PV, compared to large-scale utility PV in other countries. Such small-scale projects are relatively more expensive and require different financing vehicles compared to utility PV systems.

Unlike large-scale projects that can tap project financing vehicles, small-scale projects are typically funded with corporate loans. The nature of the solar market's development trajectory has therefore created a highly competitive business environment for industry players. Without large-scale projects that require big multinational players with strong financial backgrounds, the market is dominated by smaller players with minimal capital. Compounded by the fact the most business opportunities are distributed PV installations, this creates a largely homogenous, highly competitive, and fragmented local solar industry that competes on price. Considering this, access to financing creates another challenge for the local industry players. A typical bank loan in Singapore ranges between US$35 and 70 million, which is equivalent to a solar project size of 30–60 MW. However, distributed PV projects in Singapore have not yet exceeded even 5 MW. Consequently, solar projects are currently financed through a combination of debt and equity.

Since most solar developers in Singapore are small and medium enterprises (SMEs), the most utilized forms of bank loans are SME loans, which usually offers floating interest rates of 6–7%, and a loan term of 3–5 years. Such loans often come with a requirement to pledge certain liquid assets as collateral and are known as secured loans. Non- or limited recourse project finance may also be utilized, though this kind of finance is not readily available for smaller loan sizes. Furthermore, the lack of familiarity and standardization causes investors to demand higher returns due to higher perceived risks, while conducting more stringent due diligence that translates into higher transaction and monitoring costs. Thus, the cost of borrowing remains high for solar PV companies despite public sector credit guarantees and other forms of collateral.

The small size of projects in Singapore also limits use of innovative financial mechanisms, such as investment trusts and YieldCos due to high transaction costs and the lack of aggregation capacities. Options such as crowd lending have yet to pick up in Singapore, further limiting access to suitable investors. In this ecosystem, project developers are only left with high-cost options such as private equity.

PV Intermittency and Grid Integration. The other major challenge to large-scale adoption of PV in Singapore, as is the case elsewhere, is the intermittent and variable nature of PV output. Singapore, being situated near the equator, has a tropical climate with two monsoon seasons separated by inter-monsoonal periods (Meteorological Services Singapore 2017). Singapore receives rainfall 178 days a year on an average, and the median cloud cover is approximately 90% (mostly cloudy) throughout the year (Weatherspark 2017). Thus, despite having average annual irradiation on the order of 1580 kWh/m^2/year, PV output in Singapore can be highly intermittent. Figure 10.3 shows how highly variable solar irradiation can be over a twelve-hour period (at one-minute intervals) at a weather monitoring station in Singapore. Consequently, Singapore is concerned about the implications of high PV penetration on the stability of the power grid. Solar PV systems are thus categorized as Intermittent Generation Sources (IGS) in the electricity market.

However, the figure above does not present an accurate picture of PV output in Singapore. As mentioned in the previous sections, PV systems in Singapore will mostly be installed on rooftops distributed across the island. Thus, due to the variability smoothing effects of spatial diversity that is discussed in the literature (Marcos et al. 2011; Mills and Wiser 2010), the aggregated output of all PV systems is expected to be much less volatile as compared to the output of a single PV system. Figure 10.4 shows one-minute solar irradiation as measured at four monitoring stations across Singapore. The bold line shows the

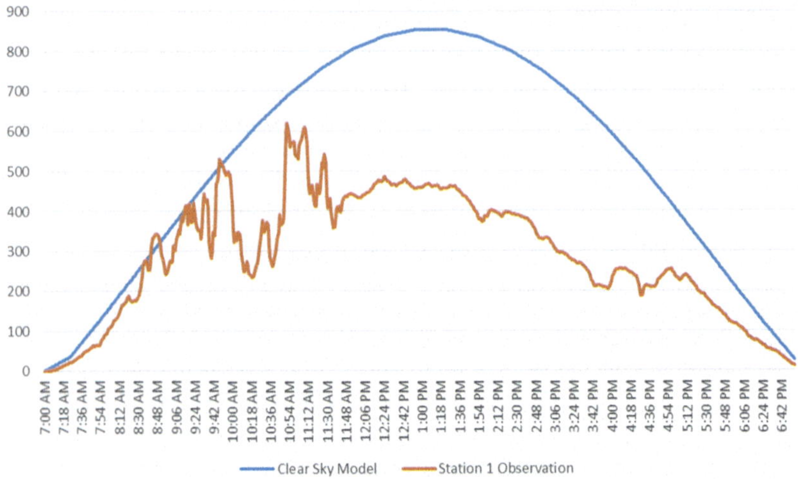

Fig. 10.3 1-minute resolution irradiation data at Singapore's monitoring station 1 in 2014 versus expected irradiation from a clear sky model (*Note* Author created figure that uses the Adnot–Bourges–Campana–Gicquel [ABCG] model as it has been observed to most precisely reflect clear sky GHI for Singapore. Based on data from Yang et al. [2011] and Weatherspark [2017])

average irradiation, which has much lower variability as compared to the individual station irradiation.

Singapore operates one of the most reliable power systems in the world. It has a System Average Interruption Duration Index (SAIDI) value of 0.56 for the year 2015–2016, which basically means that electricity consumers in Singapore experienced an average interruption in power supply of approximately 34 s during the entire year. In comparison, the SAIDI value for other major cities was much higher. For example, Tokyo had a SAIDI value of 4.00, New York's SAIDI came to 20.53, and Paris' value was slightly higher than one hour (SP Group 2018).

Thus, despite the expected smoothing effect of aggregated power output from distributed PV systems, the EMA has consistently expressed concern about the implications of high PV penetration for Singapore's grid stability and reliability. To avoid these impacts, Singapore initially imposed a cap of 350 MWac of installed PV capacity. However, in 2014 this cap was superseded by a "Dynamic Pathway" approach under

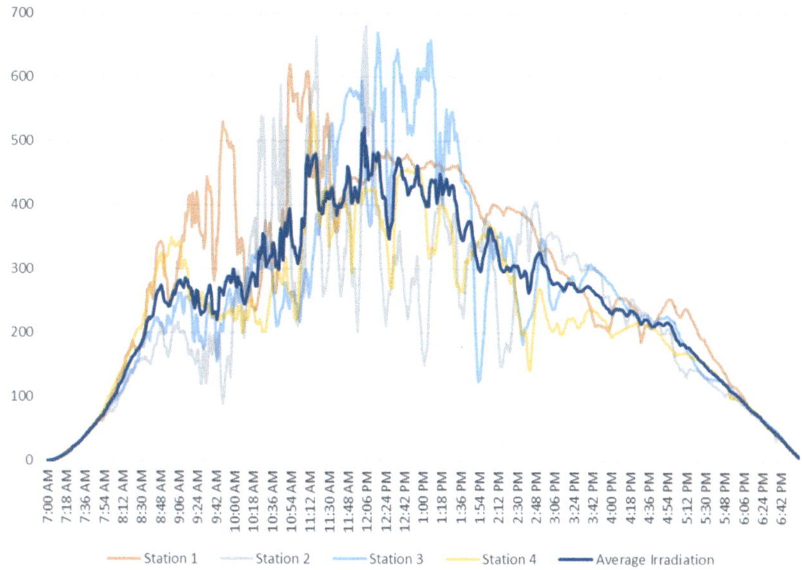

Fig. 10.4 1-minute resolution irradiation data at four monitoring stations in Singapore and their average irradiation (*Note* Author created figure based on data from Weatherspark [2017])

which frequency control reserves will be procured in increasing quantities in parallel with the growth in installed PV capacity (Energy Market Authority 2014).

Under the dynamic pathway approach, two thresholds have been defined. The first is the Intermittent Generation Threshold (IGT), which is the amount of PV capacity that can be integrated into the grid based on the existing amount of reserves. The IGT, which was equivalent to the hard cap of 350 MW set previously, was revised to 600 MWac under this pathway. The pathway includes an option to raise the IGT level further with improvements in accuracy of PV forecasting, as more accurate PV forecasts will lead to reduction in quantity of required reserves. Toward this objective, the EMA has awarded a S$6-million research grant to improve the accuracy of PV output forecasts through tools and methods from weather prediction, remote sensing, machine learning, and grid modeling. The second threshold under the dynamic pathway is called the Intermittent Generation Limit (IGL), which is the maximum amount of

PV capacity that the power system can support depending upon the total amount of reserves capacity that can be called upon at any point in time. This figure is dependent on the total quantity of reserves that can be provided by all generators and loads in Singapore's power system that have the capability to provide reserves. As of now, the IGL level has not been disclosed. Integration of PV capacity beyond the IGL levels would require Singapore to either link its electricity grid with other regional grids, or promote installation of energy storage, among other measures. However, these come with their own set of challenges.

Interconnection of power and gas grids has been a major topic of discussion among the member countries of the Association of Southeast Asian Nations (ASEAN) for several years now. An interconnected ASEAN electricity grid would theoretically result in benefits such as reduced grid costs and fuel savings, enhanced energy security and system resiliency, and increased environmental benefits—especially with the use of the vast hydropower potential in the Mekong region. However, there are a number of challenges that would need to be overcome, the most important one being the marked differences among policies and regulations for managing the energy sectors in various ASEAN countries, including the differences in electricity tariffs and subsidy levels (Andrews-Speed 2016). As a result, interconnections are currently restricted to bilateral connections between several groups of immediate neighbors.

The first attempt toward a multilateral interconnection and transfer of electricity was envisaged in 2013 in the form of the Lao PDR-Thailand-Malaysia-Singapore (LTMS) Power Integration Project (PIP). The aim of this project is to eventually allow for the transfer of up to 100 MW of power from Lao PDR to Singapore via Thailand and Malaysia through existing interconnections (Ministry of Trade and Industry 2014). Currently under the first phase of this project, Malaysia, Thailand, and Lao PDR have signed a tripartite Energy Purchase and Wheeling Agreement (EPWA) to transfer 100 MW of hydropower from Lao PDR to Malaysia. A publication by Malaysia's energy regulator claims that Singapore is expected to join the project in its second phase, possibly by 2020 (Suruhanjaya Tenaga 2017).

Regarding energy storage, Singapore opened its electricity market in 2015 to electricity storage systems, by allowing them to compete with conventional generators in offering frequency regulation services. However, unlike electricity markets in the United States, Singapore currently does not offer special tariffs for battery storage's fast response capabilities. With prices for regulation averaging at less than S$20 per MWh over the last three years (Energy Market Company 2018b), there seems to be no business case for the private sector to invest in electricity storage for this purpose.

In 2016, the EMA released a consultation paper on developing a holistic policy framework that will eventually govern technology and applications for agnostic integration of energy storage solutions in the electricity market. The objective of the framework was to answers questions such as—which energy storage applications can provide the most value in Singapore's market; are there commercially viable business models for ESS projects; and what modifications can be made to the regulatory framework or wholesale market design to accommodate certain unique characteristics of storage (Jindal 2017). While EMA's consultation papers generally lead to Final Determination Papers that contain mandatory rules or processes to be followed by relevant market participants, the result of this consultation was a "Policy Paper," which was defined by the EMA as a "living document" in the sense that it will be open to review and will continue to evolve as ESS business models and applications evolve (Energy Market Authority 2019).

In its current state, the Policy Paper recognizes the potential for electricity storage facilities to provide several services in the electricity market, to enable higher integration of intermittent PV, and to defer investment in grid infrastructure upgrading. However, the paper does not provide any special payments or compensation mechanism for ESS' superior performance as compared to thermal generators in providing certain market services (Energy Market Authority 2018d). Rather, it mentions that Singapore will continue to evaluate whether it needs new market services such as Fast Response Reserves and the use of electricity storage assets in providing these.

The Policy framework aside, two electricity storage demonstration projects are being developed in Singapore under an 18 million-dollar grant award from the EMA and Singapore Power. One is a 2.4 MW/2.4 MWh Lithium Ion battery, and the other is a 0.25 MW/2 MWh vanadium redox flow battery.

Conclusion

Solar PV is the crucial pillar of Singapore's strategy to reduce its GHG emissions and achieve its climate change targets. With the government housing agency spearheading the installation of PV systems on public housing, Singapore looks set to achieve its target of 350 MW installed capacity by 2020. However, despite the accelerated development of PV systems in Singapore under the SolarNova program and increasingly simplified net-metering rules, the sector faces several challenges to further expansion.

These challenges include unfavorable market conditions, such as low and volatile electricity prices and a small-sized yet fragmented domestic market. These are further exacerbated by high interest rates on financing and limited access to debt financing. Singapore has maintained a policy of neutral market pricing of energy by not favoring any fuel or energy technology. This is unlikely to change. The leasing model is likely to remain the main business model for the PV industry in Singapore. Thus, PV developers will have to bring in innovative strategies to carve out a unique competitive advantage. These could include models such as off-site leasing and marketing of green energy at a premium to environmentally conscious consumers.

At the same time, from the regulator's perspective, integration of intermittent PV at a rapid pace presents a potential threat to the grid stability. To maintain a high degree of reliability in the electricity supply, Singapore has had to take a guarded approach toward integrating solar PV into its electrical grid. To accommodate a higher percentage of intermittent PV generation, Singapore's electricity market will need to incorporate changes to its design in line with international best practices. Additionally, continued efforts toward improving solar forecasting, interconnecting its power system with that of neighboring countries and development of energy storage will be vital for overcoming this challenge.

Notes

1. The Reserve Margin is defined as the amount by which the total generation capacity exceeds the annual electricity peak demand, to cater to scheduled maintenance as well as forced outages of generating units in the power system. In Singapore, the minimum reserve margin to maintain system security is currently set at 30%.
2. SP PowerAssets Ltd. and SP PowerGrid Ltd. are subsidiaries of Singapore Power Ltd, which in turn is a subsidiary of Temasek Holdings (Private) Limited, a holding company owned by the Singapore government.

3. From 2014 to 2016, Singapore's Grid Emission Factor has ranged between 0.4224 and 0.4277 kg CO_2 per kWh.
4. PV system generating 1 MW of AC power at the output of the inverters.
5. Regulation is a frequency balancing service that is used to correct the generation—demand imbalances within a 30-minute trading period caused by load variability and load forecasting error.

References

Andrews-Speed, Phillip. 2016. Connecting ASEAN Through the Power Grid: Next Steps. ESI Policy Brief—11. Available at: http://esi.nus.edu.sg/docs/default-source/doc/esi-policy-brief-11—connecting-asean-through-the-power-grid-next-steps.pdf. Accessed 15 June 2018.

Energy Market Authority. 2014. Enhancements to the Regulatory Framework for Intermittent Generation Sources in the National Electricity Market of Singapore—Final Determination Paper. Available at: https://www.ema.gov.sg/cmsmedia/Renewable_Energy/Overview/53b268a6d3498Final_Determination_Intermittent_Generation_Sources_-_1_July_2014_Final_.pdf. Accessed 15 June 2018.

Energy Market Authority. 2015. Enhancements to the Regulatory Framework for Intermittent Generation Sources in the National Electricity Market of Singapore—Clarification Paper. Available at: https://www.ema.gov.sg/cmsmedia/Consultations/Electricity/Clarification%20Paper%20on%20Enhancements%20to%20the%20Regulatory%20Framework%20for%20IGS%202402015.pdf. Accessed 15 June 2018.

Energy Market Authority. 2016. Singapore Energy Statistics 2015. Available at: https://www.ema.gov.sg/cmsmedia/Publications_and_Statistics/Publications/SES2015_Finalwebsite_2mb.pdf. Accessed 15 June 2018.

Energy Market Authority. 2017a. Singapore Energy Market Outlook 2017. Available at: https://www.ema.gov.sg/cmsmedia/Singapore%20Electricity%20Market%20Outlook%20%23final%20v2.pdf. Accessed 15 June 2018.

Energy Market Authority. 2017b. Enhancement to the Central Intermediary Scheme for Embedded Generation. Available at: https://www.ema.gov.sg/cmsmedia/Enhanced%20Central%20Intermediary%20Scheme%20-%20Final%20Determination.pdf. Accessed 15 June 2018.

Energy Market Authority. 2018a. 2017 Singapore Energy Statistics. Available at: https://www.ema.gov.sg/cmsmedia/publications_and_statistics/publications/ses17/publicationsingapore_energy_statistics_2017.pdf. Accessed 1 Sept 2018.

Energy Market Authority. 2018b. Statistics—Fuel Mix for Electricity Generation. Available at: https://www.ema.gov.sg/statistic.aspx?sta_sid=20140731MocHHXHqVG5M. Accessed 1 Sept 2018.

Energy Market Authority. 2018c. Statistics-Installed Capacity of Grid-Connected Solar Photovoltaic (PV) Systems by User Type. Available at: https://www.ema.gov.sg/statistic.aspx?sta_sid=20170711hc85chOLVvWp. Accessed 1 Sept 2018.

Energy Market Authority. 2018d. Energy Storage Systems for Singapore—Policy Paper. Available at: https://www.ema.gov.sg/cmsmedia/Final%20Determination%20Paper%20-%20ESS%20vf.pdf. Accessed 1 Apr 2019.

Energy Market Authority. 2019. Energy Storage System. Available at: https://www.ema.gov.sg/Energy_Storage%20System.aspx. Accessed 1 Apr 2019.

Energy Market Company. 2018a. Market Data—Price Information. Available at: https://www.emcsg.com/MarketData/PriceInformation. Accessed 1 Apr 2019.

Energy Market Company. 2018b. Market Report 2017. Available at: https://www.emcsg.com/f279,129433/NEMS_Market_Report_2017_FINAL.pdf. Accessed 1 Apr 2019.

Housing and Development Board. 2018. HDB's Latest SolarNova Tender Brings Solar Energy to Over 2400 HDB Blocks. Available at: https://www.hdb.gov.sg/cs/infoweb/press-releases/hdbs-latest-solarnova-tender-brings-solar-energy-to-over-2400. Accessed 1 Apr 2019.

Jindal, Gautam. 2017. Singapore Looking to Develop Policy Framework for Energy Storage. *Singapore Business Review*, February 13. Available at: http://sbr.com.sg/energy-offshore/commentary/singapore-looking-develop-policy-framework-energy-storage. Accessed 15 June 2018.

Luther, Joachim, and Thomas Reindl. 2014. Solar PV Roadmap for Singapore (A Summary). Available at: https://www.nccs.gov.sg/docs/default-source/default-document-library/solar-photovoltaic-roadmap-for-singapore-a-summary.pdf. Accessed 1 Apr 2019.

Marcos, Javier, Luis Marroyo, Eduardo Lorenzo, David Alvira, and Eloisa Izco. 2011. Power Output Fluctuations in Large Scale PV Plants: One Year Observations with One Second Resolution and a Derived Analytic Model Progress. *Photovoltaics: Research and Applications* 19: 218–227.

Meteorological Services Singapore. 2017. Climate of Singapore. Available at: http://www.weather.gov.sg/climate-climate-of-singapore/. Accessed 1 Apr 2018.

Mills, Andrew, and Ryan Wiser. 2010. *Implications of Wide-Area Geographic Diversity for Short-Term Variability of Solar Power*. Berkeley, CA: Lawrence Berkeley National Laboratory. Available at: https://eta-publications.lbl.gov/sites/default/files/report-lbnl-3884e.pdf. Accessed 1 Apr 2019.

Ministry of Trade and Industry. 2014. Joint Statement of the Lao PDR, Thailand, Malaysia and Singapore Power Integration Project (LTMS PIP). Available at: https://www.mti.gov.sg/-/media/MTI/Newsroom/Press-Releases/2014/09/Joint-Statement-of-the-Lao-PDR-Thailand-Malaysia-and-Sin gapore-Power-Integration-Project-LTMS-PIP/jointpressstatementcrossborder.pdf. Accessed 1 Apr 2019.

Ministry of Trade and Industry. 2018. Economic Survey of Singapore 2017. Available at: https://www.mti.gov.sg/-/media/MTI/Resources/Economic-Survey-of-Singapore/2017/Economic-Survey-of-Singapore-2017/fullreport_aes2017.pdf. Accessed 1 Apr 2019.

National Solar Repository of Singapore. 2018. Solar Economics Handbook—Grid Parity and Economic Viability Analysis. Available at: http://www.solar-repository.sg/grid-parity-and-economic-viability-analysis. Accessed 1 Apr 2019.

Ng, Desmond, and Daniel Heng. 2018. Stepping on the Gas to Keep Singapore's Lights Burning. *Channel News Asia*, March 30. Available at: https://www.channelnewsasia.com/news/cnainsider/lng-natural-gas-electricity-singapore-energy-security-tank-10088910. Accessed 15 June 2018.

SP Group. 2018. SP Group—Year in Review 2017: Always Here for You. Available at: https://www.spgroup.com.sg/poweringthenation/always-here-for-you.html. Accessed 1 Apr 2019.

Suruhanjaya Tenaga. 2017. The ASEAN Connection. *Energy Malaysia*, 13. Available at: https://www.st.gov.my/ms/contents/publications/energyMalaysia/Energy%20Malaysia%20Volume%2013.pdf. Accessed 1 Sept 2018; Accessed 1 Apr 2019.

Weatherspark. 2017. Average Weather in Singapore. Available at: https://weatherspark.com/y/114655/Average-Weather-in-Singapore. Accessed 1 Sept 2018.

Yang, Dazhi, Panida Jirutitijaroen, and Wilfred Walsh. 2011. The Estimation of Clear Sky Global Horizontal Irradiance at the Equator. *Energy Procedia* 25: 141–148.

CHAPTER 11

Renewable Energy Policy in Vietnam

Nam Hoai Nguyen, Binh Van Doan, Huyen Van Bui, and Quyen Le Luu

Overview

Vietnam's economic transition has led to a high economic growth rate and remarkable poverty reduction in recent decades. This transformed Vietnam from one of the world's poorest countries into a lower middle-income country by 2010. The country's high economic growth rate has led to an increasing demand for energy, with an average growth rate of 11.8%. Electricity demand is expected to increase during the current decade, driven mainly by further industrialization, increased urbanization, and economic growth. The country's energy policies have primarily focused on building up electricity generating capacity through investment in coal, gas, and medium and large-scale hydropower. Vietnam's domestic

N. H. Nguyen (✉) · B. Van Doan · Q. Le Luu
Institute of Energy Science, Vietnam Academy of Science and Technology, Hanoi, Vietnam
e-mail: nhnam@ies.vast.vn

H. Van Bui
Institute of Economics, Ho Chi Minh National Politics Academy, Hanoi, Vietnam

P. Midford and E. Moe (eds.), *New Challenges and Solutions for Renewable Energy*, International Political Economy Series,
https://doi.org/10.1007/978-3-030-54514-7_11

energy resources have contributed greatly to meeting the increasing demand so far, but it is very likely that imported fuels will be dominating the country's energy context well after 2020.

Vietnam is privileged regarding its untapped potential for renewable energy. Adoption of energy efficiency and demand-side management measures have not been fully developed. More efficient use of the renewable energy potential would make Vietnam's electricity mix more diverse and therefore more resilient, flexible, and sustainable. It would also decrease Vietnam's future need for imported energy sources and electricity from other countries in the region. The current National Renewable Energy Strategy (RES) sets goals for each renewable energy source, and for the first time, solar power was taken into account. The targets of the RES were also mainstreamed in the revised National Master Power Development Plan (PDPr), reflecting a more concrete and stronger political commitment for energy transition in Vietnam.

Energy Context

Over the last two decades, Vietnam has achieved remarkable improvements in energy development (see Fig. 11.1). The country has remained largely self-sufficient, with substantial proven reserves of fossil fuels and renewable energy resources, including hydropower potential. Vietnam has been an important coal, oil, and natural gas producer in Southeast Asia, while being a net importer of oil products. The country's ongoing exploration activities, especially offshore, could increase this figure in the

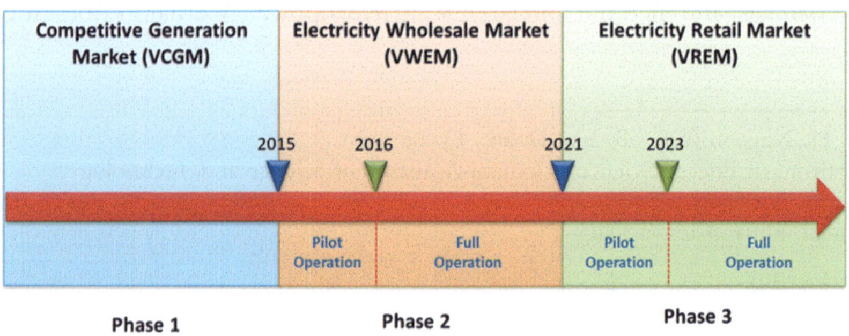

Fig. 11.1 Different levels of electricity market liberalization in Vietnam (*Source* Authors' created figure compiled with data from MOIT 2015c)

future. Vietnam's total power capacity has increased by approximately 12% per annum over the past decade, reaching 46.673 GW of installed capacity by end of August 2018. Hydropower, coal, and natural gas are the most important primary energy sources for electricity production. Hydropower accounted for the largest share of capacity (40%), followed by coal (38%), and gas and oil (18.7%).

Vietnam's power mix was mainly dominated by large hydropower, followed by gas power and coal power. It appears that the intention of the Vietnamese government is to retain these primary sources for electricity production over the coming decades. However, nearly all large and medium hydro-resource potential has already been fully exploited, while domestic coal is currently insufficient to supply existing power plants. New coal power plants require higher-quality coal to generate high-pressure steam, and while low-quality coal is cheaper, it is not usable for newly built plants. This makes the Levelized Cost of Energy (LCOE) for coal-fired plants less competitive than new renewable energy power plants such as solar PV. As a result, in the PDPr has set the goals of 8000 MW of installed wind capacity, and 12,000 MW of solar capacity, by 2030.

Renewable Energy Potential

Vietnam has excellent potential for the development of renewable energy generation in the context of Southeast Asia, and it is on the right path for translating this potential into a developed renewable energy sector. Our review of the current progress of the RES suggests that most forms of renewable energy in Vietnam are at an early state of unlocking their market potentials. We focus on wind, solar, and biomass.

Solar Energy Potential. The theoretical potential of solar power in Vietnam is high, at about 43.9 billion Tons of Oil Equivalent (TOEs) per year (Dang and Le 2005; To 2012; Trinh 2012). Most of the potential is concentrated in the Central and Southern regions, with an average of over 300 sunny days per year, and a solar radiation intensity of 5 kWh/m^2. In the North, those numbers are 250–280 days per year and 2.4–5.6 kWh/m^2, respectively (VUSTA 2007). Before 2016, solar energy was mainly utilized on a very small scale, and mostly deployed at household and commercial scale, including at hotels, restaurants, hospitals, military bases, and services centers (To 2012). The total installed capacity of PV solar power in June 2019 was 4440 MW, most of which was commissioned before the end of that month with a Feed-in Tariff

(FIT) of US$0.0935 US/kWh. The locations of the large-scale solar PV farms are mostly concentrated in the South due to its high solar radiation (Table 11.1).

Wind Energy. According to the World Bank's wind potential assessment conducted in 2000, Vietnam's wind energy potential is estimated at 513,360 MW. The most promising areas for wind power development are in the coastal areas and highlands of south-central and southern Vietnam. 8.6% of Vietnam's land area has wind energy potential ranging from "high" to "very high" (i.e., wind speeds above 7 m/s), which is suitable for large-scale wind turbine deployment (World Bank 2010). In 2012, the Ministry of Industry and Trade (MOIT) in cooperation with the German Agency for Development Cooperation (GIZ) conducted wind assessment at 10 sites in the Central Highlands and Central Coast at altitudes ranging from 40 to 80 m. The potential of wind energy at an altitude of 80 m is presented in Table 11.2.

Wind energy has been used in Vietnam for many years, for water pumping and power generation in rural, remote, and isolated areas. Since the 1990s, the Institute of Energy (IE) has designed and installed off-grid wind turbines in the capacity range of 50–500 W. More recently, Vietnam's wind power technology has progressed with the transfer of foreign technology and the installation of large-scale wind power equipment (IE 2012). By June 2019, the total installed wind power capacity in Vietnam reached nearly 200 MW, mainly composed of large-scale wind farms along the coast of the Southern region.

Biomass Energy

As a country with a large agricultural sector, Vietnam has great potential for biomass. The main types of biomass are wood, crop residue and by-products, livestock manure, municipal waste, and other organic waste. The sustainable production of biomass for energy production in Vietnam amounts to 100–150 million tons per year. Some types of biomass can be exploited for electricity generation or energy (electricity and heat) production, including crop residues (rice husks, straw, bagasse, etc.), municipal solid waste, livestock manure, and other organic wastes from agro-forestry-fishery processing industries (Zwebe 2012; Nguyen 2012).

The distribution of crop residues varies among localities. Taking rice straw as an example, volume is highest in the provinces of An Giang, Kien Giang, Dong Thap, and Long An. Notably, the amount of straw in An

Table 11.1 Solar radiation variation by month in selected provinces in Vietnam

	Average daily solar radiation by months (Wh/m²/day)												*Yearly figure kWh/m²/year*
	JAN	***FEB***	***MAR***	***APR***	*MAY*	*JUN*	*JUL*	***AUG***	*SEPT*	***OCT***	*NOV*	*DEC*	
Cao Bang	2391	2487	2964	3591	4752	4983	5331	5382	5016	3930	3276	2745	1425
Sapa	3708	4053	4293	4776	4488	4230	4371	4251	3675	3111	3141	3406	1445
Lai Chau	3072	3753	4347	4707	5079	4686	4935	5031	5013	4326	3210	2934	1554
Thanh Haa	2802	2370	2883	3834	5853	5748	5928	5364	5280	4563	3555	3921	1597
Nghe An	2559	2313	2646	4110	5673	5610	6186	5202	4557	386	2967	2616	1468
Da Nang	3564	4221	5100	5760	6312	5964	6480	5886	5148	4140	3027	2454	1766
Nha Trang	5028	5983	6961	6458	5900	6370	5960	6210	6520	5620	4350	4360	2120
Pleiku	4210	5150	5540	5630	5500	4990	4820	4570	4490	4500	3850	3780	1740
Binh Thuan	3591	4232	5215	6764	6712	5864	6381	5886	5158	4160	3031	2474	1792
Ninh Thuan	5972	6812	6895	5925	4893	4091	5464	4681	4328	4269	4611	5275	1980
Tan Son Nhat	4180	4860	5270	5120	4450	4330	4410	4490	4170	4110	3940	3960	1621
Da Lat	5934	6702	6885	5820	4872	4080	5361	4683	4306	4257	4566	5184	1906

Note Wh/m² is Watt hours per square meter. *Source* Measurements data of Ministry of Natural Resource and Environment

Table 11.2 Potential of wind energy in Vietnam at a height of 80 m

Average wind speed	*<4 m/s*	*4–5 m/s*	*5–6 m/s*	*6–7 m/s*	*7–8 m/s*	*8–9 m/s*	*>9 m/s*
Square area (km^2)	99,916	70,868	40,473	2345	220	20	1
Percentage of area (%)	45.7	33.8	19.3	1.2	0.1	0.01	<0.01
Potential (MW)	956,161	708,678	404,732	24,351	2202	200	10

Source GIZ (2012)

Giang is 2.2 million tons/year, an amount equivalent to Malaysia's total domestic straw production (Yano et al. 2009). Meanwhile, 87% of the straw is removed or burned directly in the field which causes post-harvest air pollution (Dang et al. 2010). A small portion is used for livestock or mushroom cultivation. The currently burnt straw would be a good source of energy for the area, once biomass energy is developed.

By 2018, 20 biomass power plants had been constructed, with rice husks as the feedstock. These plants are located in five provinces: An Giang, Kien Giang, Hau Giang, Dong Thap, and Can Tho in the Mekong Delta. The total installed capacity of these plants is 200 MW. According to calculations, a plant with a capacity of 10 MW will consume approximately 85,000 tons of rice husks a year (SNV 2013). Besides, bagasse is also a form of biomass energy with many advantages for electricity generation. There are about 40 biomass cogeneration plants with total capacity of 150 MW in Vietnam (MOIT2015c) (Table 11.3).

Electricity Grid. Transmission and distribution infrastructure system in Vietnam is structured into 500, 220, and 110 kV grid. The 500 and 220 kV transmission grids are managed by the National Power Transmission Corporation; the distribution grid from 6 to 110 kV is under the management of the local electricity corporations. The 500 kV transmission network is the backbone of Vietnam's electricity system, with a total length of 7346 km from the North to the South. This system plays a very important role in the national energy balance and has a great influence on the reliability of electricity supply in each region. The transformer capacity for 500 kV lines reached 26,100 megavolt amperes (MVA) in 2016.

The 220–110 kV grid is responsible for ensuring the safe and continuous power supply of the regions and localities. In 2016, the length of

Table 11.3 Potential of renewable energy for power generation in Vietnam

Renewable energy	*Economic potential*	*Development plan*
Hydropower	84 TWh/yr	
Hydropower (>50 MW)	18–20 GW	16.6 GW by 2020
Small hydropower (<30 MW)	2–4 GW	2.5–3.2 GW
Micro hydropower (<1 MW)	100 MW	
Storage hydropower	10.2 GW	10.2 GW
Geothermal	1.4 GW	300–400 MW by 2020
Wind energy	45.5 GW	500 MW
Solar energy	53 GW	2–3 MW
Energy from rice husk	250 MW	500 MW
Energy from rice straw	550 MW	
Energy from bagasse	200 MW	
Energy from forestry residue	100 MW	

220 and 110 kV transmission lines was 35,100 km, and the capacity of substations was 95,096 MVA. According to the revised PDPr, the transmission and distribution system will be developed in five areas as spelled out in the roadmap. First, transmission expansion is required to meet technical requirements of the transmission system; by 2020, the transmission system will meet the reliability standard N-1 for main devices, and other quality standards. Second, to avoid overload, congestion, low power quality of the transmission system, and ensure the supply of power it is necessary to improve reliability for load centers. Third, it is necessary to further develop the 500 kV transmission grid to transmit power from large generating centers to load centers, linking regional power grids and transmitting power to neighboring countries. Fourth, it is necessary to develop the 220 kV transmission grid according to dual-loop configuration; the substations in large demand regions will be designed according to suitable configurations for flexible operations. Fifth, the roadmap calls for conducting research for constructing GIS substations, 220/22 kW

substations, underground stations, anonymous stations in load centers, and introducing smart grids in power transmission.

Regarding interconnections with neighboring countries, the roadmap specifies four goals. First, it calls for continuing to collaboratively study power connection development with neighboring countries in ASEAN and the Greater Mekong Subregion. Second, it advocates connecting to Laos' power grid through 220 kV transmission lines to import electricity from hydropower plants in Southern and Central Laos. Third, it specifies maintaining the connection with Cambodia through the existing 220 kV transmission lines and conducting research on improving the existing interconnection between Vietnam and Cambodia through multilateral and bilateral programs. Fourth, it advocates maintaining the connection with China through the existing 220 and 110 kV transmission lines, conducting research on asynchronously harmonizing the power system through single and alternative substations, and conducting research on exchanging power with China through 500 kV transmission lines.

Electricity Import and Export

Vietnam currently imports electricity from Laos, especially from Xekaman 1 and 3 hydropower plants, which have a capacity of 540 MW. It is expected that by 2020, Vietnam will import electricity from Xekaman plants 1, 3, and 4, and the Xekaman Sanxay hydropower plant, which comprises a total installed capacity of 635 MW, and annual electricity generation of 2 billion kWh. By 2030, Vietnam expects to import electricity from hydropower plants with a capacity of approximately 2000 MW at Xekong 1 and 2, plus from some small hydropower plants.

Vietnam also imports electricity from China at a maximum capacity of 700 MW. By 2020, the total electricity generation of 3.5 billion kWh will be annually imported through the 220 kV transmission line in the North. By 2025, Vietnam starts importing electricity from China through 500 kV transmission lines, and by 2030, the power will be directly transmitted to the South at a maximum capacity of 3000 MW. Vietnam exports electricity to Cambodia through 220 kV transmission lines with a maximum capacity of 200 MW. Vietnam continues to export through 220 kV transmission lines. There is no plan to import electricity from Cambodia in the near future.

Grid Connection of Renewable Power Projects

Grid connection of renewable power projects come under the same regulations as conventional power projects. These regulations include:

- Circular No. 12/2010/TT-BCT, dated 15 April 2010 of the Ministry of Industry and Trade on regulations on the transmission power system.
- Circular No. 39//2015/TT-BCT, dated 18 November 2015 of the Ministry of Industry and Trade on regulations on the distribution power system.

The first is applicable for power projects being connected to the 220 kV and higher voltage transmission lines, while the latter is applicable for the lower voltage distribution lines (under 220 kV lines). Additionally, for power projects above 220 kV, developers need to conclude an agreement with the National Power Transmission Corporation (NPT) as per Decision No. 1449/QD-EVNNPT, dated 17 June 2015, which covers regulations and procedures for connecting power plants to the transmission grid.

Considering the technical differences between renewable energy sources and conventional electricity generation sources, the connection of renewable power projects raised several challenges to grid operation such as predictability, non-synchronous generation, congestion, and decentralization. As the output of renewable power is volatile over time, the amount generated consequently can be uncertain. Therefore, predicting renewable power generation is crucial for ensuring the reliability and quality of electricity, and the security of the power system.

Besides, renewable energy sources are often deployed in remote areas, and sometimes at low voltage. The capacity of the grid needs to be enhanced to transfer the amount of renewable power without encountering congestion due to voltage surges or overload. In case the renewable energy sources are connected to the grid at low voltage, the distribution as well as transmission grid need to be enhanced. At the same time, the agencies managing different voltage-level grids need to efficiently coordinate to ensure the smooth operation of the power system.

Electricity Market Liberalization

In 2006, Decision No. 26/2006/QD-TTg was issued to regulate and layout the roadmap toward creating a competitive electricity market in three phases (Fig. 11.2).

Phase 1—Competitive Generation Market: Establish a Market Operator and single buyer under the national electric company, Electricity of Vietnam (EVN). On that basis, the competitive electricity generation market has been built and operated, implementing the spot market mechanism so that all generators can participate in the market.

Phase 2—Competitive Wholesale Market: At this level, the demand side will be expanded with an organized power market. The model of "single buyer" is replaced by an agreement whereby consumers can negotiate to purchase electricity directly from the generation entities regardless of the single buyer. From 2015 to 2016, a pilot competitive wholesale market was implemented. From 2017 to 2021, a fully competitive wholesale market is being implemented.

Phase 3—Competitive Retail Market: At this level, the electricity distribution company will be separated by their functions of grid management and retail from those of electricity generation (unbundling). End users who are served by distribution companies can then freely choose their suppliers (retailers). From 2023 onward, the fully competitive retail market will be implemented.

Competition in electricity generation has been prepared since 2009, and officially operated since 2012 under the name of "Competitive Generation Market" in Vietnam. The competitive generation market was open for power investors outside EVN, thereby diversifying the ownership of electricity production. Participants in the competitive generation market include all power plants with installed capacity of more than 30 MW, and which are directly connected to the national grid (except for BOT power plants, wind power plants, geothermal power plants). Since 2015, hydropower plants with a capacity of up to 30 MW, connected to a grid of 110 kV or higher, and meeting all requirements on infrastructure, are entitled to participate in the electricity market.

The Vietnam competitive wholesale market was designed and commissioned from 2016 and planned to officially begin operations by the end of 2017, according to the Ministry of Industry and Trade's Decision No. 8266/QD-BCT of 2015, which approved a detailed design for the competitive wholesale market. Participants in the electricity wholesale market are classified into three groups: (i) power sellers; (ii) power buyers; and (iii) service providers. The power sellers include power generation facilities and entities setting prices and selling power on behalf of power generation units. The power buyers include five electricity corporations, large power consumers, and eligible power buyers. An overview of participants in the competitive wholesale market is depicted in Fig. 11.2.

The power market has been transforming from a monopoly market into a competitive market. Presently, the power market has become competitive at the stage of power generation (competitive generation market) and at the stage of the wholesale market.

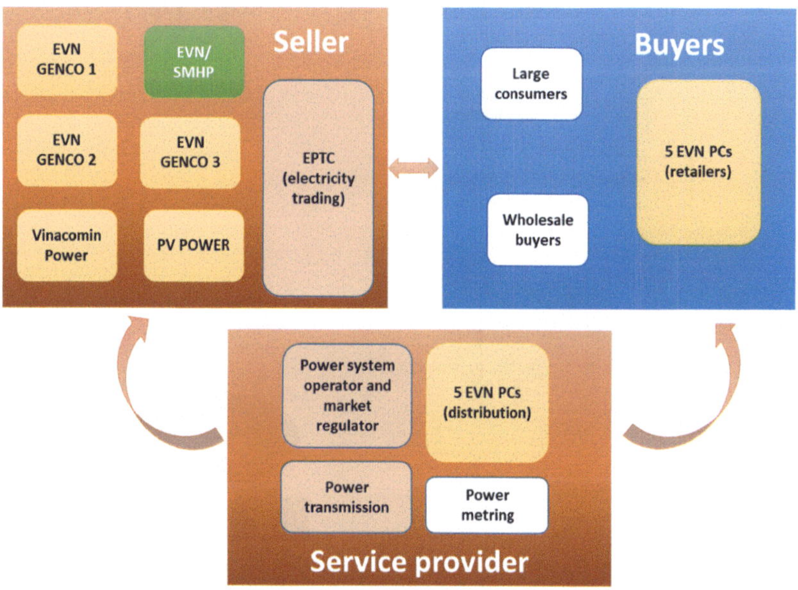

Fig. 11.2 Competitive wholesale power market (*Source* Authors' created figure compiled with data from MOIT 2015c)

Institutional and Legislative Framework

Key Stakeholders. Governance of the electricity supply industry is primarily carried out by MOIT via the Electricity and Renewable Energy Authority, the Department of Energy Efficiency and Sustainable Development, and the Department of Oil, Gas and Coal. The Electricity and Renewable Energy Authority is the agency for planning, appraisal, and approval of electricity projects, including renewable power projects. Governance of the electricity market is exerted by the Electricity Regulatory Authority of Vietnam (ERAV), a quasi-independent regulator. This is the agency responsible for managing the demand side of the electricity sector, including electricity price setting, electricity purchase mechanisms, and setting the roadmap toward a competitive electricity market. The institutional structure of electricity governance is illustrated in Fig. 11.3.

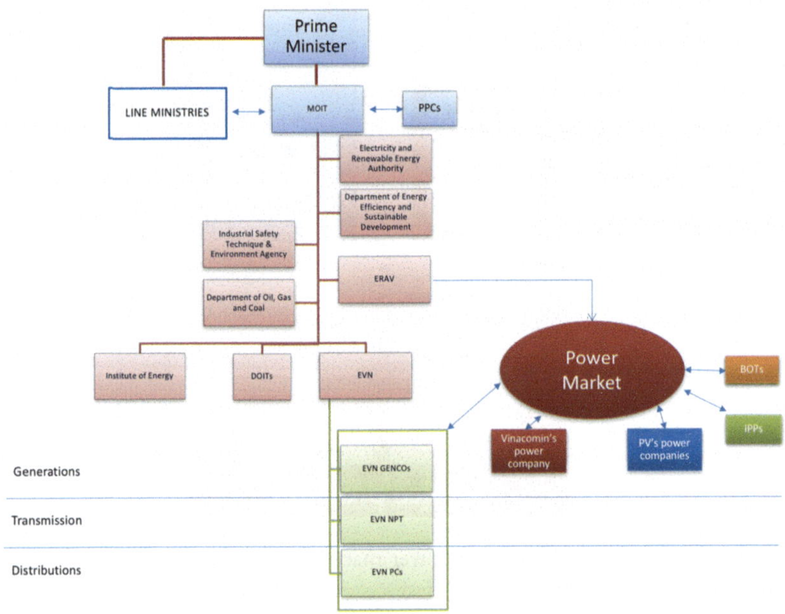

Fig. 11.3 Electricity supply industry governance in Vietnam (*Source* Authors' created figured)

MOIT manages and regulates energy development and utilization, promulgates circulars which guide implementation of Government and Prime Minister Decrees and decisions, issues standards, technical-economic norms under its authority, and organizes and verifies power master plan development. In addition to the MOIT, as for the other sub-sectoral policy, the other main institutions involved in renewable energy are the MPI and MoF as the bodies in charge of coordinating planning and allocating resources for policies. They are important stakeholders that review and appraise renewable energy sourcing programs[1] in accordance with their mandates and duties in planning, financing, and resource allocation. EREA/MOIT assists in the implementation of MOIT state management functions. It is the most important policy maker within the field of renewable energy planning, with significant authority over national and provincial renewable energy planning.

The Provincial People's Committees (PPCs) are in charge of implementing at the local level renewable energy policies in close coordination with MOIT, particularly for wind power planning in Southern and coastal provinces, and small hydropower planning in Northern and Central and Central Highland provinces. PPCs play a crucial role in renewable energy projects at the local level as they are entitled to appraise Development Proposals and certify the Investment Certificate for developers, issue land-use permits, provide local incentives (if any) for renewable energy project investors who are developing projects in their territories.

Vietnam has a vertically integrated electricity sector whereby EVN is the single buyer of electricity and maintains control over generation, transmission, distribution, dispatch, market operations, and ancillary services. It is the third-largest state-owned enterprise in Vietnam and, despite partial unbundling into subsidiary companies, exerts firm monopoly control over most of the electricity supply industry in Vietnam. PetroVietnam (PV) is now the single largest producer of biofuels in Vietnam. Currently, PV owns three bioethanol facilities across the country with a capacity of approximately 100,000 cubic meters per year.

Renewable Energy Policy Formulation

Renewable energy policy is formed through consultations among relevant ministries, provincial-level governments, state-owned companies in the energy sector such as EVN, Petro Vietnam, Vinacomin, and selected private partners. Private stakeholders could be engaged in some policy-formation steps, but normally energy policy development is a top-down process and private stakeholders' involvement remains limited.[2] PPCs are responsible for provincial small-scale hydropower policies and planning. These plans must include a consulting phase with MOIT before adoption.

The monitoring and evaluation of renewable energy planning is mainly left to the MOIT in terms of state management at the highest level and to PPCs at the local level. For small hydropower plants and wind power, PPCs will have roles to develop provincial renewable energy Master Plans, to monitor progress and report on renewable energy development to MOIT and the Prime Minister. Master Plans are prepared for wind, biomass, and small hydropower (in provinces with highest potential). The monitoring and evaluation are mostly conducted on a yearly basis, and this information is used to review the planning. The monitoring and evaluation frameworks are aligned with domestic processes and conform to local accountability mechanisms. Performance evaluation, financial assessment, and audit procedures are undertaken in accordance with relevant regulations.

Renewable Energy Targets and Strategies

In relation to renewable energy promotion, several legislative documents, strategies, and action plans serve as the key components that define the country's renewable energy policy framework. These documents are outlined below.

Electricity Law 24/2012/QH13. This electricity law was introduced in 2004, and the most updated version of the electricity law came into force in 2013. The Law provides the framework for the introduction of a competitive power market, restructuring of the electricity sector and related regulations in national and provincial Master Power Plans. The Law also provides that Power Master Plans must conform to a National Master Energy Plan (NMEP). This must be developed in order to ensure the consistent, comprehensive, systematic, and integrated character of energy resource planning.

National Energy Development Strategy to 2020 and 2050. The National Energy Development Strategy (NEDS) to 2020 and 2050 was issued in Decision 1855/QD-TTg by the Government of Vietnam in 2007. The coordinating government agency for NEDS is MOIT. This is the central framework for the entire energy sector in Vietnam, encompassing all sub-sectors such as coal, oil, electricity, renewable energy, and power market reform. The strategy identifies the need for a sufficient supply of energy to service projected developments in the national socioeconomic plan. It proposes the gradual formation of a state-oriented energy market with diversified ownership and business. NEDS goals include strengthening national energy security, promoting improvements in energy efficiency, and increasing the share of renewable energy in the energy mix. NEDS also includes power market reform and electrification of remaining off-grid rural areas. NEDS impacts all stakeholders in Vietnam's energy sector, from state agencies to public utilities, household energy consumers, and project developers.

NEDS requires that the exploitation of all energy resources must be efficient and mitigate environmental impact in conformity with regional and international environmental standards to limit emissions and pollution. NEDS includes the following targets for renewable energy and energy efficiency:

- Renewable energy as a share of total primary energy supply is targeted to reach 3, 5, and 11% by 2010, 2020, and 2050, respectively.[3]
- The share of households using renewable energy for cooking is targeted to reach 50% by 2010 and 80% by 2020.

National Renewable Energy Development Strategy (REDS)

The National Renewable Energy Development Strategy to 2030 with outlook up to 2050 was approved by the Prime Minister in Decision 2068/QD-TTg dated 25/11/2015. The Strategy creates foundations for more specific and sub-sectoral renewable energy planning by providing targets for biomass, biogas, wind, and solar regarding energy and electricity uses. The country will promote onshore wind power until 2030 and study potential offshore developments as of 2030. Solar power development will also be focused on islands, borders, and remote areas where no electricity supply is possible. MOIT will be the focal agency to implement the strategy that is expected to most benefit electricity customers, renewable energy investors, and the renewable energy industry.

In order to reach the targets included in the REDS, the government of Vietnam included a number of incentive mechanisms such as net metering, FITs (established separately from REDS), support for local manufacturing of renewable energy technologies and equipment, power purchase agreements with EVN, and the development of an renewable energy market. One of the key upcoming activities will be the development of a RPS for Vietnam (see the subsequent discussion on this).

National Master Power Development Plan 2011–2020, 2030 Revised (PDP7r)

The National Master Power Development Plan 2011–2020, 2030 was revised (PDP7r) as per Decision 428/QĐ-TTg in 2016. The main objectives of the PDP7s are to define high-level strategies for national power development in four main areas: (i) the development of power sources; (ii) the development of the power transmission grid; (iii) interconnection of Vietnam's power networks with neighboring countries; and (iv) electricity supply to rural, mountainous, and island areas.

- Among the specific targets of PDP7r is developing renewable energy sources (including storage, medium, and large hydropower) for electricity production, increasing the percentage of electricity produced from these energy sources from 7% in 2020 to 10% in 2030;
- Increasing the total capacity of hydroelectric power from 17,000 MW in 2015 to 21,600 MW by 2020, 24,600 MW by 2025, and 27,800 MW by 2030, increases of 29.5, 20.5, and 15.5%, respectively;
- Increasing total wind power capacity from 140 MW in 2015 to 800 MW by 2020, 2000 MW by 2025, and 6000 MW by 2030, amounts equivalent to 0.8, 1, and 2.1% of total electricity production, respectively;
- Developing biomass power, and power generation in sugar mills, and cogeneration plants to provide 1% of electricity production by 2020, 1.2% by 2025, and 2.1% by 2030; and
- Accelerating the development of solar power in the forms of mega solar and rooftop solar by increasing capacity of solar electricity generation to 850 MW by 2020, 4000 MW by 2025, and

12,000 MW by 2030, accounting for 0.5, 1.6, and 3.3% of total electricity production by 2020, 2025, and 2030, respectively.

Apart from these comprehensive strategies and plans on renewable energy, there are several master and regional plans for specific types of renewable energy such as biofuel and biomass power, including those listed below.

Renewable Energy Master Plan for Northern Region

The Renewable Energy Master Plan for the Northern Region to 2020, with a vision to 2030 (Decision No. 8217/QD-BCT, dated 28 December 2012 of the Minister of Industry and Trade) provides targets that are monitored annually for renewable energy development as follows:

- 2013–2020: the target for grid connection of renewable energy is 325.7 MW, including small hydropower plants, biogas, waste incineration, sugar cane, and rice husk biomass.
- 2020–2030: the target for grid connection of renewable energy is 1230.5 MW, including small hydropower plants, biogas, waste incineration, sugar cane, and rice husk biomass (Table 11.4).

Master Plan for Biomass Power Generation in Mekong Delta

The Master Plan for Biomass Power Development in the Mekong Delta region until 2020, including a vision to 2030, was endorsed by Decision 9486 in 2013 by MOIT. The Plan aims to develop biomass power generation in this region to reach 214 MW of installed capacity by 2020, and 304 MW of installed capacity by 2030. Power generation will be fed mostly by bagasse, rice husk, woodchip, and rice straw biomass (Table 11.5).

The main principles of Vietnam's energy policy in the field of renewable energy aim to:

Table 11.4 RES master plan for northern region to 2030

Year	*Grid connected RE*	*Off-grid RE*	*RE for heating and biofuel*
2013–2020	325.7 MW, in which: • Small hydro: 269.2 MW • Biogas power: 5.5 MW • Waste incineration power: 26 MW • Bagasse power: 15 MW • Rice husk: 10 MW	For rural electrification: 22,899 HHs, totaling at 7801 kWp For farms: 2,421,000 m^3 of biogas systems	521 million m^2 of solar water heating 6.46 million m^3 of biogas system 316,000 biomass gasification cook stoves 198 million liters of ethanol
2020–2030	1230,5 MW, in which • Small hydro: 939.8 MW • Biogas power: 5.5 MW • Waste incineration power: 50.5 MW • Biomass power: 111 MW • Wind power: 110 MW	For farms: 463,000 m^3 of biogas systems	1158 million m^2 of solar water heating 7.98 million m^3 of biogas system 672,000 biomass gasification cook stoves 198 million liters of ethanol

Source Table created by authors and compiled from Renewable Energy Master Plan for Northern Region of Vietnam

Table 11.5 Installed power generation capacity from biomass in Mekong Delta region (MW)

Targets (MW)	*From 2013 to 2020*	*From 2021 to 2030*
Sugar cane	50	30
Rice husk	140	150
Wood	24	44
Rice straw	0	80
Total	214	304

Source Compiled from biomass power master plan in Mekong Delta region

- Increase shares of renewable energy resources in energy production and consumption, especially in electricity generation;
- Develop the most promising renewable energy sources in Vietnam such as wind, solar, biomass, and waste; and

- Gradually reduce reliance on fossil fuels, diversify the energy mix, increase national energy security, and protect the environment while ensuring sustainable development.

More specifically, the Energy Policy of Vietnam in terms of renewable energy focuses on achieving the progressive expansion of renewable energy in the national energy mix:

- The share of renewable energy should be 38% of total electricity generation in 2020, 32% in 2030, and 43% in 2050,[4] (Table 11.2); and
- The shares of renewable energy sources (excluding storage, medium and large hydropower) in electricity generation should increase to 7% by 2020, and 10% by 2030[5] (Table 11.3) (Tables 11.6 and 11.7).

Table 11.6 RES renewable energy for power production targets

Electricity generated from different sources of RE	*2020*		*2030*		*2050*	
	Amount (TWh)	*Share (%)*	*Amount (TWh)*	*Share (%)*	*Amount (TWh)*	*Share (%)*
Hydropower (including large hydropower plants)	90	33.5	96	17	104	9.9
Biomass	7.8	3.0	37	6.3	85	8.1
Wind	2.5	1.0	16	2.7	53	5.0
Solar	1.4	0.5	35.4	6.0	210	20

Source Decision 2068/QĐ-TTg issued by prime minister in 2015

Table 11.7 PDP7r renewable energy targets for installed capacity and share of power generation

Targets of RE for power generation	*2020*		*2025*		*2030*	
	Install capacity (MW)	*Share in power generation (%)*	*Install capacity (MW)*	*Share in power generation (%)*	*Install capacity (MW)*	*Share in power generation (%)*
Biomass	–	1.0	–	1.2	–	2.1
Wind	800	0.8	2000	1.0	6000	2.1
Solar	850	0.5	4000	1.6	12,000	3.3

Source Data from Decision No. 428/QD-TTg issued by prime minister in 2016

Promotion Measures

In order to reach these targets, the government is introducing considerable incentives and mechanisms, such as funding and loans for renewable energy projects, tax exemptions, net metering, FITs, support for local development of renewable energy technologies and manufacturing of equipment, power purchase agreements with EVN, the RPS mechanism, and the renewable energy market mechanism. These measures can be categorized into three main types: (1) financial incentives designed to improve the financing of projects; (2) fiscal incentives such as tax exemptions and other tax-related subsidies; and (3) market price support mechanisms and favorable regulation.

Financial Incentives. Decree No. *75/2011/ND-CP* mandated that projects utilizing any renewable energy source (wind, solar, geothermal, bioenergy, small hydropower, etc.) will be supported with investment and export credits covering construction costs for investment: loans up to 70% of the total investment capital of the project (excluding working capital), terms up to 12 years with an interest rate equivalent to government bond interest rates. Developers of renewable power project are responsible for any investment from the connection point back to the power plant, and the grid operators are responsible for any investment from the connection point to the grid (either local or national grid under their control). Developers of renewable power projects do not bear any fees for grid connection. Renewable energy scientific research and technology development programs also receive financial support from the government.

Fiscal Incentives. The Vietnamese government has introduced several incentive measures for renewable energy developers including tax incentives, financing supports, and licensing. Detailed description on incentives for wind, biomass power, and solar PV is provided below.

Incentives for Wind Power. For wind development, import tax exemptions for equipment which is not domestically manufactured can be initiated. Developers can also qualify for a lower corporate income tax rate, with either a 10% tax rate for the first 15 years with a possible extension up to 30 years, or an exemption for the first 4 years and a 50% reduction for the next 9 years. A fast depreciation mechanism, 1.5 times faster than normal projects, can be granted, followed by an exemption of land-use tax, charges, and environmental protection fees.

Incentives for Biomass Power and Solar PV. On 26 December 2013, the Government promulgated Decree No. 218/2013/ND-CP, which provided detailed regulations and guidelines for the implementation of the Law on Corporate Income Tax. According to this Decree, newly established enterprises working on renewable energy projects can qualify for an exemption and a reduction in corporate income tax. These enterprises enjoy a preferential tax rate of 10% (instead of the normal tax rate of 20% applicable since January 1, 2006). On top of that, these enterprises enjoy additional tax incentives including tax exemption for the first 4 years and 50% tax reduction for the next 9 years. The corporate income tax rate for renewable energy-based power generation enterprises can be summarized as follows: 0% for years 1–4; 5% for years 5–13; 10% for years 14–15; and 20% from year 16 onwards.

Biomass-based power projects are exempted from import tax on imported goods that would become fixed assets of the project and goods used as raw materials, input or semi-finished products that are not available on the domestic market for the project operation. This is based on the Law on Import and Export Taxes and other regulations on import and export duties.

Market Price Support and Regulation

Net Metering. Electricity-end-users purchasing power from the national electricity system and producing electricity (generated from renewable sources) for their own demand may apply for the net metering mechanism. The payment for power will be calculated on the principle of net metering at suitable prices. The power generated from renewable sources of the end user will be credited to the RPS of the power distributing and trading units.

Feed-in Tariffs. Decisions endorsed by the prime minister on avoided costs/FITs for power generation development from wind, biomass, solid waste incineration, and small hydropower are:

- Decision No. 18/QD-BCT, dated 18/8/2008 on promulgation of an avoided cost-based tariffs schedule and standard Power Purchase Agreement (PPA).
- Joint circular 58/2008/TTLT-BTC-BTN&MT: Tariff support for CDM renewable energy projects (including: wind, solar, geothermal, tidal, or recovered methane gas projects) from the Vietnam Environment Protection Fund.
- Decision No. 37/2011/QD-TTg, dated 29/06/2011 by the prime minister on a wind power price subsidy mechanism (currently under review and being updated): (i) standard PPA; (ii) grid-connected electricity price subsidy of 7.8 US cents/kWh with 1 USD cent coming from the Environmental Protection Fund.
- Decision No. 24/2014/QD-TTg, dated 24/3/2014 by the prime minister on the biomass power price subsidy mechanism: (i) the electricity price of biomass power projects connected to the grid set at 5.8 US cents/kWh.
- Decision No. 31/2014/QD-TTg, dated 5/5/2014 by the prime minister on the solid waste power price subsidy mechanism: purchase price of 10.05 US cents/kWh for incineration technology and 7.28 US cents/kWh for burial technology.
- Decision No. 11/2017/QD-TTg, dated 11/4/2017 by the prime minister on the mechanism to encourage the development of solar PV projects in Vietnam: The purchasing price at agreed locations is 9.35 US cents/kWh. This power price is applied for on-grid solar PV projects with solar-cell efficiency of more than 16%, or module efficiency of more than 15%. For rooftop solar PV with the net metering system, the purchase price of excess power is the same as that for on-grid solar PV projects, being paid at the end of the year or at the completion of the PPA. The power purchasing price is annually adjustable in accordance with exchange rate of USD/VND. Since introduction of the solar FIT, installed solar PV capacity has skyrocketed from almost zero to more than 4500 MW at various locations by June 2019. This even surpassed the targets set by the PDP7r.

- Decision No. 39/2018/QD-TTg, dated 10/9/2018 by the prime minister on revision, and supplements for some articles of the Decision number 37/2011/QD-TTg, dated 29/6/2011 on the wind power price subsidy mechanism. This Decision revised the grid-connected electricity selling price to 8.5 US cents/kWh for onshore wind power and 9.8 US cents/kWh for offshore wind power. This decision became immediately effective on November 1, 2018. For projects that generated power before the effective date of the Decision No. 39/2018/QD-TTg, they can also benefit from the new FITs from November 1, 2018. This decision applies to a component or the whole grid-connected power plant being commercially operated before November 1, 2021; and remains in effect for 20 years since the date of commencing commercial operation.

Power Purchase Agreement (PPA). Standard PPAs have been published for different types of renewable energy, including wind power, solar power, biomass energy, and waste to energy.

- Decision No. 18/QD-BCT, dated 18/8/2008 on the promulgation of an avoided cost-based tariffs schedule and a standard power purchase agreement;
- Decision No. 37/2011/QD-TTg, dated 29/06/2011 by the prime minister on the wind power price subsidy mechanism;
- Circular 32/2012/TT-BTC: Project development procedures and the standard PPA for wind power;
- Circular 44/2015/TT-BCT: Project development procedures and the standard PPA for biomass energy;
- Circular 32/2015/TT-BCT: Project development and the standard PPA for solid waste to energy conversion; and
- Circular 16/2017/TT-BCT, dated 12/9/2017: Project development and the standard PPA for solar power projects.

Renewable Portfolio Standard (RPS). All power distributing, trading, and generating units need to meet the RPS. Power generating units with an installed capacity of more than 1 MW (excluding the BOT power project) must acquire a share of electricity from renewable energy sources (excluding hydropower plants with a capacity exceeding 30 MW): at least 3% by 2020, 10% by 2030, and 20% by 2050. The power distributing units that produce or purchase electricity and the electricity end users who produce electricity by themselves must acquire a share of electricity from renewable sources (excluding hydropower sources with capacity exceeding 30 MW): at least 5% by 2020, 10% by 2030, and 20% by 2050. However, detailed guidelines and policies for RPS have not yet been developed (Fig. 11.4).

Fig. 11.4 Evolution of financing and other measures to promote renewable energy (*Source* Compiled by authors)

Opportunities

Availability of Supporting Policy Framework. The policy framework to promote the development of renewable energy is now in place, with a range of laws, strategies, action plans, and incentives. The most three important documents include the National Energy Development Strategies, the National Renewable Energy Development Strategy, and the Power Development Plan 7 revised. The national master plans are prepared for the sub-sectors of coal, oil, gas, and electricity every 10 years. It is mandated in the 2013 Electricity Law that sub-sector energy master plans must be developed in accordance with the National Energy Development Strategy. This will ensure a degree of consistency and integration between the sub-sectors of energy, including renewable energy sources.

The existing strategies and action plans set both general and specific objectives. Clear targets set by year, or several year intervals, are defined and come together with monitoring and evaluation frameworks. For PDP7 revised, a National Steering Committee has been established where the Deputy Prime Minister is Chairman, and Minister of MOIT serves as Vice Chairman of the Committee. MOIT is the coordinating body for the National Steering Committee. The Committee evaluates the progress of the PDP's annual implementation and reports directly to the Prime Minister. With specific reference to renewable energy, the coordinating and implementing agency is MOIT. MOIT is also in charge of evaluations and reporting to the prime minister. The Ministry of Agriculture and Rural Development (MARD) and Ministry of Natural Resources and Environment (MONRE) are also key agencies in biomass, biogas, and biofuel feedstocks, environmental management, and climate change.

Moreover, there exists a set of measures defined in these strategies, plans, and follow-up decisions to promote renewable energy, including tax exemptions, FITs, power purchase agreements, the renewable portfolio standard, net metering, etc. Tax exemptions are applicable for all renewable energy sources, including import tax, corporate income tax, land-use tax, and environmental protection fees. FITs vary among renewable energy sources (see **Promotion Measures: Fiscal Incentives**). Although there are several concerns regarding the low levels of current FITs, they are nonetheless expected to somehow promote the development of renewable energy, assuming the continued fast development of renewable energy technology, the reduction of renewable energy technology prices, and the integration of the global renewable-energy technology market.

Challenges and Barriers

Institutional Barriers. Several bodies are involved in the implementation of policy. Among the main players (MOIT-Electricity and Renewable Energy Authority, MARD, MONRE, EVN, PPC, various financial institutions and private developers), the level of capacity to implement policy varies significantly. MOIT-Electricity and Renewable Energy Authority is responsible for the sectors of power, new and renewable energy. It is responsible for drafting laws and resolutions on electricity and renewable energy, developing strategy, planning, including investment plans for electricity and renewable energy, developing mechanisms and policies to encourage electrification and renewable energy; preparing national technical standards on electricity and renewable energy, etc.

MARD mainly serves as an agency overseeing feedstock for biomass and biofuels. MARD's current performance indicates that they are capable and experienced as seen through the success of the biogas program, and even in the role of donor, through the constant presence of technical assistance from SNV, which has been a key point. However, the development of policies promoting sustainable feedstock for biomass will require more efforts and capacities from this institution.

MONRE is the agency in charge of planning, monitoring, evaluating, and reporting on environment and climate change policies. The institutional arrangement of this environment climate change body has been well established and considerably supported by donors in recent years. MONRE has set up separate bodies, specifically the General Directorate of Environment, and the National Committee for Responding to Climate Change, which demonstrates strong institutional commitment and credible performance. However, waste to energy promotion policies that are under the jurisdiction of MONRE have realized insufficient success, with only limited and inconsistent supporting policies developed.

EVN has been involved in selected renewable energy development projects (e.g., the Phu Lac Wind Project) as shareholder. EVN seems to have capacity to develop renewable energy projects, especially grid-connected power generation. However, renewable energy development does not seem to be a priority for EVN, especially off-grid renewable energy projects and power purchase from renewable power projects. PPCs have more authority in local energy planning, implementing, and monitoring. Nevertheless, close coordination and consultation with MOIT needs to be sustained to ensure their proper performance in the renewable energy sector. Domestic banks have demonstrated only a limited capacity to provide capital sufficient to finance renewable energy projects on a large scale. More generally, they lack banking experience regarding renewable energy projects and the capacity to effectively evaluate renewable energy investments.

Policy Challenges

Unbeneficial Market Framework. As mentioned in the National Renewable Energy Development Strategy (NREDS), the renewable energy market will be established and promoted by prioritizing the investment and use of renewable energy. Organizations and individuals with different forms of ownership (private/public) will be encouraged to participate in the development and use of renewable energy. Their legitimate rights and interests will be protected under laws and regulations. However, the market for renewable energy has not been established, and the support mechanisms are not sufficient to attract investors to developing renewable energy.

Moreover, a ESCO (Energy Service Company) system in Vietnam is immature and there is no mechanism to promote the development of ESCOs. Currently, the number of ESCOs is small, with only 6 ESCOs established by 2015. Despite the important role of ESCOs in facilitating the deregulation of the energy market, especially in terms of finance, the development of ESCOs has not been clearly encouraged, and it would take several years to promote ESCOs.

Feed-in Tariffs Alone Cannot Unlock Market Potential. **FIT** introductions have varying impacts on different forms of renewable energy. The solar PV market has skyrocketed with an increase of installed capacity from nearly zero before 2018 to 4500 MW by June 2019. This is attributed to Decision 11/2017/QD-TTg, which was approved by the prime minister in April 2017, mandating the same FIT for solar farms and rooftop solar at 9.35 US cents/kWh. The second promising renewable energy to enjoy subsidies is wind power with new a FIT launched in 2019, and projects consisting of thousands of megawatts of new capacity now awaiting approval by MOIT from 2020 onward. However, these FITs are being passed through to consumers as they pay electricity bills. The dilemma here is that the electricity tariff is relatively low, at 7–8 US cents/kWh, and highly regulated. Consequently, these FITs are unlikely to continue for a prolonged period, and this reality creates an opportunity for more market-driven mechanisms to be introduced. These may include auctions, and Direct Power Purchase agreements that allow renewable electricity sellers and buyers to trade directly with lower transaction and network costs.

Conclusion

Vietnam is endowed with excellent potential for renewable energy, especially solar and wind resources. It is expected that the power mix will transition from fossil-fuel reliance, with renewable-energy technologies requiring more space and a concrete roadmap to improve their contributions to the national energy mix. Several strategies and master plans have been introduced to promote the development of solar, wind, and biomass energy. These strategies and master plans establish the development targets for each renewable energy technology as well as the policy frameworks required to achieve these targets. The success of solar power in Vietnam, with installed capacity increasing from a few MWs to 4500 MW over a mere 1.5 years, demonstrates how a strong FIT could unlock the huge potential of solar deployment and private investment in the power sector.

However, instead of subsidies like the FIT, more market-based mechanisms are needed to sustainably accelerate renewable energy development in Vietnam. Low electricity tariffs are hindering the liberalization of the electricity market and longer-term engagement from private stakeholders. Possible options such as Direct Power Purchase Agreements and solar and wind power auctions are being considered at the political level, and are expected to fully realize Vietnam's renewable energy potential as one of the key contributors to national energy security, and a sustainable energy mix.

Notes

1. Applicable for programs utilizing the state budget.
2. In most cases, nationally recognized and competent consultant firms have been involved in developing and appraising the energy plans. Recent PDPs were all developed by the public think tank Institute of Energy. For the latest PDP, new procedures were requested by the deputy prime minister related the involvement of an international expert to review the revised PDP.

3. The share of renewable energy target for the 2010 milestone was not satisfactory.
4. Decision 2068/QD-TTg (2015) issued by Vietnamese Prime Minister (National Renewable Energy Development Strategy).
5. Decision No. 428/QD-TTg issued by the prime minister in 2016, Power Development Plan 2011–2020, 2030 revised (PDP7 revised).

References

Dang, Dinh Thong, and Danh Lien Le. 2005. *New and Renewable Energy Foundation*. Hanoi: Science and Technology Press (in Vietnamese).

Dang, Thanh Tu, Osamu Saito, Yugo Yamamoto, and A. Tokai. 2010. Scenarios for Sustainable Biomass Use in the Mekong. *Journal of Sustainable Energy and Environment* 1: 137–148.

Electricity Law 24/2012/QH13. 2012. On Revising and Supplementing Some Articles of the Electricity Law, November 20.

GIZ. 2012. Status of Wind Power Development and Financing of These Projects in Vietnam.

Government of Vietnam. 2007b. Decision 1855/QD-TTg, Dated 27 December 2007 of the Prime Minister on Approving the National Energy Development Strategy to 2020, with Vision to 2050.

Government of Vietnam. 2011a. Decision 37/2011/QD-TTg, Dated June 29 of the Prime Minister on Wind Power Price Subsidy Mechanism in Vietnam.

Government of Vietnam. 2011b. Decree 75/2011/ND-CP, Dated August 30, of the Prime Minister on Investment Credits and Export Credits of the State.

Government of Vietnam. 2013. Decree 218/2013/ND-CP, Dated December 26 of the Prime Minister on Regulation and Guidelines on Corporation Income Tax.

Government of Vietnam. 2014. Decision 31/2014/ QD-TTg, Dated May 5 of the Prime Minister on Solid Waste Power Price Subsidy Mechanism in Vietnam.

Government of Vietnam. 2015. Decision 2068/QD-TTg, Dated November 25 of the Prime Minister on Approving the National Renewable Energy Development Strategy to 2030, with vision to 2050.

Government of Vietnam. 2016. Decision 428/QĐ-TTg, Dated March 18, of the Prime Minister on the Revised National Power Development Plan in the Period 2011–2020, with Consideration to 2030.

Government of Vietnam. 2017. Decision 11/2017/QD-TTg, Dated April 11, of the Prime Minister on Mechanism to Encourage the Development of Solar PV Projects in Vietnam.

Institute of Energy (IE). 2012. Master Plan of Renewable Energy Development in Red River Delta and Northern Mountainous Areas.

Joint Circular 58/2008/TTLT-BTC-BTN&MT. 2008. Tariff Support for CDM Renewable Energy Projects (Including: Wind, Solar, Geothermal, Tidal or Recovered Methane Gas Projects) from Vietnam. Environment Protection Fund.

MOIT. 2008. Decision 18/QD-BCT: The Minister of Industry and Trade on Promulgation of Avoided Cost Based Tariffs Schedule and Standard Power Purchase Agreement, August 18.

MOIT. 2012a. Circular 32/2012/TT-BTC: The Minister of Industry and Trade on Project Development Procedure and Standard Power Purchase Agreement for Wind Power, November 12.

MOIT. 2012b. Decision 8217/QD-BCT: The Minister of Industry and Trade on Approving The Renewable Energy Development Plan in North Region to 2020, with Vision to 2030, December 28.

MOIT. 2013. Decision 9486/2013/QD-BCT in 2013 by Ministry of Industry and Trade on Approving the Biomass Power Development in Mekong Delta Region to 2020, with Consideration to 2030.

MOIT (Ministry of Industry and Trade). 2015a. Circular 32/2015/TT-BCT: The Minister of Industry and Trade on Project Development and Standard PPA for Solid Waste to Energy, October 8.

MOIT. 2015b. Circular 44/2015/TT-BCT: The Minister of Industry and Trade on Project Development Procedure and Standard PPA for Biomass Energy, December 9.

MOIT. 2015c. Renewable Energy, Presentation by Mr. Pham Trong Thuc, General Director, Renewable Energy Department, General Directorate of Energy, Ministry of Industry and Trade.

Nguyen. 2012. Overview on Status and Trends of Renewable Energy Market of Vietnam.

To, Quoc Tru. 2012. Review and Recommendations on Sustainable Development of Vietnam Renewable Energy. *Journal of Vietnam Energy* (in Vietnamese). Accessed at: http://nangluongvietnam.vn/news/vn/nhan-dinh-phan-bien-kien-nghi/phan-bien-kien-nghi-giai-phap-phat-trien-ben-vung-nguon-nang-luong-tai-tao-viet-nam-(ky-2).html (June 1, 2020).

Trinh, Quang Dung. 2012. Solar Radiation and Current Application in Vietnam. *Journal of Vietnam Energy* (in Vietnamese). Accessed at: http://nangluongvietnam.vn/news/vn/dien-hat-nhan-nang-luong-tai-tao/nang-luong-tai-tao/buc-xa-mat-troi-va-thuc-trang-ung-dung-o-viet-nam.html.

SNV. 2013. Off-Grid Opportunities and Challenges in Vietnam. Netherlands Development Organisation and United States Agency for International Development.

VUSTA. 2007. *Evaluation on Vietnam Power Development Plan*. Vietnam Union of Science and Technology Association.

World Bank. 2010. Wind Energy Resource Atlas of Southeast Asia, July.

Yano, Shinichi, et al. 2009. Potential of Ethanol Production from Major Agricultural Residues in Southeast Asia. *International Energy Journal* 10: 209–214.

Zwebe, Dagmar. 2012. Biomass Business Opportunities Viet Nam, SNV Netherlands Development Organisation.

PART III

New Challenges and Opportunities in Norden

CHAPTER 12

Why Norway as a Green Battery for Europe Is Still to Happen, and Probably Will Not

Espen Moe, Susanne Therese Hansen, and Eirik Hovland Kjær

Introduction

The notion of Norway as a green battery goes back approximately a decade. In 2011, an International Energy Agency (IEA) report discussed the potential role of Norway as a green battery for Germany, Great Britain, and the Netherlands. In Germany, *Der Spiegel* (2010) had earlier written about how chancellor Angela Merkel's green revolution would benefit from Norway as a green battery. When Germany in 2011 decided to phase out nuclear power, imports of clean Norwegian energy gained attraction. Merkel demonstrated her interest by getting personally involved in securing an agreement on the construction of subsea cables from Norway to Germany (Aftenposten 2012). The interest was however not confined to Germany; in 2011, then Norwegian Prime Minister Jens

E. Moe (✉) · S. T. Hansen
Norwegian University of Science and Technology (NTNU), Trondheim, Norway
e-mail: Espen.Moe@ntnu.no

E. H. Kjær
Oslo, Norway

P. Midford and E. Moe (eds.), *New Challenges and Solutions for Renewable Energy*, International Political Economy Series,
https://doi.org/10.1007/978-3-030-54514-7_12

Stoltenberg (Labor Party) waxed lyrical about the vision of Norway as a rechargeable battery for Europe (TV2 2011). The term rapidly became an established part of the Norwegian energy policy vernacular, but also came to signify different things to different actors. As this chapter demonstrates, some embrace it for contributing to an energy transition, others believe that it distorts the debate by suggesting that Norway has unlimited amounts of power to spare, whereas others again object that industrial actors actively use it to greenwash strategies that they would have pursued regardless.

From a climate perspective, the green battery metaphor looks tremendously attractive. The idea is one of utilizing Norwegian hydropower and wind resources to contribute to a European energy transition. Because renewable energy is intermittent, storage capacity is necessary for when the sun does not shine, and the wind does not blow. Norway's reservoirs hold roughly half of Europe's hydro storage capacity. The country has some of the greatest renewable energy resources in Europe, both existing (hydropower) and potential (wind power, especially offshore) and domestic consumption of electricity are derived almost solely from hydropower. In a climate perspective, it would be akin to environmental folly not to exploit the resources that exist on Norwegian territory.

There are two obvious hurdles to be overcome for the battery scenario to be realized. First, the potential of Norway as a green battery can only come to fruition if Norway integrates more closely with the European Continent, i.e., builds interconnectors in the shape of subsea cables for electricity exports. Second, the wind power potential must be harnessed, so that Norwegian renewable electricity production can be increased.

The political feasibility of Norway as a green battery was explored by Anne Therese Gullberg in 2013. She found that while Norwegian politicians were by and large positively inclined to selling surplus renewable energy at a profit, there was no willingness to invest in the infrastructure required to transform the country from a minor electricity exporter to a green battery. Rather, renewable electricity exports would be a logical response to electricity demand from the continent, and subsea cables would only be built if deemed socially profitable.

Since 2013, a lot has however changed. Take Germany: in 2013, Germany derived 13% of its electricity from wind and solar power, against well above 30% today. Globally, wind power capacity has more than doubled. The EU's renewable energy consumption has increased, and the Union has intensified its efforts to create an energy union with an internal electricity market. Moreover, there has been a tremendous surge in the focus on climate. Thus, the physical demand for a green battery to draw on is far greater now than in 2013, and the overall demand for Norwegian electricity has increased. Meanwhile, at the time of writing, two Norwegian subsea cables are close to completion (NordLink to Germany, due in 2020, and North Sea Link to Britain, due in 2021). In sum, many of the underlying conditions for greater renewable integration between Norway and Europe have changed since 2013. Thus, time is overdue to re-explore the feasibility of Norway as a green battery for Europe.

In many countries, vested interest actors such as grid and electric utility companies have been formidable incumbents, resisting grid expansion and keeping new renewables (primarily wind and solar) at arm's length (e.g., Moe 2012, 2015). However, as we return to later, when it comes to subsea cables, these same actors may easily push for change instead of resisting it (e.g., Kelsey 2018). Indeed, the two cables currently under construction were enabled by a grand vested interest compromise between the power sector and the energy-intensive industry, whereby the power sector was granted cables for exports in return for wind- and hydropower installations generating cheap electricity for the industry. Thus, with respect to wind power, conditions have also changed. The 2012 introduction of a green certificate scheme has boosted installations, and wind power costs have come dramatically down, dropping by an average of 70% over a decade and 45% since 2012 (Lazard 2019). This has eroded previous resistance from the Ministry of Finance (MOF) and the Ministry of Petroleum and Energy (MOPE). Additionally, climate change has acquired ever more prominence in Norwegian politics. With less incumbent aversion to change, a more cost-competitive wind power industry, and a generally greater focus on climate, one could reasonably suspect Norwegian politicians—whether through suasion or conviction—to have become more favorably inclined toward the green battery idea since 2013.

Despite this, to a major extent we reiterate the conclusions of Gullberg. Norwegian politicians have consistently been eager and willing to export electricity, but are reluctant to build the necessary cable infrastructure, or to phase in enough renewable energy for large-scale electricity sales to occur. Rather than green battery, we find that the idea that is embraced by most Norwegian actors is that of *increased power exchange* with other states, i.e., Norway as a provider of balancing power, including some net exports of electricity. Furthermore, while we suggest that a long-term future with (some) new cables and more wind power is likely, the short-term prospects for the green battery look worse now than in 2013. The vested interest compromise between the power sector and the energy-intensive industry no longer stands. Instead, industry interests now seem distinctly more influential than power sector interests, and many Norwegian political parties have become deeply skeptical both about new cables and more wind power. One consequence is that in March 2020 a potential further cable (NorthConnect) was put indefinitely on hold by MOPE.

Climate arguments are part of the rhetoric of most of the involved actors. However, it is a distinctly secondary—and oftentimes not even that—argument. Instead, like in 2013, the concern of Norwegian politicians primarily resides in how to utilize Norwegian energy resources for maximum social profitability. While on the European Continent, the green battery idea is unambiguously perceived as environmentally progressive, there is no united embrace of the concept in the Norwegian environmental movement. Many nature conservationists, including several environmental organizations, ardently resist it on the grounds that more wind power leads to the destruction of large tracts of pristine nature for marginal climate gain. With a few exceptions, most Norwegian actors mainly think of Norwegian renewable energy in national rather than broader European or global terms. Hence, Norway as a green battery for Europe is still a long way from happening.

This case study relies on literature reviews, official reports, as well as the data from 15 interviews with Norwegian national politicians, environmental organizations, and industrial actors involved in the decision-making process behind the building of subsea cables and wind power installations. Where our interviewees are cited in the text, we have translated their statements from Norwegian to English.

Theory

Previous research (e.g., Kjær 2018; Moe 2014, 2015, 2016) has demonstrated the importance of vested interests in explaining Norwegian renewable energy policy. Most notably, wind power had to rise in the shadow of a strong petroleum sector, whose interests to a large extent have been embedded within the MOPE. The insistence of both MOPE and MOF on cost-effectiveness and industrial neutrality as a compass for industrial and energy policy, and the existence of large reservoirs of hydropower, meant that wind power stood little chance of expanding.

Transforming the energy system from fossil-based to primarily renewable energy-based—which the green battery idea requires—constitutes an enormous structural economic change. Structural change, however, creates winners and losers and is typically fought by the losers, which have a vested interest in the perpetuation of the system. New and rising industries, like wind power, have not had the time to garner the support of powerful interests, not yet organized to form strong special interest organizations, and not yet accumulated powerful political allies. In contrast, incumbents have had ample time to organize, have access to political networks, and have succeeded in forming beneficial institutional structures. Politically, economically, and institutionally, the old established vested interests have all the advantages.

Furthermore, decisions with large redistributive consequences are politically costly. Politicians rarely make such decisions without major political support across party lines or in the populace. Thus, the safe bet for a politician is to back established interests, preserve the institutional structure, eschew major structural change, and in essence shelter the most important vested interests.

There are numerous vested interests in any political economy, but energy interests are among the most powerful. Once locked in, they are not easily replaced. Today's oil, coal, and energy companies are the biggest industrial giants on the planet, part of what Unruh (2000) labels a techno-institutional complex that perpetuates a fossil fuel-based infrastructure, exacerbated by government subsidies and institutions, resulting in "carbon lock-in".

Theoretically, the above logic rests on Joseph Schumpeter (1942, 1983) and Mancur Olson (1982). It blends Schumpeter's evolutionary economics and "waves of creative destruction" (a renewable energy transition being one such potential wave) with Olson's focus on vested interests. Schumpeterian economics emphasizes structural change. Industries rise because they provide society with new and more efficient ways of doing things, and they fall because what was once technologically revolutionary becomes commonplace and obsolescent as society invents other technologies by which to solve its problems. Hence, there is a steady rise and fall of industries, linked to technological progress: waves of innovation, waves of stagnation, and "waves of creative destruction".

From Olson (1982) flows the observation that rigidities gradually silt up in the economy. Without major shake-ups, economies become ever more inefficient. This is because vested interests wrest power away from elected policymakers, occasionally accumulating enough influence to reduce politics to propagating the interests of the most powerful vested interest groups. Thus, even if time is ripe for renewable energy, structural change will not just happen by itself. It is only when political decision-makers have the autonomy to pursue policies of structural change that they may realistically go against vested interest groups to actively pursue an energy transformation away from fossil fuels.

The petroleum bias of Norwegian energy policy in combination with abundant hydropower explains the until recently very slow emergence of wind power installations in Norway in a manner that dovetails with the framework sketched out above; incumbents were blocking change and Norway did not particularly need wind power. However, regarding subsea cables, Norway does not easily conform to this framework, since here, incumbents in the power sector—as in grid (Statnett) and utility companies (Statkraft the biggest among these)—are not blocking change, but actively *pushing* it. Here, the existing vested interests are harbingers of Schumpeterian creative destruction rather than stability and status quo, completely unlike the role they have played in many other countries, where grid companies have resisted wind power because of concerns with intermittency whereas electric utility companies have been reluctant because of costs (e.g., Moe 2015, 2017).

To explain when incumbents opt for change over stability, we propose expanding the above-described vested interest framework with Kelsey's (2018) division of energy-political actors into four groups: winners, losers, convertibles, and management. Winners and losers are easily defined. They are the actors of the regular vested interest framework, and either gain from change or indisputably lose.

The management category is not particularly relevant for our purpose.[1] Instead, the interesting category is the *convertibles*. According to Kelsey (2018, p. 620), "[c]onvertible industries are industries that make polluting products but do have the capability to switch to non-polluting products." Here, among others, we find the grid and utility companies. The key issue for these is that they generate profits and run a stable electricity system, not whether the energy is derived from coal, petroleum, nuclear, or renewables. They have (in theory) no ideological allegiances to any specific type of energy.[2] Thus, if renewable energy is the better solution, they will switch from resisters to promoters. Ultimately, breaking down the interests of incumbents into winners, losers, and convertibles (and management) enables us to explain why some defend the status quo, whereas others push for change. First, it enables explanations as to why Statnett has lobbied for subsea cables, as more European renewable energy increases the need for power exchange with countries that have energy storage (i.e., Norway). It also increases Statnett's potential for bottleneck income: Norwegian renewable electricity can be stored (hydropower) and can be sold when the price is high. In contrast, wind and solar power from the continent cannot be stored and must be exported when there is a surplus, i.e., when the price is at its lowest. Second, breaking down incumbent interests also enables explanations as to why wind power has become more attractive, with the introduction of the green certificates and the dramatic cost reductions for wind power.

Why would we reasonably anticipate a different result than Gullberg (2013)? First, the climate threat and thus the need for renewable energy has only grown since 2013. Second, as suggested above, the strength of the vested interests that favor the green battery idea can reasonably be assumed to have increased. Thus, one might expect that politicians to a greater extent have come to favor the green battery idea and increasingly use climate arguments to legitimize this. As we shall see, this has not to any great extent happened.

Is Norway Already a Green Battery?

In one sense, Norway has been a green battery for many years already, with decades of power trading between the Nordic countries. Nord Pool was founded in 1996 by Norway and Sweden as a Nordic power exchange, with Denmark and Finland joining by 2000, and power trading with Sweden goes back to 1960. The basic idea was to have a reliable power exchange between neighboring countries with rather complementary power situations: Norway has large reservoirs of hydropower, Denmark has wind power, Sweden has hydropower, wind power, biomass, and nuclear power, and Finland also has biomass and nuclear power. The Norwegian rhetoric surrounding Nord Pool has however not contained notions of green battery. Instead, Nord Pool has enhanced the security of supply for Norway by allowing power exchange between the Nordic countries (Energifaktanorge 2020a; TU 2019a).

Yet, this power exchange has allowed Norway to become a green battery of sorts for Denmark. In 2019, Denmark derived 47% of its electricity consumption from wind power, the highest figure of any industrialized country. The Danish success however relies heavily on four subsea cables (Skagerrak 1–4) from Norway, constructed in 1976, 1977, 1993, and 2014, with a total capacity of 1.7 GW (Statnett 2011; TU 2019a). On windy days, when the Danish grid does not have the capacity to accept all the wind energy produced domestically, the surplus is sold to Norway. The price is low, because the surplus of Danish wind pushes the power price down, allowing Norwegian producers to import cheap wind power to cover domestic consumption while saving the water stored in Norwegian hydropower reservoirs. Norwegian hydropower is instead sold to Denmark on days with less wind and higher power prices. This illustrates the gain from tying different types of complementary renewables together.

Norway also has eight interconnectors with Sweden, the first of which dates from 1960, at a total capacity of nearly 4 GW. In 2018, 40% of Norwegian energy exchange was with Sweden, 24% with Denmark. The total power exchange amounts to around 30 TWh (Bysveen 2018, interview). To enjoy the same security of supply without this power exchange, Norway would hence need to generate an extra 30 TWh domestically, i.e., approximately 20–25% of what hydropower generates in a normal year.

Norway has been a net exporter of electricity in 17 of the past 25 years. However, at an average of 8 TWh per year, with a maximum of almost 15 TWh in 2015, this hardly constitutes a green battery for Europe, let alone Germany. Compared to other European countries, Norway is nonetheless abundant in interconnector capacity. In addition to the interconnectors to Denmark and Sweden (and a minor line to Finland), in 2008 NorNed, with a capacity of 0.7 GW, was built to the Netherlands. Current power trading capacity with foreign countries is 6.2 GW, increasing to 9 GW as NordLink and North Sea Link are completed in 2020 and 2021, extending the number of Norwegian foreign interconnectors to 17. If approved, the cable NorthConnect (to Scotland) would have added another 1.4 GW of interconnector capacity by 2024. The cable is however politically highly contentious and is now put on hold by MOPE. Statnett itself favors another interconnector to Sweden over NorthConnect, *inter alia* because of the possible future decommissioning of Swedish nuclear power, which could remove as much as 40 TWh from the Swedish grid. Skagerrak cables 1 and 2 to Denmark will also have to be replaced within a decade (Energifaktanorge 2020b; NVE 2019; Statnett et al. 2019; TU 2019a).

The Norwegian Renewable Electricity Potential

The idea of Norway as a green battery draws on the utilization of Norwegian renewable energy resources. How substantial are these resources?

The Norwegian energy situation is privileged. Norway may be known as a petroleum country, but the petroleum is for exports, not domestic consumption.[3] 91% of all electricity is generated by hydropower; 98–99% is renewable (NVE 2020a). Abundant hydroelectric power was Norway's original key to energy, industrialization, and wealth. Norwegian industrialization, from the early twentieth century onward, rested on the exploitation of waterfalls for hydroelectric power. By the 1960s, between a third and a half of all Norwegian waterways were regulated (Jansen and Mydske 1998). The 1950s and 1960s saw the rise of a hydro-industrial complex, resulting from the political prioritization of abundant electricity at reasonable prices for households and Norway's energy-intensive industries (Christiansen 2002; Midttun 1988).

Consequently, Norway has the world's seventh-largest hydroelectric production. In a "normal" year, hydropower yields 130–140 TWh, with a record of 149 TWh from 2016 (Energifaktanorge 2020c). Compared to Denmark's much-lauded wind power, Norwegian hydropower generates almost ten times as much electricity. Norway also holds roughly 50% of Europe's hydro storage capacity and a total of 32 GW of hydropower (NVE 2020a; Statkraft 2016). Head of Statnett, Auke Lont, in 2014 predicted that by 2030 as much as 20–25% of German electricity might come from Norway (TU 2014). In other words, Norway's contribution would be significant, but hardly enough for a green battery for all of Europe. The Norwegian Water Resources and Energy Directorate (NVE) estimates the economic potential of new hydropower to 23.2 TWh, with 14.6 TWh deemed realistic (NVE 2020b). The technical potential is far higher, at 600 TWh, but this is neither economically nor politically feasible. In 2001, Prime Minister Jens Stoltenberg (Labor Party) stated that the time for major hydropower developments had passed, a statement reiterated by succeeding Prime Minister Kjell Magne Bondevik (Christian Democratic Party) (Moe 2016; Regjeringen.no 2001). Since then, electricity demand has however kept increasing, requiring Norway to find solutions beyond hydropower.

The wind power potential is also huge, both on land and offshore, with a technical potential of 1400 TWh (NVE 2020b), but it is unclear how much is politically and economically feasible. In 2006, Buen (2006) presented an estimate for land-based wind of 76 TWh. NVE (2019) operates with a far more conservative 26 TWh. For many years, the rate of land-based wind power installations was slow, with lukewarm government support. The notion that MOPE should play a major role in promoting renewable energy seemed distant. In 2006, as feed-in tariffs were discussed, the impression of Norwea (the organization representing the interests of renewable energy producers in Norway) was one of facing a wall of MOPE bureaucrats with no belief in wind power. Upon asking a senior MOPE bureaucrat if a change of government would make a difference, Øyvind Isachsen of Norwea was told that it would not, as the bureaucrats ensured that policies remained stable (Hager 2014). However, in 2012 a green certificate system was introduced, partly in response to the EU renewable energy directive. Here, Norway and Sweden committed to financing a total of 28.4 TWh of renewable energy by 2020.[4] In combination with a dramatic fall in wind power costs, this has increased the pace of Norwegian wind power installations. At the end of 2020, capacity was 3.7 GW. Europe's largest onshore wind power project is being installed in Central Norway, with six wind parks, 278 turbines, and a capacity of more than 1 GW, resulting in 3.4 TWh of annual electricity (Otte et al. 2018). Near 4 GW by 2020 implies more than a quadrupling of capacity in only four years, which however also suggests how slow the pace was until 2016.[5] By the end of 2021, wind power might generate 16 TWh of electricity, or roughly 10% of total electricity generation, up from less than 3 TWh and 1.9% as late as 2017 (Vindportalen 2020).

How about offshore wind power? In 2007, the media labeled the government's tottery offshore wind initiatives a second lunar landing (after Carbon Capture and Storage). In 2009, Petroleum and Energy Minister Terje Riis-Johansen (Center Party) presented a framework for another "future industrial and energy adventure", namely offshore wind power (Moe 2015; Regjeringen.no 2009). In 2009, Equinor's Hywind project became the world's first full-scale floating offshore wind turbine, triggering expectations of a future offshore wind boom (Bellona 2009; TU 2016, 2017). However, successive Petroleum and Energy Ministers were ever less encouraging, until Ola Borten Moe (Center Party) in 2011 explicitly denounced offshore wind as too expensive, backtracking on expectations created by his predecessors (TU 2011a). Costs are however falling far faster within offshore wind than land-based, and Equinor is seemingly more eagerly diversifying away from petroleum than before.

Furthermore, in 2019 the government committed to granting Equinor NOK2.3 billion (roughly €230 million) to Hywind Tampen to electrify the oilfields Gullfaks and Snorre by 2022 through an 88 MW floating offshore wind farm (E24 2019; Regjeringen.no 2019).

While these are interesting developments, they hold few immediate implications for the green battery idea. Conservative estimates state that there is an 18–44 TWh potential from offshore wind (NVE 2020b). WindEurope (2019) states that by 2050 it is realistic for Norway to have 30 GW of offshore capacity in the North Sea, half of which should be exported to Europe. This would however require expensive grid investments.[6]

For now, Norwegian petroleum platforms will be used as testbeds for Norwegian floating wind turbine technology, but the electricity is too expensive to be competitive domestically. Equinor primarily looks for markets with electricity prices high enough for offshore wind to be competitive. Despite its sizeable potential, offshore wind based on *Norwegian* renewable energy resources will therefore not contribute to any green battery soon.

Based on this, it is feasible for Norway to export 30–40 TWh a year to the Continent, which might be expanded to 70 TWh by modernizing hydropower storage capacity (Røynesdal et al. 2015). 11 TWh of increased power generation is currently in the pipeline (Aftenposten 2019). This however fades in comparison with the fact that in 2014 petroleum exports accounted for the equivalent of 1774 TWh. Norway will also need more electricity itself. Electrifying the transport sector would add 13 TWh to the existing need for electricity; 23 TWh if implementing a fully electric solution involving hydrogen. The electrification of the Norwegian Continental Shelf adds another 12 TWh, 23 TWh if also including land-based industry, and a massive 54 TWh for a fully electric solution with hydrogen. Including buildings and construction into this mix adds another 3 TWh. The grand total is a consumption that will have increased by 30–50 TWh plus another 40 TWh if hydrogen is included to make the energy system completely carbon-free (Statnett 2019).

Increasing Norwegian electricity generation to cope with increased domestic demand is technically possible. What this however makes evidently clear is that without significant expansion in terms of hydropower, wind power, and offshore wind, as well as more subsea cables (and considerable progress on energy efficiency), Norway is far from able to supply an entire continent with renewable power, although it can serve as a balancer. Such expansion however requires political support.

Will Norway Become a Green Battery?

The crucial question is whether the expansion necessary for Norway to become a green battery will occur. Gullberg in 2013 concluded that most Norwegian politicians welcomed selling surplus energy at a profit, but shied away from heavy, export-oriented investments in transmission capacity not deemed socially profitable or cost-effective. Policy conformed to regular Norwegian energy policy, market-based and aimed at cost-effectiveness. The willingness to pursue a course of action that put Norway on a trajectory to become a green battery was modest.

When it comes to exporting electricity, there are a few hurdles. In principle, the building of interconnectors is determined by social profit considerations. However, there are major conflicting vested interests in the Norwegian energy sector. Chief among these are the electricity *producers* and the electricity *consumers*. The electricity producers, including the transmission system operator, Statnett, are in favor of electricity exports as the electricity price abroad typically is higher than in Norway.

Energy consumers instead fear rising electricity prices at home. First, the costs associated with building interconnectors and grid upgrades are passed on to consumers. Second, exporting electricity abroad reduces the supply at home, which drives the electricity price up. This concern is shared by the Norwegian energy-intensive industries, which account for as much as 7% of GDP. Their international competitiveness is based on cheap, carbon-free energy derived from Norwegian hydropower (Szulecki et al. 2016). Norwegian politicians know that higher electricity prices are unpopular, and that it is hard to sell the idea that Norwegian electricity prices need to increase to provide Europe with green electricity (Gullberg et al. 2014). Cheap and abundant electricity is considered a Norwegian birthright.

Environmentalists constitute another block of interests. While not really a vested interest—this sector consists of unorganized or loosely organized individuals as much as environmental organizations—Norway has a strong conservationist strand that is generally against both hydropower and wind power. Here, we find interests supporting energy conservation over energy production, but also organizations actively advocating Norway as a green battery.

In the next subsections, we synthesize the empirical material from the interviews and secondary sources from four loosely identified groups or actors, environmentalists, industry (representing electricity consumers), the power sector (electricity producers), and finally the political decision-makers. As we shall see, the discussion around Norwegian climate and energy policy is overwhelmingly nationally centered, and not significantly about contributing to international climate solutions. To the extent that the international level is relevant, it is through Europeanization from EU regulations that impact national decision-making (e.g., the renewable energy directive).

The Environmentalists

Among the actors that distinctly think of Norway in global terms, we find environmental organizations like Bellona and Zero. Wind power installations and cable construction are applauded by Bellona and Zero for contributing to a transition away from fossil fuels toward an energy system based on renewable energy. Hence, Norway can contribute to a more stable European grid that can more rapidly phase in renewable energy. Einar Wilhelmsen of Zero (2018, interview) states that "we need to decarbonize the entire power sector, both in Norway and in the world. To accomplish that, we need large amounts of new renewable production." While both organizations regret the local environmental cost, they favor wind power expansion if the electricity is used *not* to increase Norwegian consumption, but to fuel the electrification of the transport sector and the Norwegian Continental Shelf, in conjunction with subsea cables to other countries. The sum of these measures would make Norwegian wind power (and hydropower) a part of the European energy transition, and Norway a green battery of sorts. Wilhelmsen emphasizes that Norway has a lot of space and has the possibility to install wind power at a low environmental cost.

The environmental movement is however not united. Large parts of the environmental movement abhor wind power for its destruction of ecosystems, habitat, and scenery. One of the largest environmental organizations—Friends of the Earth Norway (Naturvernforbundet)—articulated a phrase that has become ubiquitous among wind power opponents: "We will not save the climate by destroying the nature," advocating energy efficiency and circular economy over energy production (Lundberg and Esmark 2019). Oddvin Lund (2018, interview) of the Norwegian Trekking Association (Turistforeningen) accepts that there is substance to the green battery idea since Norway to some extent can contribute electricity to Europe, and because a flexible, well-functioning power exchange may lessen the need for more wind power within Norway. Yet, he also argues that the power sector "pushes this as a climate action [even if] it is interested in it primarily for economic reasons." He stresses that they are not against cables *per se*, but against building cables without understanding the consequences, and that the power sector's wind power ambitions are far beyond what the Trekking Association supports. He sees politicians having uncritically bought into the power sector's story of wind power as a climate solution, while ignoring environmental costs. The conflict also permeates the Green Party, which during its 2019 national congress ended up with a tenuous compromise saying that wind turbines should preferably be reserved for areas with infrastructure and already existing interventions in nature (VG 2019a).

Public opinion, especially through local nature-protection agencies, has actively lobbied against wind power for arguably ruining large tracts of pristine wilderness. The lacking social and local acceptance of wind power in Norway however dates far back (Buen 2006; IEA 2005; Jerpåsen and Larsen 2011). The political onus on cost-effectiveness has meant that power supply has taken precedence over siting. Thus, Norway has ended up with wind farms in windy and exposed, but scenic areas rather than in visually less-intrusive locations (Buen 2006; Karlstrøm and Ryghaug 2014; Lund 2018, interview). This, in combination with the perception that local opinion is neglected, has invariably caused major popular resistance. There have been cases of physical resistance and sabotage in at least four current construction areas,[7] and demonstrations against many more. In Frøya, construction machines have been sabotaged, an excavator set on fire and there have been two attempts at forcing company cars off the road. The entrepreneur publicly admits that he is afraid of his employees' lives and well-being (Adresseavisen 2020).

Several polls have suggested considerable or increasing resistance against wind power if it leads to the destruction of nature: 60% of the population opposes wind power when the result is major loss of pristine nature, 71% is against if it leads to major disturbances for animal life (DNT 2019). Another poll showed 40% in favor of land-based wind power and 36% against, with 70% in favor if it is offshore (NRK 2019b). Polls however also show that despite a very polarized debate, two out of three are indecisive or not feeling particularly strongly about the subject (Linnerud et al. 2019).

In 2019, NVE published its national plan for wind power, to identify suitable and non-suitable areas for wind power development. The intention was to replace regional wind power plans, reducing the conflict level among sector interests and increasing the predictability of decision-making. However, of the 101 municipalities singled out as suitable for wind power, only 5 said yes. 78 said outright no. The government ultimately shelved the plan (Inderberg et al. 2019; TU 2019b).

Thus, faced with the necessary evils of the green battery scenario—wind farms and cables—major parts of the environmental movement and the overall population favor domestic nature conservation over increased renewable energy production and exports. Although the environmental organizations are fully aware that climate change is a global challenge, domestically it often does not trump the stronghold of nature conservation. Some environmental organizations certainly advocate the green battery argument, but there is no unanimity.

The Industry

The first prerequisite for the green battery scenario, installing more wind power to generate a greater power surplus, meets little resistance from industry as more wind power yields lower energy prices and hence benefits industry. Building subsea cables is more contentious. The fear in the energy-intensive industry, where energy may account for 45–50% of the total costs (Løfsnes 2018, interview), is that more energy exports lead to higher domestic electricity prices. For the energy-intensive industry, cheap electricity is one of its main competitive advantages. Industry interests argue that it would be folly to let the greenest energy-intensive industries in Europe suffer higher power prices and lose competitiveness compared to foreign competitors and far less green energy-intensive industries elsewhere. Accounting for 7% of GDP, energy-intensive industry is a potent political stakeholder, with strong ties to the Labor Party and the far left.

The Norwegian Confederation of Trade Unions (LO), where most of the energy-intensive industry is organized, has of approved the construction of the current subsea cables. Others are less supportive. The far-left think tank Manifest Tankesmie sees the green battery idea as pushed by the power sector, which has strong economic interests in subsea cables and the importation of European power prices to Norway,[8] employing self-serving green rhetoric and actively searching for green arguments for wind power and cable construction, among other things by funding research.[9] Instead, Manifest Tankesmie argues that the electricity contributes more to the climate if kept on domestic hands, as a cheap, carbon-free input for Norwegian energy-intensive industries. The carbon footprint from Chinese aluminum is for instance eight times higher than that of Norwegian aluminum (Manifest Tankesmie 2019; Marsdal 2019). Hence, by helping Norwegian industry, we help the climate. Several of our interviewees reiterate this logic. Leif Sande (2018, interview) of Industri Energi—the union that has been the most critical of cables—argues that Norway exports 100 times more energy to Europe through gas than electricity, which for most European countries counts as a climate measure. Furthermore, Norway exports manufacturing goods using renewable energy, which is a climate bonus contingent on being able to provide cheap electricity to Norwegian industry. Eilertsen (2018), also of Energi Industri, moreover maintains that the climate contribution of Norwegian cables is negligible. Arguably, cables are only about the interests of the power sector, whereas the potential damage to Norwegian industry is potentially severe. He argues that electricity prices will not increase by 3–6 øre/kWh as predicted by Statnett (E24 2017), but by 10–40 øre.

Industrial interests are however not as unambiguously negative as above. Ole Løfsnæs (2018, interview) of Norsk Industri agrees that the climate contribution from Norwegian cables is marginal. Yet, Løfnsæs downplays the interest battle between the two main vested interests, the industry and the power sector, stressing that they are really in the same boat. Industry does not want the power companies to go bankrupt over low power prices, and the power sector does not want the industry flagging out because of high power prices. There are no indications that the power sector would opt for unlimited amounts of cables if given the choice. There is a far greater harmony of interest between these two vested interests than expected.

In summary, the industry's overall attitude is industry first. The industry generally favors wind power installations but is wary about cables. Furthermore, the green battery concept is viewed with skepticism, and primarily as a way for the power sector to advocate its own interest. Arguably, the climate is better served through the green manufacturing of energy-intensive products, accomplishable by few countries beyond Norway.

The Power Sector

The electricity producers, as well as the transmission system operator, Statnett, are in favor of exporting electricity to markets where the electricity price is higher than in Norway. With an inherent interest in wind power installations and cable construction, the power sector is therefore an expected advocate of the green battery. Indeed, Steinar Bysveen (2018, interview) of state-owned Statkraft—the largest Norwegian power producer and the largest producer of renewable electricity in Europe, most of this hydropower—even claims that Statkraft *invented* the green battery idea (or at least used it a lot), as they realized that the German *Energiewende* generated major need for backup capacity and power exchange in Germany.

However, beyond this, what we see from all our power sector interviewees (as well as many of the other interviewees) is an emphasis on how the green battery idea distorts the debate by making it all about Norway *supplying* Europe with massive amounts of hydropower. The reality is that Norway is making good money from *power exchange*. Subsea cables are about linking (and thus enlarging) markets to exchange power and utilize renewable energy better. The need and potential for power exchange increases with more renewable energy capacity on the Continent. Hence, notwithstanding seasonal fluctuations[10] and differences between cables,[11] in the long run this is more about power exchange than about exports.

To the extent that the power sector focuses beyond the state, it is on the Nordic region as much as the EU. As Statnett director of customer and public affairs, Erik Skjelbred (2018) told us, the focus for future cable construction may shift toward the Nordic countries as EU processes are extremely slow and heavy. But MOPE-owned Statnett—responsible for the Norwegian power grid (the stem net), interconnectors, and subsea cables abroad—has social profits as its mandate. Thus, subsea cables will be built whenever socially profitable. This does however not mean that Statnett is a voice for more cables regardless. The two cables already under construction will, according to Skjelbred, change the interconnector landscape sufficiently for Statnett to need to gain experiences with the two new cables before further cables are built.

While Statnett does not need to run a profit, climate is not included in their social profitability mandate. It is an important bonus, but Skjelbred emphasized that they have no tools by which to incorporate climate concerns into their calculations. He however also stated that while reliability of energy supply is priority number one, there really is a triangle of priorities: reliability of supply is at the top, and at the two bottom corners we find value creation for consumers and producers, and environment and climate. Magne Fauli (2018, interview), industrial policy adviser of Energi Norge—an interest organization comprising 275 member companies in the power sector—added his take on the climate rationale behind subsea cables: "We built cables and power exchange with the Danes long before it was relevant to include climate concerns in this." While stressing that climate is indeed important, it would not alone have garnered support for cables.

Not surprisingly, the main interest of the power sector is to increase the Norwegian transmission and renewable energy capacity. Climate is certainly used as an argument but is clearly secondary. Several of our interviewees were however frustrated with the trade union complaint that the power sector is sacrificing Norwegian industry to sell electricity to Europe. Among others, the Electrician and IT workers union (EL og IT) (Andersen, Heien, Hennum, 2018, interview) labeled this complaint nonsensical. It would be completely counterproductive for the power sector to let any green battery idea go at the expense of Norwegian industry as Norwegian industry is a far bigger customer than foreign countries will ever be, meaning the dependence goes both ways. Arguably, the rise in electricity prices from cable construction will merely bring prices back to where they used to be before the recent influx of wind power, leaving industry with little to worry about (E24 2017).

In sum, the power sector holds the real value of Norway's "battery" to lie in power exchange, not exports, and the fears of increased prices are exaggerated. However, social and monetary profits are the key rationales for the power sector. Climate is certainly also used to argue for wind power and cables but is clearly secondary, or simply a bonus.

The Politicians

It is not clear who in Norway was the originator of the term green battery, but among the politicians, in 2011 PM Jens Stoltenberg and present-day Labor Party leader Jonas Gahr Støre were early adopters. It was the

2005–2013 left-center coalition government of Stoltenberg (Labor Party, Center Party, Socialist Left Party) that in 2012 approved the construction of two subsea cables to Germany and Britain—NordLink and North Sea Link—with support from the opposition.

Although part of the government coalition, the Center Party was never as sanguine about the green battery idea as the Labor Party. Illustratively, when then Petroleum and Energy Minister Ola Borten Moe (Center Party) in 2011 gave the official approval to NordLink and North Sea Link, he supported renewable energy installations to lower the electricity price, but generally warned against subsea cables (Statnett then had plans for five) (TU 2011b). His opinion was that cables could be built only if socially profitable *and* if strengthening the reliability of energy supply. His aim was to improve the *power exchange* (TU 2012). Borten Moe was criticized by environmentalists for preventing the realization of Norway as a green battery, and claims surfaced that German politicians were unhappy with perceived Norwegian backpedaling on earlier pledges (Dagens Perspektiv 2011; TV2 2011). However, despite Stoltenberg's earlier green battery rhetoric, by 2012 Stoltenberg *and* Borten Moe were adamant that subsidizing Norwegian renewable energy to sell it cheaply to Europe was a horrible business idea. One telling quote from Stoltenberg's autobiography, referring to the green certificate system eventually introduced in 2012, was "I have nothing against Germans, but I cannot understand why we should build expensive electricity in Norway only to sell it cheaply to German power companies?" (Stoltenberg 2016, p. 420). To Stoltenberg, the green certificate system constituted erroneous investments of billions of Norwegian kroner in transfers to foreign countries (Stoltenberg 2016). For similar reasons, in 2013 Borten Moe put the partially private-owned cable NorthConnect on indefinite hold. Stoltenberg emphasized that the government would not privatize cable construction, arguing that this might hurt Norwegian industry (TU 2013).

Thus, despite lofty early rhetoric about contributing to a European energy transition, it is conspicuous how the focus soon shifted from the international to the national, and from green battery to power exchange. Cables would improve the flexibility of the grid and increase the reliability of supply. Norway would profit from power exchange since electricity prices abroad are generally higher than in Norway. Simultaneously, the increased generation of renewable energy would lower electricity bills for households and industry.

When interviewing politicians today, most of the reasoning from the Stoltenberg administration is echoed both in the opposition and in the conservative-led government with Erna Solberg (Conservative Party) as prime minister of a shifting coalition of parties on the Norwegian right. Several of our interviewees particularly stress how the green battery term erroneously suggests that Norway can simply serve the Continent in a one-way-street manner, whereas the figures rather enable Norway to become a balancing power. Industrial policy spokesman for the Labor Party, MP Terje Aasland (2018, interview), even claimed that the green battery idea has primarily been pushed by the interests that benefit the most from it, predominantly the power sector.

If there is little support for the green battery idea, does that also mean that climate concerns are unimportant for Norwegian politicians with respect to cable construction? The answers vary. Our interviews with MPs Kjetil Kjenseth (Liberal Party) and Lars Haltbrekken (Socialist Left Party) suggest that climate and environmental issues were very important for their parties' acceptance of NordLink and North Sea Link.

For the three biggest Norwegian parties over the past few decades—the Labor Party, the Conservative Party and the Progress Party—the emphasis is far more clearly on what benefits Norway economically, especially from a social profit point of view, and on how best to manage Norwegian energy resources and provide energy security for Norway. These correspond with long-standing Norwegian energy policy goals. For these parties, any climate benefit is a major bonus, but no independent reason for building cables. Beyond this, our interviews uncover an ideological nuance. The Labor Party places substantial weight on utilizing Norwegian resources in a manner that explicitly supports Norwegian industry. Thus, renewable energy expansion should not simply be export oriented (Aasland 2018, interview). With the two parties on the political right, we rather see a tendency toward emphasizing the power sector (interviews with Odd Henriksen, Tord Lien, and Oskar Jarle Grimstad, 2018). Since Norway has a power surplus, not building cables for exports would lead to falling power prices and an erosion of profits for the power sector, subsequently also lowering social profits for the country.

Apart from the party Red, there has been political consensus around the building of two subsea cables, and on the expansion of domestic wind power. The way forward may however not hold the same consensus on these core prerequisites for the green battery. The potential construction of a third cable, NorthConnect, illustrates the withering consensus.

Some of the parties on the right are positively inclined. Stefan Heggelund of the Conservative Party for instance argues that NorthConnect will be socially profitable *and* annually cut 2 million tons of CO_2 (NRK 2020). However, the government is bound by a 2018 agreement with the Labor Party and the Green Party that no new cables can be built until Norway has gained necessary experience from the two cables under construction and thorough analyses have been completed (*and* that any new cable must be socially profitable) (Høyre 2018). This compromise preserves some consensus about the future, but one that is far weaker than the past consensus.

The Conservative Party and the Liberal Party (and conceivably the Labor Party) will approve of more cables. Several parties are, however, unambiguously opposed. The Progress Party (since leaving the government in January 2020) will not allow new cables that will produce more expensive electricity for households, the Socialist Left Party will not allow private cables, the Red Party argues that Norway should prioritize providing energy-intensive industry with cheap electricity over exporting electricity, whereas the Center Party also emphasizes that cables are a matter of critical infrastructure, over which Norway needs to preserve national control. These all seek the outright rejection of NorthConnect and a prohibitive stance on future cables (Dagbladet 2019b; Filter Nyheter 2019b; NRK 2020; Stortinget 2019). In line with the compromise between the government and the Labor Party and the Green Party, MOPE has however neither rejected nor approved the concession application, but as of March 2020 put it on indefinite hold, stating that it currently has no parliamentary mandate to make a decision in either direction (ABC Nyheter 2019; E24 2020).

Underlying the different political stances are miscellaneous rationales. The Labor Party and the Conservatives arguably represent the Norwegian political mainstream, and their stance, whether they support industry or the power sector, is ultimately one of utilizing Norwegian natural resources in a manner that maximizes social profits. Climate is not unimportant, but clearly secondary. The dissenting voices advocate arguments against higher electricity prices and private cables and embrace the importance of national control over natural resources and energy policy. Furthermore, notwithstanding their particular stance on Norway as a green battery, every Norwegian political party stresses the importance of a well-functioning power exchange with the neighboring countries for energy security reasons.

Arguably, the most strongly dissenting voice belongs to the far-left party Red and their leader Bjørnar Moxnes, who echoes the arguments made by the industry and Manifest Tankesmie by arguing that the battery argument is greenwashing by the power industry and that the real climate effort Norway can make is to produce aluminum, silicon, and other metals and minerals cleaner than any other country. Further, he argues that the green battery argument is akin to trying to run a Tesla on a flashlight battery. Building more cables will vacuum the country for electricity and hurt the climate by eroding the competitiveness of Norwegian energy-intensive industry (Dagbladet 2019a; Filter Nyheter 2019b).

The other prerequisite for the green battery is wind power installations. Here, the short version is that for years, when Norway saw little wind power installed, every political party was *in principle* in favor of wind power. It was however not a serious political priority, and the support system was poor, until the 2012 introduction of the green certificate system, largely a consequence of the EU renewable energy directive (e.g., Moe 2014, 2015, 2016). The green certificate system has led to a significant increase in wind power installations, especially over the final few years. However, it has brought with it increasing popular opposition against wind power installations, with captions such as "wind power rebellion" and "wind power opposition" featuring increasingly in national media.

Following this uptick in popular opposition, Norwegian politicians have grown considerably more lukewarm towards wind power. PM Solberg has expressed sympathy with the opposition against turbines, acknowledging that enjoying nature is central to Norwegians' daily lives (VG 2019b). Numerous caveats are expressed by other parties: The Labor Party wants wind power, provided installations are located in places with lots of wind, where conservation concerns have been accommodated, and where there is local support and local financial gain (Arbeiderpartiet 2020; Filter Nyheter 2019a). The Socialist Left Party wants all wind power license applications halted until there is a comprehensive, national plan for wind power (NRK 2019a), and the Green Party only supports wind power installed away from pristine nature (VG 2019a). Several parties (Center Party, Green Party, Socialist Left, Progress Party, Christian Democratic Party) advocate stronger local veto power, and legally enshrining the right for the municipalities to reject license applications (Stortinget 2019; TU 2020; VG 2019c).

Moreover, beyond marginal voices, wind power has never been part of a green battery/climate strategy. Rather, wind power has historically been perceived as a useful complement to the existing, hydropower-based system, to prevent dry years from producing electricity deficits (Moe 2014, 2015). However, in the context of the present "wind power rebellion", Norwegian politicians have become extremely wary of supporting further installations. Meanwhile, the end of the certificate system to some extent will result in an automatic drop in installations. The new mantra is instead renewed support for offshore wind, as exemplified by Enova funding for Equinor's floating offshore wind project (e.g., Arbeiderpartiet 2020).

Vested Interest Influence?

In the theory section, we suggested that Norwegian energy policy routinely has been dominated by vested interests. The main vested interests of relevance to the green battery are the power sector, organized through the Confederation of Norwegian Enterprise (NHO) and the energy-intensive industry, organized through LO. Additionally, while not constituting a uniform or strongly institutionalized vested interest *per se*, there is no doubt that environmental organizations have engaged in lobbying activity. This section explores the extent to which these interests have influenced policymaking, and whether this has changed since 2013.

Regarding subsea cables, the vested interest influence is easily identified. Until 2018, there was near unanimity among two different government coalitions on constructing two subsea cables. Several interviewees pointed out that this consensus was founded on a big compromise between the two main vested interests, preceding the MOPE concession approval: LO agreed to the power sector's demand for surplus renewable electricity to be exported through new subsea cables. In return, new wind power (and hydropower) installations would ensure that the industry could still rely on cheap and abundant electricity. One interviewee stated that this compromise to a significant extent was initiated by Energi Norge, which recognized the importance of getting ahead of the politicians to forge an agreement between the most important interests before the license application process started. Within LO the agreement was contentious, and some LO sub-branches strongly opposed it, but the overall stamp of approval from LO made it far easier, especially for the Labor Party, to support cables.

Wilhelmsen (2018, interview) stated that an analysis by Zero showed major industry lobbying toward the Labor Party, which was in power at the time, resulting in the two-cable compromise. While there was consensus in the parliament that this was a good solution, Wilhelmsen states that political decision-makers really had no clear plan. Rather, projects were initiated bottom-up. Statnett, while not actively part of the compromise, had for instance already initiated talks with Germany over cables. NorthConnect was another driver for cables. Lund of the Trekking Association (2018, interview) states that the power sector was also actively and successfully lobbying politicians.

The compromise meant that in some ways there was less lobbying pressure on politicians than would otherwise have been the case. The two main vested interests had already forged an agreement, and this removed a lot of pressure especially from the Labor Party, which could now approve subsea cables with the official backing of the industry. Thus, in one way, politicians did not experience major vested interest pressure, as confirmed by several of our interviewees. Instead, they had the necessary autonomy to make the decision they wanted with respect to cables. Another reading of the situation is however that little pressure was felt because the main vested interests had already practically made the decision on behalf of elected politicians. One of the politicians we interviewed was clear that any alliance between LO and NHO carries a lot of weight on both the left and the right, and that the compromise between the industry and the power sector made it far easier for politicians—especially those on the left—to accept cables.

Has this changed? With NorthConnect, there was no similar compromise. Odd Øygarden of NorthConnect (2018, interview) was already in 2018 worried that while the government seemed in favor, the Labor Party was under severe pressure from the unions to reject NorthConnect. While the power sector has lobbied for the cable, far more interests, including a practically unanimous LO, opposed its construction, arguing that NorthConnect will produce higher electricity prices, harming households and industry, while only marginally improving Norway's security of supply. Parts of NHO, i.e., Norsk Industri, are also opposed, and aluminum giant Norsk Hydro—one of the largest companies of the Norwegian energy-intensive industry—favors postponing the cable, as does Statnett (E24 2020; FriFagbevegelse 2019, 2020; VG 2019d). Thus, instead of any vested interest compromise, with NorthConnect, industry interests are clearly stronger. Consequently, it has been put on hold by MOPE.

While we cannot positively confirm that vested interests swung the decision against approving NorthConnect, the industry arguments advocated by LO (and others) were very much used by all the parties that demanded the rejection of the cable (Red Party, Socialist Left Party, Center Party, Progress Party). However, LO has also argued that it is important to gain experiences from NordLink and North Sea Link before considering new cables. This is the compromise stance of the Labor Party, quite conceivably a consequence of the cross-pressure the party is under on this matter (ABC Nyheter 2019; E24 2020; FriFagbevegelse 2020). However, unlike with NordLink and North Sea Link, the grand compromise here is not a vested interest compromise, but a *political* compromise, between the Labor Party (and the Green Party) and the government. It ensures that the Labor Party has the continued backing of LO by for now closing the door on new cables, while simultaneously—together with the government parties—keeping the door for future cables open (ABC Nyheter 2019; Arbeiderpartiet 2018). A plausible interpretation is that while politicians often bow to vested interest pressure, they are also actors in their own right. The political compromise appears to be an example that illustrates this.

While industry interests weighed heavily with NorthConnect, Terje Aasland (2018, interview) suggests the existence of a counteracting, embedded institutional bias. Energy policy sorts under MOPE, which is inclined toward exports. If instead it had sorted under the Ministry of Trade, Industry and Fisheries, greater weight would have been put on industry, both as in protecting existing energy-intensive industries, and in promoting new industries like floating offshore wind. Thus, the institutional structure arguably pulls in the direction of the power sector over industry. While this will not make Norway into a green battery, it is likely that the result is a slow long-term increase in power exports. For now, however, for the cable prerequisite for the green battery, the situation is less promising than in 2013. The cables under construction will be completed, but in the short term, no new cables will be built.

Regarding wind power, it is not equally clear that vested interests were influential. Vested interests obviously had no influence on costs dropping by 70% over a decade (Lazard 2019). The other main development was the introduction of green certificates. Prominent among those lobbying hard for certificates, believing this would provide cheap electricity for industry, was Norsk Hydro, backed by both LO and NHO, and Labor Party mayors from industrial municipalities. They were also supported by most of the environmental movement (DN 2013). In his autobiography, Prime Minister Stoltenberg (2016, p. 429) states that he experienced massive lobbying pressure from what he labeled an enviro-industrial complex, consisting of the environmental movement, LO, NHO, and the industry, in cooperation with several Norwegian political parties, including his coalition partners, the Center Party and the Socialist Left Party. These had, he says, enormous power. However, while there was heavy vested interest pressure behind the certificates, they were not considered a victory for wind power. The warning in 2012 was that certificates would favor hydropower, threatening the last remaining natural Norwegian river systems, while suppressing wind power, which was still not cost-competitive (ABC Nyheter 2012; FriFagbevegelse 2013).

It is however no longer obvious that there is strong vested interest pressure for continued wind power expansion. The certificates expire in 2021, no new support system will replace it, and popular resistance against wind power installations has made politicians wary. Thus, also for this prerequisite for the green battery the short-term prospects seem if anything worse than in 2013.

In conclusion, the vested interest story is ambiguous. The vested interest constellation that produced a cable compromise and a green certificate system no longer exists. Instead, industry interests have grown stronger and environmentalist support for renewable energy has to a major extent been replaced by nature conservation concerns about the pace of wind power expansion. Norwegian politicians are still influenced by lobbying, but these are lobbying efforts that for now have made the green battery less likely.

Conclusions

Norway is an interesting case. With its abundance of actual and potential renewable energy resources, it is an antidote to a world where most countries are energy scarce. It is also a country with a strong environmental identity. If not Norway, what other countries could credibly be able and willing to serve as a green battery for Europe?

In one way, Norway is already a green battery—for Denmark—and has for decades engaged in power exchange through Nord Pool. Thus, Norway has a long history of exporting and importing electricity. Power exchange is a source of income and it improves Norwegian energy security by providing easy access to electricity during dry years when Norwegian reservoirs have little water. If Norway does not approve any cables beyond NordLink and North Sea Link, the capacity to act as a green battery will however be limited to 9 GW. Increased energy production would be used within Norway, leading to falling electricity prices rather than European emissions reductions. Granted, there will still be need for more electricity—the two new cables increase the export potential from 6.2 to 9 GW, and more power is required to electrify sectors like transportation, construction, and land-based industry. Thus, Norway will arguably see more wind power installed, but the rate of expansion will slow down, and the additional power will be used domestically. It does not bring about a green battery.

If our vantage point is instead that with an almost fully renewable electricity sector and the best renewable resources in Europe, Norway's ability to contribute to solving global climate problems is through the deployment and sale of renewable energy to Europe, and to act as a balancer of intermittent renewable power on the Continent, the green battery idea makes more sense. If so, then Norway should install more wind power and build more cables.

Grid and utility companies profit from electricity exports. They prefer more cables. However, the threat of higher electricity prices for industry and households is a major hurdle for strategies aimed at exporting more electricity. How big this cost is, today and in an unknown future, is both contentious and politicized. For politicians, higher electricity prices are unpopular with voters. For industry, it makes little sense—economically and environmentally—to relinquish one of Norway's main competitive advantages, i.e., cheap and carbon-free electricity for a large energy-intensive industrial sector. Thus, for Norway to become a green battery, significant obstacles must be overcome, and a balance struck between the two main vested interests, the power sector and the industry. Meanwhile, conservation concerns and rising popular opposition against wind turbines must be more seriously taken into account.

How do Norwegian politicians relate to the green battery concept and the broader climate discussion surrounding it? First, what was striking was that very few had any enthusiasm for the green battery idea. Many expressed how the term obscured what was the real issue, namely power exchange and Norway providing balancing power to Europe. While Norway has Europe's largest hydro storage capacity, Germany could alone easily consume all the electricity that Norway generates. Thus, the potential Norwegian contribution, while significant both from a power and a climate perspective, is limited.

Second, also striking was how few actors thought of energy exports primarily as a climate contribution and as climate politics. Except for the most ardent opponents of cables and wind power, nearly all interviewees recognized the potential climate contribution to a European energy transition through renewable energy exports. However, beyond environmental organizations like Zero and Bellona, and the most climate-conscious political parties, the belief that Norwegian renewable energy should be utilized to as great an extent as possible to assist a European energy transition is entirely absent. Instead, the dominant perspective among political decision-makers is that energy exports are about utilizing natural resources in as socially profitable a way as possible, without unduly reducing the competitiveness of Norwegian industry.

On the one hand, considerations of social profitability clearly weigh more heavily than climate considerations. On the other, several interviewees talked about climate gain by other means than electricity exports, arguing that Norway's most important climate contribution is one of manufacturing and exporting energy-intensive products (needed among other things for an energy transition) with a lower carbon footprint than any other country on the planet. The job was made easier for the politicians by the two major vested interests—the power sector and the industry. These came together and carved out a compromise whereby more wind power would be installed, alleviating the energy-intensive industry's fear of increasing electricity prices, in return for subsea cables being built for renewable energy exports to Europe. With the political right tending toward the power sector and the left toward the industry, this grand bargain alleviated important political tension, especially easing the pressure on the Labor Party. It is also a bargain that points towards the future. The strength of the power sector and the industry suggests that their interests to a major extent will be accommodated by politicians also in the future. Policy swings are likely to be moderate rather than dramatic.

Furthermore, very few Norwegian politicians think of renewable energy exports as part of a green battery vision but as belonging to the realm of energy politics. They are preoccupied with how Norwegian natural resources can be utilized to maximize social profits. Despite the global nature of man-made climate change, and despite ever closer integration with the EU energy union, Norwegian decision-making is characterized by national over international considerations.

Unlike what we proposed earlier, the green battery is not edging any closer to fruition. True, in the long term, our prediction is a future with more cables (but not many), primarily because the Labor Party only supports a pause. Also, as wind power costs keep falling, wind power will become competitive even without green certificates. Hence, Norway will install more wind turbines, conceivably paving the way for larger exports. Thus, a somewhat more comprehensive and extensive power exchange is a likely future outcome. However, in the short term, developments are edging in the opposite direction. There will be no new cables in the immediate future, while post-2021 the pace of wind power installations will slow down because of the expiration of the green certificates and popular opposition based primarily on nature conservation. Hence, neither prerequisite for the green battery is currently being fulfilled.

Notes

1. Industries that have an indirect stake in pollution regulation because they make inputs or generate wastes linked to pollution (Kelsey 2018).
2. This is not always true. The Japanese utility companies have for instance traditionally had strong allegiances to nuclear power (Moe 2012).
3. Norway exports almost eight times more energy than it consumes (Godzimirski 2014).
4. The commitment is to the joint financing of 28.4 TWh. This goal was reached in May 2019: 20 TWh in Sweden and 8.4 TWh in Norway, more than half of the Norwegian contribution from hydropower (Kraftnytt 2019).
5. In comparison, Denmark has over 6 GW, Sweden nearly 9 GW, and Germany above 60 GW (IRENA 2020).
6. WindEurope's (2019) projections are based on Europe meeting 30% of its 2050 electricity demand by offshore wind. Norway's projected share is 30 GW from a total of 450 GW.
7. Frøya, Flatanger, Sandnes, Kvinesdal (Kommunal Rapport 2019).
8. Manifest Tankesmie (2019) claims that for every 10 øre (€0.01) price increase per kWh, the power companies gain an extra NOK13.5 billion (€1.35 billion).
9. The research program CEDREN was funded by the Norwegian Research Council and 15 Norwegian power companies, with Statkraft having the chairman of the board (Manifest Tankesmie 2019).
10. In the summer, Norway tends to have a rather large surplus, being a season with high production and low consumption. The opposite is the case in the winter.

11. North Sea Link will predominantly be export-oriented, and NordLink primarily about balancing power.

References

ABC Nyheter. 2012. SV-politiker angrer på grønne sertifikater, August 24. Available at: https://www.abcnyheter.no/nyheter/2012/08/24/157671/sv-politiker-angrer-pa-gronne-sertifikater. Accessed 11 Apr 2020.

ABC Nyheter. 2019. Arbeiderpartiet vil ikke være med på vedtak mot ny utenlandskabel, June 14. Available at: https://www.abcnyheter.no/nyheter/politikk/2019/06/14/195586081/arbeiderpartiet-vil-ikke-vaere-med-pa-vedtak-mot-ny-utenlandskabel. Accessed 10 Apr 2020.

Adresseavisen. 2020. Frykter for ansattes liv og helse, January 24. Available at: https://www.adressa.no/pluss/nyheter/2020/01/24/Frykter-for-ansattes-liv-og-helse-20908794.ece?rs48753158240750901 5&t=1. Accessed 13 Apr 2020.

Aftenposten. 2012. Strekker ny el-kabel mellom Tyskland og Norge, June 21. Available at: https://www.aftenposten.no/okonomi/i/P3Kye/strekker-ny-el-kabel-mellom-tyskland-og-norge. Accessed 13 Apr 2020.

Aftenposten. 2019. Norge kan halvere klimautslipp med flere vindmøller, March 25. Available at: https://www.aftenposten.no/okonomi/i/7lEMg4/norge-kan-halvere-klimautslipp-med-flere-vindmoeller. Accessed 13 Apr 2020.

Arbeiderpartiet. 2018. EUs tredje energimarkedspakke (ACER). Available at: https://www.arbeiderpartiet.no/politikken/eus-tredje-energimarkedspakke/. Accessed 11 Apr 2020.

Arbeiderpartiet. 2020. Mer fornybar energi samtidig som vi tar vare på naturen. Available at: https://www.arbeiderpartiet.no/politikken/vindkraft/. Accessed 11 Apr 2020.

Bellona. 2009. Verdens første flytemølle åpnet, August 8. Available at: https://bellona.no/nyheter/energi/fornybar-energi/2009-09-verdens-forste-flytemolle-apnet. Accessed 13 Apr 2020.

Buen, Jørund. 2006. Danish and Norwegian Wind Industry. *Energy Policy* 34: 3887–3897.

Christiansen, Atle C. 2002. New Renewable Energy Developments and the Climate Change Issue. *Energy Policy* 30: 235–243.

Dagbladet. 2019a. Strømbløff, March 4. Available at: https://www.dagbladet.no/nyheter/strombloff/70826052. Accessed 13 Apr 2020.

Dagbladet. 2019b. Ikke gi EU mer kontroll over norsk energi, August 29. Available at: https://www.dagbladet.no/kultur/ikke-gi-eu-mer-kontroll-over-norsk-energi/71533233. Accessed 10 Apr 2020.

Dagens Perspektiv. 2011. Borten Moe vil fornye Tyskland med fossil gass, October 30. Available at: https://www.dagensperspektiv.no/borten-moe-vil-fornye-tyskland-med-fossil-gass. Accessed 13 Apr 2020.

Der Spiegel. 2010. Will High Costs Kill Merkel's Green Revolution? September 23. Available at: https://www.spiegel.de/international/germany/the-expensive-dream-of-clean-energy-will-high-costs-kill-merkel-s-green-revolution-a-718951.html. Accessed 13 Apr 2020.

DN [Dagens Næringsliv]. 2013. Det store kraftkuppet, March 9. Available at: https://www.dn.no/det-store-kraftkuppet/1-1-1099565. Accessed 11 Apr 2020.

DNT. 2019. Folket setter naturen fremfor vindkraft, May 27. Available at: https://www.dnt.no/artikler/nyheter/17191-folket-setter-naturen-fremfor-vindkraft/. Accessed 13 Apr 2020.

E24. 2017. Denne kabelen øker strømprisen din, March 26. Available at: https://e24.no/privatoekonomi/i/Ql6M5J/denne-kabelen-oeker-stroemprisen-din. Accessed 13 Apr 2020.

E24. 2019. Hywind Tampen-prosjektet får 2,3 milliarder i Enova-støtte, August 22. Available at: https://e24.no/olje-og-energi/i/VbedX6/hywind-tampen-prosjektet-faar-23-milliarder-i-enova-stoette. Accessed 13 Apr 2020.

E24. 2020. Setter den omstridte NorthConnect-kabelen på vent, March 25. Available at: https://e24.no/olje-og-energi/i/Op7r11/setter-den-omstridte-northconnect-kabelen-paa-vent. Accessed 10 Apr 2020.

Eilertsen, Roar. 2018. EUs energiunion og norsk tilknytning til ACER? *De Facto*, Rapport 1/2018.

Energifaktanorge. 2020a. The Power Market. Available at: https://energifaktanorge.no/en/norsk-energiforsyning/kraftmarkedet/. Accessed 13 Apr 2020.

Energifaktanorge. 2020b. The Electricity Grid. Available at: https://energifaktanorge.no/en/norsk-energiforsyning/kraftnett/. Accessed 13 Apr 2020.

Energifaktanorge. 2020c. Kraftproduksjon. Available at: https://energifaktanorge.no/norsk-energiforsyning/kraftforsyningen/. Accessed 13 Apr 2020.

Filter Nyheter. 2019a. Støre om: • Vindkraft-veto • «Teslaskatt» • Trumps velgere, December 4. Available at: https://filternyheter.no/partienes-klimaplan-jonas-gahr-store/. Accessed 13 Apr 2020.

Filter Nyheter. 2019b. Moxnes: – Det hadde vært enklere å si «drill, baby, drill» og cashe inn oljebransjen, December 19. Available at: https://filternyheter.no/moxnes-det-hadde-vaert-enklere-a-si-drill-baby-drill-og-cashe-inn-oljebransjen/. Accessed 13 Apr 2020.

FriFagbevegelse. 2013. Frykter press på vannkraft, December 16. Available at: https://frifagbevegelse.no/article-6.158.64586.3c3d1102ea. Accessed 11 Apr 2020.

FriFagbevegelse. 2019. Disse LO-tillitsvalgte vil ikke at Norge skal bygge flere strømkabler til utlandet, August 30. Available at: https://frifagbevegelse.no/

nyheter/disse-lotillitsvalgte-vil-ikke-at-norge-skal-bygge-flere-stromkabler-til-utlandet-6.158.646729.419b101c3e. Accessed 10 Apr 2020.

FriFagbevegelse. 2020. LO sier nei til NorthConnect-kabelen til Skottland, March 12. Available at: https://frifagbevegelse.no/lo-sier-nei-til-northconnectkabelen-til-skottland-6.158.685028.005efafde8. Accessed 10 Apr 2020.

Godzimirski, Jakub. 2014. The Norwegian Energy Security Debate. In *The Political Economy of Renewable Energy and Energy Security*, ed. Espen Moe and Paul Midford, 116–136. Houndsmills: Palgrave Macmillan.

Gullberg, Anne Therese. 2013. The Political Feasibility of Norway as the 'Green Battery' of Europe. *Energy Policy* 57: 615–623.

Gullberg, Anne Therese, Dörte Ohlhorst, and Miranda Schreurs. 2014. Towards a Low Carbon Energy Future—Renewable Energy Cooperation Between Germany and Norway. *Renewable Energy* 68: 214–222.

Hager, Caroline P. 2014. Skiftende bris? En casestudie av de største endringene i norsk vindkraftpolitikk i perioden 1990 til 2012. *FNI Report*, 6/2014. Oslo: Fridtjof Nansens Institutt.

Høyre. 2018. Enighet om EUs tredje energimarkedspakke, March 20. Available at: https://hoyre.no/aktuelt/nyheter/2018/enighet-om-eus-tredje-energimarkedspakke/. Accessed 13 Apr 2020.

IEA. 2005. *Energy Policies of IEA Countries: Norway 2005 Review*. Paris: OECD/IEA.

Inderberg, Tor Håkon Jackson, Helga Rognstad, Inger-Lise Saglie, and Lars H. Gulbrandsen. 2019. Who Influences Windpower Licensing Decisions in Norway? *Energy Research & Social Science* 52: 181–191.

IRENA. 2020. *Renewable Capacity Statistics 2020*. Abu Dhabi: International Renewable Energy Agency (IRENA).

Jansen, Alf-Inge, and Per Kristian Mydske. 1998. Norway: Balancing Environmental Quality and Interest in Oil. In *Governance and Environment in Western Europe*, ed. Kenneth Hanf and Alf-Inge Jansen, 181–207. London: Longman.

Jerpåsen, Gro B., and Kari C. Larsen. 2011. Visual Impact of Wind Farms on Cultural Heritage: A Norwegian Case Study. *Environmental Impact Assessment Review* 31: 206–215.

Karlstrøm, Henrik, and Marianne Ryghaug. 2014. Public Attitudes Towards Renewable Energy Technologies in Norway. *Energy Policy* 67: 656–663.

Kelsey, Nina. 2018. Industry Type and Environmental Policy. *Business and Politics* 20 (4): 615–642.

Kjær, Eirik Hovland. 2018. Mellomlandskabler og fornybar energi. Master thesis, Department of Sociology and Political Science, Norwegian University of Science and Technology (NTNU).

Kommunal Rapport. 2019. Vindkraftopprør i emning, May 9. Available at: https://www.kommunal-rapport.no/kommentar/vindkraftoppror-i-emning/110833. Accessed 13 Apr 2020.

Kraftnytt. 2019. 28,4 TWh ny kraftprodusjon siden 2012, May 31. Available at: https://kraftnytt.no/2019/05/31/284-twh-ny-kraftproduksjon-siden-2012/. Accessed 5 Apr 2020.

Lazard. 2019. Lazard's Levelized Cost of Energy Analysis—Version 13.0, November 2019. Available at: https://www.lazard.com/media/451086/lazards-levelized-cost-of-energy-version-130-vf.pdf?fbclid=IwAR3vv8ADl2XB3zErld9JF0BuoOCgTLIdCYhBrDI9tqoNvRzjW0Eau7IRsTI. Accessed 13 Apr 2020.

Linnerud, Kristin, Marianne Aasen, and Merethe Dotterud Leiren. 2019. Åtte råd for mindre strid om vindkraft, August 20. Available at: https://www.dn.no/innlegg/energi/fornybar-energi/vindkraft/atte-rad-for-mindre-strid-om-vindkraft/2-1-656196. Accessed 13 Apr 2020.

Lundberg, Silje Ask, and Maren Esmark. 2019. Vi må redde både naturen og klimaet. *Dagens Næringsliv*, August 7. Available at: https://www.dn.no/innlegg/klima/miljo/naturvern/vi-ma-redde-bade-naturen-og-klimaet/2-1-649243. Accessed 13 Apr 2020.

Manifest Tankesmie. 2019. *For grønt til å være sant?* Industrinotat nummer 1–2019. Available at: http://manifesttankesmie.no/wp-content/uploads/sites/2/2019/03/For-gr%C3%B8nt-til-a%CC%8A-v%C3%A6re-sant-Industrinotat-1-2019.pdf. Accessed 13 Apr 2020.

Marsdal, Magnus E. 2019. Klimakampen trenger grønn industri, March 29. Available at: https://www.dagbladet.no/kultur/klimakampen-trenger-gronn-industri/70925673. Accessed 13 Apr 2020.

Midttun, Atle. 1988. The Negotiated Political Economy of a Heavy Industrial Sector. *Scandinavian Political Studies* 11 (2): 115–143.

Moe, Espen. 2012. Vested Interests, Energy Efficiency and Renewables in Japan. *Energy Policy* 40: 260–273.

Moe, Espen. 2014. Energy Policy and Renewables in Japan, China, Norway and Denmark. In *The Political Economy of Renewable Energy and Energy Security*, ed. Espen Moe and Paul Midford, 276–317. Houndsmills: Palgrave Macmillan.

Moe, Espen. 2015. *Renewable Energy Transformation or Fossil Fuel Backlash.* Houndsmills: Palgrave Macmillan.

Moe, Espen. 2016. 11 Environmental Issues in Norway. In *Small States, Big Challenges*, ed. Kristen Ringdal, Sabrina Ramet, and Danica Fink-Hafner, 271–288. Nomos: Baden-Baden.

Moe, Espen. 2017. Does Politics Matter? Explaining Swings in Wind Power Installations. *AIMS Energy* 5 (3): 341–373.

NRK. 2019a. Utsetter kampen mot vindkraft på land, March 31. Available at: https://www.nrk.no/norge/sv-utsetter-kampen-mot-vindkraft-pa-land-1.14496922. Accessed 13 Apr 2020.

NRK. 2019b. Motstanden mot vindkraft på land aukar, September 17. Available at: https://www.nrk.no/hordaland/motstanden-mot-vindkraft-pa-land-aukar-1.14705126. Accessed 13 Apr 2020.

NRK. 2020. Listhaug rett i opposisjon – lovar omkamp om straum, January 22. Available at: https://www.nrk.no/norge/sylvi-listhaug-vil-prove-a-stoppe-north-connect-i-opposisjon-1.14870712. Accessed 13 Apr 2020.

NVE. 2019. National Report 2019, 33/2019. Available at: http://publikasjoner.nve.no/rapport/2019/rapport2019_33.pdf. Accessed 10 Apr 2020.

NVE. 2020a. Kraftproduksjon. Available at: https://www.nve.no/energiforsyning/kraftproduksjon/?ref=mainmenu. Accessed Apr 2020.

NVE. 2020b. Ressursgrunnlag. Available at: https://www.nve.no/energiforsyning/ressursgrunnlag/?ref=mainmenu. Accessed 10 Apr 2020.

Olson, Mancur. 1982. *The Rise and Decline of Nations*. London: Yale University Press.

Otte, Pia, Katrina Rønningen, and Espen Moe. 2018. Contested Wind Energy: Discourses on Energy Impacts and Their Significance for Energy Justice in Fosen. In *Energy, Resource Extraction and Society*, ed. Anna Szolucha, 140–158. Routledge: London.

Regjeringen.no. 2001. *Statsministerens nyttårstale 2001*, January 1. Available at: https://www.regjeringen.no/no/dokumentarkiv/Regjeringen-Stoltenberg-I/smk/Taler-og-artikler-arkivert-individuelt/2001/statsministerens_nyttarstale_2001/id264461/. Accessed 13 Apr 2020.

Regjeringen.no. 2009. Havenergilova – tilrettelegging for havvindmøller, June 26. https://www.regjeringen.no/no/dokumentarkiv/stoltenberg-ii/oed/Nyheter-og-pressemeldinger/pressemeldinger/2009/havenergilova–tilrettelegging-for-havvi/id570220/. Accessed 13 Apr 2020.

Regjeringen.no. 2019. Enova Supports Equinor's Pilot Project for Floating Offshore Wind, August 29. Available at: https://www.regjeringen.no/en/aktuelt/enova-stotter-equinors-demonstrasjonsprosjekt-for-flytende-havvind/id2666182/. Accessed 13 Apr 2020.

Røynesdal, Jostein, Alexander Strøm Arnesen, and Terje Omland. 2015. Krafteksport en dårlig forretningsidé. *Energi og klima*, December 22. Available at: https://energiogklima.no/kommentar/krafteksport-en-darlig-forretningside/. Accessed 13 Apr 2020.

Schumpeter, Joseph A. 1942. *Capitalism, Socialism and Democracy*. New York: Harper Torchbooks.

Schumpeter, Joseph A. 1983. *The Theory of Economic Development*. London: Transaction Publishers [1934].

Statkraft. 2016. Globale energitrender og norske muligheter. Available at: https://www.statkraft.com/globalassets/1-statkraft-public/_news-2016/statkrafts_lavutslippsscenario_2016.pdf. Accessed 10 Apr 2020.

Statnett. 2011. The Skagerrak 4-Interconnector—Cable Contracts Signed, July 1. Available at: https://www.statnett.no/en/about-statnett/news-and-press-releases/News-archive-2011/the-skagerrak-4-interconnector—cable-contracts-signed/. Accessed 13 Apr 2020.

Statnett. 2019. Et elektrisk Norge – fra fossilt til strøm. Available at: https://www.statnett.no/globalassets/for-aktorer-i-kraftsystemet/planer-og-analyser/et-elektrisk-norge–fra-fossilt-til-strom.pdf. Accessed 13 Apr 2020.

Statnett, Fingrid, Energinet, and Svenska kraftnät. 2019. Nordic Grid Development Plan 2019. https://www.statnett.no/contentassets/61e33bec85804310a0feef41387da2c0/nordic-grid-development-plan-2019-for-web.pdf. Accessed 13 Apr 2020.

Stoltenberg, Jens. 2016. *Min historie*. Trondheim: Gyldendal.

Stortinget. 2019. Innstilling fra energi- og miljøkomiteen om Representantforslag fra stortingsrepresentant Bjørnar Moxnes om å sikre ren energi til miljøvennlig industri gjennom å stoppe NorthConnect, November 26. Available at: https://www.stortinget.no/no/Saker-og-publikasjoner/Publikasjoner/Innstillinger/Stortinget/2019-2020/inns-201920-069s/?all=true. Accessed 12 Apr 2020.

Szulecki, Kacper, Severin Fischer, Anne Therese Gullberg, and Oliver Sartor. 2016. Shaping the 'Energy Union'. *Climate Policy* 16 (5): 548–567.

TU [Teknisk Ukeblad]. 2011a. For dyrt med vindmøller til havs, March 17. Available at: https://www.tu.no/artikler/for-dyrt-med-vindmoller-til-havs/238153. Accessed 13 Apr 2020.

TU [Teknisk Ukeblad]. 2011b. Ola Borten Moe vil ha billigere strøm, March 31. Available at: https://www.tu.no/artikler/ola-borten-moe-vil-ha-billigere-strom/245960. Accessed 13 Apr 2020.

TU [Teknisk Ukeblad]. 2012. Bygg ut fornybar kraft nå! October 18. Available at: https://www.tu.no/artikler/bygg-ut-fornybar-kraft-na/245361. Accessed 13 Apr 2020.

TU [Teknisk Ukeblad]. 2013. Krafteksport kan skade industrien, January 9. Available at: https://www.tu.no/artikler/krafteksport-kan-skade-industrien/233997. Accessed 13 Apr 2020.

TU [Teknisk Ukeblad]. 2014. 25 prosent av kraften i Tyskland kan komme fra Norge, spådde Statnett-sjefen, November 21. Available at: https://www.tu.no/artikler/25-prosent-av-kraften-i-tyskland-kan-komme-fra-norge-spadde-statnett-sjefen/230586. Accessed 13 Apr 2020.

TU [Teknisk Ukeblad]. 2016. Derfor tok ikke offshore vind-industrien av i Norge, October 21. Available at: https://www.tu.no/artikler/derfor-tok-ikke-offshore-vind-industrien-av-i-norge/359574. Accessed 13 Apr 2020.

TU [Teknisk Ukeblad]. 2017. Mange har vært villige til å ta på seg farskapet – her er historien om Hywinds trange fødsel, October 21. Available at: https://www.tu.no/artikler/mange-har-vaert-villige-til-a-ta-pa-seg-farskapet-her-er-historien-om-hywinds-trange-fodsel/410309#cxrecs_s. Accessed 13 Apr 2020.

TU [Teknisk Ukeblad]. 2019a. Her er Norges strømkabler til utlandet, February 21. Available at: https://www.tu.no/artikler/her-er-norges-stromkabler-til-utlandet/457474?key=j5Qb4Alc&fbclid=IwAR0SpQoDguJIMleiCT6T1ggsf3iNta9HCfssCLtskJ8zHeNQZlBIQ4mU_2c. Accessed 13 Apr 2020.

TU [Teknisk Ukeblad]. 2019b. Massiv motstand mot vindkraft: kun 5 av 101 kommuner sier ja, September 17. Available at: https://www.tu.no/artikler/massiv-motstand-mot-vindkraft-kun-5-av-101-kommuner-sier-ja/474164. Accessed 13 Apr 2020.

TU [Teknisk Ukeblad]. 2020. Listhaug lover ny kamp mot vindkraft og kraftkabler, January 20. Available at: https://www.tu.no/artikler/listhaug-lover-ny-kamp-mot-vindkraft-og-kraftkabler/483283. Accessed 13 Apr 2020.

TV2. 2011. Vil gjøre Norge til Europas grønne batteri, January 20. https://www.tv2.no/a/3393390/. Accessed 13 Apr 2020.

Unruh, Gregory C. 2000. Understanding Carbon Lock-in. *Energy Policy* 28: 817–830.

VG. 2019a. Kompromiss i bitter vindmøllestrid i MDG, May 25. Available at: https://www.vg.no/nyheter/innenriks/i/vQxoWL/kompromiss-i-bitter-vindmoelle-strid-i-mdg. Accessed 13 Apr 2020.

VG. 2019b. Erna Solberg: –Jeg forstår mølle-motstanden, July 24. Available at: https://www.vg.no/nyheter/innenriks/i/Jo3xom/erna-solberg-jeg-forstaar-moelle-motstanden. Accessed 13 Apr 2020.

VG. 2019c. KrF vil gi kommuner vetorett i omstridt vindkraftplan, August 5. Available at: https://www.vg.no/nyheter/innenriks/i/P9EJP0/krf-vil-gi-kommuner-vetorett-i-omstridt-vindkraftplan. Accessed 13 Apr 2020.

VG. 2019d. Nå vil de selge ut strømmen vår, December 11. Available at: https://www.vg.no/nyheter/meninger/i/mREA1l/naa-vil-de-selge-ut-stroemmen-vaar. Accessed 10 Apr 2020.

Vindportalen. 2021. Vindkraft i Norge. Available at: https://www.vindportalen.no/. Accessed 6 Jan 2020.

WindEurope. 2019. Our Energy, Our Future: How Offshore Wind Will Help Europe Go Carbon-Neutral. Available at: https://windeurope.org/wp-content/uploads/files/about-wind/reports/WindEurope-Our-Energy-Our-Future.pdf. Accessed 13 Apr 2020.

Interviews

Terje Aasland, Labor Party, Parliamentary Standing Committee on Energy and the Environment.

Jan Olav Andersen, General Secretary, EL og IT Forbundet.

Steinar Bysveen, Executive Vice President, Statkraft.

Magne Fauli, Industrial Policy Adviser, Energi Norge.

Oskar Jarle Grimstad, Progress Party, Parliamentary Standing Committee on Energy and the Environment.

Lars Haltbrekken, Socialist Left Party, Parliamentary Standing Committee on Energy and the Environment.

Karianne Hansen Heien, Union Branch Secretary and Energy Adviser, EL og IT Forbundet.

Vidar Hennum, Vice-chairman, EL og IT Forbundet.

Odd Henriksen, Conservative Party, Former Parliamentary Standing Committee on Energy and the Environment Member.

Kjetil Kjenseth, Liberal Party, Leader of the Parliamentary Standing Committee on Energy and the Environment.

Tord Lien, Regional Director Confederation of Norwegian Enterprise (NHO) Trøndelag, Former Petroleum and Energy Minister (2013–16, Progress Party).

Ole Løfsnæs, Technical Director, Norsk industri.

Oddvin Lund, Head of Nature Management, The Norwegian Trekking Association (Turistforeningen).

Odd Øygarden, Director and Chairman, NorthConnect.

Leif Sande, former Union Leader, Industri og energi.

Erik Skjelbred, Director of Customer and Public Affairs, Statnett.

Einar Wilhelmsen, Former Head of Energy Systems, Zero.

CHAPTER 13

Beyond Wind: New Challenges to the Expansion of Renewables in Denmark

Luis Boscán, Brooks A. Kaiser, and Lars Ravn-Jonsen

Introduction

Denmark functions as a world-class example in the transformation of energy systems and in the integration of variable renewable energy sources into electricity systems. Over four decades, the country has transitioned from a fossil-fuel dependent energy economy to a renewable powerhouse. The share of renewables in Denmark's gross final energy consumption has gone from 1.4% in 1972 to 32.6% at the end of 2017 (Danish Energy Agency 2018d).

The views and opinions expressed in this article are those of the authors and do not necessarily reflect the official position of any agency of the Danish government.

L. Boscán (✉)
Danish Utility Regulator (DUR), Frederiksværk, Denmark
e-mail: boscan@sam.sdu.dk

L. Boscán · B. A. Kaiser · L. Ravn-Jonsen
University of Southern Denmark (SDU), Esbjerg, Denmark

P. Midford and E. Moe (eds.), *New Challenges and Solutions for Renewable Energy*, International Political Economy Series,
https://doi.org/10.1007/978-3-030-54514-7_13

Danish wind power development, which started in 1977, is a visible facet of this success. At the end of 2018, installed wind power capacity was 6.1 GW, accounting for more than one-third of installed electricity generation capacity (Danish Energy Agency 2018c, 2019). At the end of 2017, wind power production accounted for 43.2% of gross domestic electricity supply, setting a new record; this represents an increase of more than 20 percentage points compared to 2010 (Danish Energy Agency 2018c). Growth continued in 2019 with another increase to 46.9%. Windmills now out-produce domestic electricity demand frequently. In 2018, the highest ratio of wind power production to domestic electricity demand for an hour was 149%, reached at 4:00 (UTC + 1) on September 22, 2018 (Energinet 2019). Improved electricity interconnections enabled excess production exports to Denmark's neighbors (Norway, Sweden, and Germany).

However, beyond wind, there are several less visible aspects at the core of Denmark's success. These include energy planning and organized community participation. Market and infrastructure integration with neighboring countries, which possess fuel and technology mixes that complement Denmark's, also contribute. Similarly, natural resource development, including oil and gas, has enabled sustainable and secure supply.

When the first oil crisis broke out (October 1973), Denmark had recently begun its first production of oil and gas (the Dan field) and was still reliant on imported oil. Domestic natural gas production was insignificant and only used for own consumption in the North Sea platforms, which were at the exploration stage. Yet in 1972, oil represented 92% of gross final energy consumption, 78% of fuels used in electricity production, and 75% of final energy consumption for heating purposes. In contrast, at the end of 2017, oil accounted for 37% of gross final energy consumption and only 1.4% of all fuels used in electricity production (Danish Energy Agency 2018d). The Danish government's first energy plan (1976) responded directly to the crisis. It proposed a shift to coal and nuclear. The plan laid out the foundations for a country-wide natural gas network and established a national heating planning system at the municipality level, with a focus on energy efficiency and socio-economic viability. Popular rejection of nuclear generation led to a decision in 1985 to forego it (Danish Energy Agency 2016; Meyer 2000), but otherwise these plans came into action and shaped Denmark's long-term strategy toward independence from fossil fuels.

Here we focus our attention on Denmark as a case study of energy transformation. To this end, we first present a concise and recent history of Denmark's energy economy and policy. Our history reveals that over the years, Denmark has successfully combined centralized planning of the energy sector together with market integration in an ad hoc manner. Then we identify and investigate four of the most critical challenges for the present and future of the ongoing Danish transition. These are: greening the Danish transportation sector, reforming the Danish district heating sector while increasing the share of renewables in heat production, removing the cost barriers in the usage of electric transportation and heating, and finally, meeting the increasing flexibility requirements in conjunction with dwindling domestic backup capacity. In the third section, we identify both transferrable and non-transferable lessons for other countries. Finally, we conclude with a series of policy recommendations useful for Denmark and other countries aiming at transforming their energy systems.

A Concise and Recent History of Denmark's Energy Economy and Policy

In 1972, Denmark's gross fuel consumption was highly concentrated: 92% was oil. It was the most important fuel source for space heating (78%), electricity generation (75%), and transportation (100%). This meant that the first oil crisis (October 1973), which began when the Organization of Petroleum Exporting Countries (OPEC) declared an oil embargo, hit Denmark particularly hard (Sovacool 2013). Average international crude oil prices (non-deflated) more than tripled in a very short period, going from 3.29 USD/barrel in 1973 to 11.58 USD/barrel in 1974 (BP 2018). As an immediate reaction, Denmark decreed car-free Sundays, imposed stricter speed limit restrictions on motorists, ordered crude oil savings for industries and transport, and promoted energy conservation during the crisis' winter months.

The crisis gave momentum to the energy policy debate in Denmark. For the first time, there was an initiative to define an energy plan with clear goals over a two-decade timeframe. Shifting from oil to coal and developing nuclear power were leading components of the Danish government's first energy plan (1976). In 1974, the government had announced nine possible nuclear sites in the country, proposing to deploy 6.2 GW between 1985 and 1999. Another initiative of the first plan was developing a nationwide network for North Sea natural gas. In 1962, A.P. Møller Mærsk obtained an exclusive concession to explore and produce oil and gas in Denmark, leading to a first discovery in 1966 (Kraka field). Crude oil production began in 1972 (Dan field), leading to the first exports in 1977. In 1976, the state-owned Danish Oil and Gas (DONG) company gained pre-emption rights to North Sea output, and in 1979 the parliament passed a law governing the domestic sale and export of natural gas, which began in 1984 (Danish Energy Agency 2016; Meyer 2000; Mortensen and Overgaard 1992). Boosted by North Sea oil and gas exports, Denmark became self-sufficient in energy for the first time in 1997. Oil self-sufficiency and natural gas self-sufficiency peaked in 2004 and 2008, respectively (see Fig. 13.1). Presently, as a net oil exporter, Denmark has a stockholding obligation as an EU Member State for energy security purposes (International Energy Agency 2017).

A third element was implementing heat planning in all Danish municipalities, with the goal of reducing dependence on oil and increasing

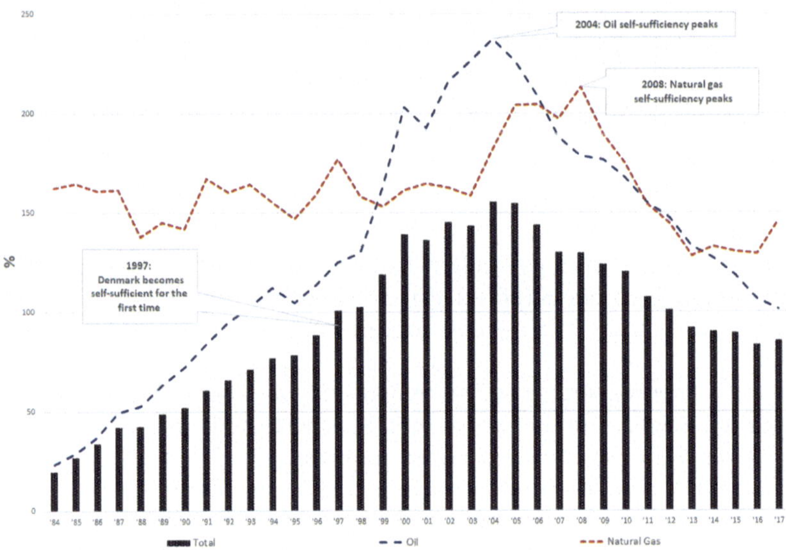

Fig. 13.1 Denmark's self-sufficiency in oil and gas (primary energy production/gross energy consumption), 1975–2017 (*Source* Author created figures based on data from Danish Energy Agency [2018d])

energy efficiency. The heat planning goals were legally operationalized in 1979. This initiated a radical transition from individual usage of oil to collective supply systems based on natural gas and district heating. According to the first energy plan estimates, 800 thousand individual boilers needed to switch to natural gas and coal (Mortensen and Overgaard 1992; Danish Energy Agency 2016). The law mandated that municipalities, in cooperation with locally owned heating utilities, estimate and plan local heat requirements and methods that would substitute oil and could harness surplus heat from different activities. Zoning and planning occurred between 1981 and 1982 and constituted the basis for district heating systems; these now cover 64% of Danish households (see Fig. 13.2).

The 1970s ended in Denmark with policies and institutions that would shape the following decades. Taxation of electricity and other energy products was introduced during this period (1977), sending price signals to incentivize energy conservation and accruing funds for Research and Development (R&D) in alternative sources of energy. Once oil

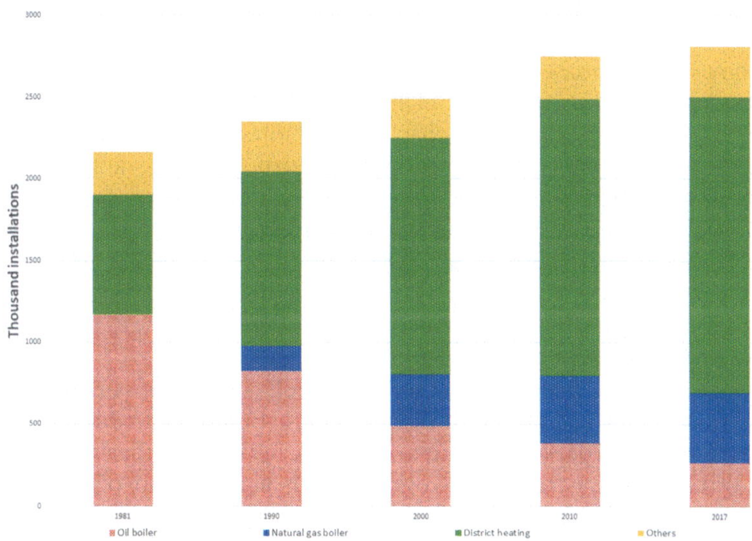

Fig. 13.2 Number and type of heating installations in Danish dwellings (*Source* Author created figure based on data from Danish Energy Agency [2018c] [primary statistics originally produced by Statistics Denmark])

prices dropped in the 1980s, Danish policymakers responded by further increasing taxes to consolidate the transition away from oil (Sovacool et al. 2008; Sovacool 2013). The creation of the Danish Energy Agency (in 1976) and the Energy Ministry (in 1979) institutionalized energy policy-making. Except for the reconsidered nuclear power, all the initiatives put forward in the first energy plan were set in motion within a decade's time.

The second energy plan (1981) set goals of securing cost-efficient energy supplies for society, increasing security of supply, and decoupling energy consumption from economic growth (Meyer 2000). From there, the Danish economy has grown while reducing energy consumption, carbon emissions, and increasing efficiency. Today Denmark's energy intensity and carbon intensity are among the lowest of the 30 states of the International Energy Agency (IEA) (International Energy Agency 2017).

Among other things, the second energy plan proposed increasing the extraction rate with extraction by the Danish state in North Sea oil and gas, increasing energy efficiency standards in new building projects, and securing resources for R&D. The plan continued with increasing district heating and promoted the expansion of Combined Heat and Power (CHP) plants, including decentralized CHP. Later, in 1985, the Danish parliament introduced adjustments to the initial heat planning act: domestic resources would have priority in heat planning. Natural gas, surplus heat from electricity plants, and biomass (straw, waste, and wood) would be prioritized over coal. Furthermore, a still present feature of Danish energy policy was introduced at that time: all new power plants must include the possibility of cogenerating heat and power. Only by exception, the government would authorize building condensation (i.e., electricity-only) thermal units (Mortensen and Overgaard 1992). The Danish government also prohibited electric heating in 1988 in areas served by district heating and natural gas networks, effectively granting monopoly power to district heating suppliers.

Cogeneration has been a key factor in the achievement of energy efficiency in Denmark and the country is, by far, a European leader in this respect (Sovacool 2013). At the end of 2016, 69% of all thermal electricity production was produced together with heating, and 65% of all heating production was simultaneous with thermal electricity production. Similarly, energy efficiency measures in buildings (insulation, new heating installations) have increased the heated area in Danish dwellings with a simultaneous decrease of final energy consumption per square meter. Final energy consumption for heating purposes (space and water heating) has also remained relatively stable.

The second energy plan also promoted wind power and electricity interconnections with neighboring countries and continued the existing investment subsidy schemes for wind turbines, available since 1979. The government initially covered 30% of the capital costs of new wind turbines, which had to be certified by a national test center (International Renewable Energy Agency 2013; Hvelplund et al. 2017). These grants were progressively reduced to 20%, then to 10%, and eliminated in the late 1980s. They supported the Danish wind industry during the collapse of the California market for wind turbines during the mid-1980s (Hvelplund et al. 2017). As early as 1981, Denmark introduced feed-in tariffs for wind power output. These were fixed at a discount relative to the electricity retail price paid by a 20,000 kWh/year consumer (Sovacool et al. 2008; Hvelplund et al. 2017). By the late 1970s, local wind cooperatives, where organized community groups—e.g., farmers—invested together in wind turbines, were emerging (International Renewable Energy Agency 2013). The number of cooperatives involved in community ownership exceeded two-thousand by the late 1990s (Oteman et al. 2014).

Nonetheless, the government's stance remained favorable for nuclear power. The Danish public debate on nuclear power (1974–1985) gained considerable attention and exemplifies how Danish grassroots political movements succeeded in reversing an initially favorable consensus for the nuclear option in the Danish parliament. In line with Denmark's democratic tradition of organizing public participation and allowing dissent, the Danish government sponsored an information campaign where both camps could raise their arguments. The technical and scientific community was divided between enthusiasts and moderates, who advocated for nuclear power based on the possibility of having long-term clean energy supplies. The main arguments raised against nuclear power were the risk of irreversible accidents, the lack of safe radioactive waste disposal facilities, and the risk of fissile material ending in terrorists' hands. Later, underestimated decommissioning costs were also brought up as relevant issues against nuclear power (Danielsen 2005; Meyer 2000). The nuclear debate also resulted in renewable energy sources such as wind and solar becoming increasingly present in the public opinion.

Once the question mark about nuclear power was removed in 1985, the Danish government embraced wind power development in a more systematic manner. One early measure was the "100 MW agreement" of 1985 between the Energy Ministry and electric utilities, which aimed at developing 100 MW between 1986 and 1990. New agreements came later: in 1990 for 100 MW onshore wind capacity to be added before 1994, and in 1996 for 200 MW onshore wind capacity by the year 2000. Denmark also had the first offshore wind farm in the world, which began operating in Vindeby, Denmark in 1991 (Danish Energy Agency 2016). Figure 13.3 shows the evolution of wind power in Denmark from 1977 to 2018.

Since the early days of wind power development in Denmark, governmental support for renewables has aimed at providing certainty for investors and reducing cost barriers, while facilitating community ownership and participation. In this respect, Oteman et al. (2014) identify three types of community initiatives in Denmark. First, community district heating employs municipalities or consumer cooperatives of distinct size to invest and manage heating networks on a not-for-profit basis. Second, the wind power cooperatives are usually formed by farmers, household owners, and local investors. This bottom-up approach has proven crucial to public acceptance. Third, there is support for a diverse array of community, experimental-stage initiatives in which people organize around the development of community-backed investments for energy efficiency and innovative technologies.

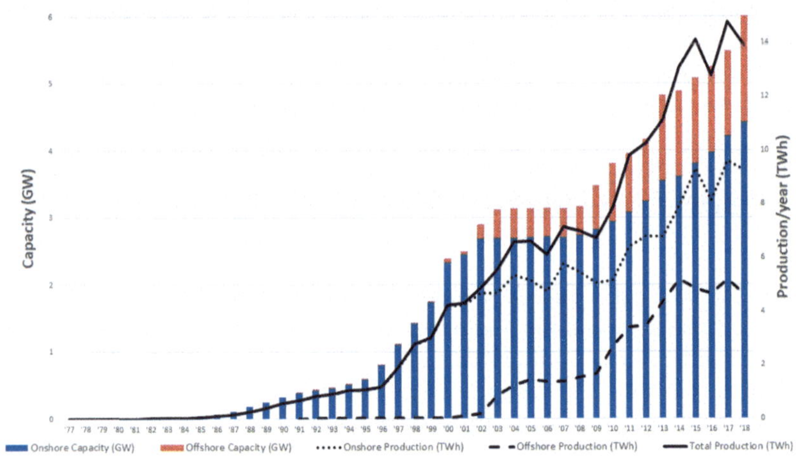

Fig. 13.3 Trends in installed wind power capacity (GW) and annual output (TWh/year), 1977–2018 (*Source* Author created figure based on data from Danish Energy Agency [2019])

Community participation in wind power development has become enshrined into law in Denmark (Oteman et al. 2014). The current law on renewable energy (*VE-loven*) requires that project developers must offer a 20% ownership share to residents in the area where the wind project is developed. Members of the local community are also granted protection against household valuation losses, in case the value of the household falls more than 1% because of siting windmills. Local developers can obtain state loans to finance feasibility studies for the siting of wind projects. Finally, municipalities that host wind power projects obtain a grant per installed MW, which must be geared toward cultural initiatives and "green" projects.

Electricity market reform was first implemented in Denmark in 1999. Key pre-market reform economic incentives for producers included a combination of power purchase agreements, production subsidies, tax credits, and carbon taxation refunds. Thus, during the first two decades of Danish wind power development, producers obtained an almost fixed feed-in tariff without exposure to market conditions, in which approximately half of the payments were governmental subsidies. The high feed-in tariff meant it was quite profitable to establish wind turbines in Denmark (Meyer 2004; Morthorst 2000).

Following electricity market reform, Denmark proposed to adopt a tradeable certificate market, together with renewable quotas imposed on selected consumers. Wind producers' revenue would be the sum of electricity prices plus the selling price of the tradeable certificate. However, the system was never put in practice. An EU-level decision determined that feed-in tariffs were not in conflict with EU state aid rules or the EU internal market in electricity. Instead, a series of transitional renewable support schemes were put in place, creating considerable uncertainty among investors (Meyer 2004). In fact, the number of onshore wind turbines peaked in Denmark in 2001 and capacity additions stagnated between 2003 and 2008, with a slight decrease in installed capacity happening between 2006 and 2007—see Fig. 13.3 (Danish Energy Agency 2016). In parallel, both the number of offshore wind installations and installed capacity kept growing during the 2000s.

One possible explanation for the relatively better performance of offshore wind development from the late 1990s onwards is the simplification and transparency of siting, tender processing, and permitting. According to the Danish framework, there is a one-stop-shop concept, where the Danish Energy Agency is the main point of contact for investors. Following the political decision to build an offshore wind farm, in consultation with municipalities regarding siting, a technical dialogue begins with developers. Second, pre-qualification is granted. Third, an environmental assessment is performed. Fourth, the tendering process takes place and winning bids, in which producers obtain a Contract-for-Difference over the lifetime of the project, are determined competitively. Fifth, the Danish Energy Agency acts as coordinator with remaining state institutions. Finally, the production stage begins (State of Green 2017).

With the advent of a new Danish energy plan put forward in 2007 and a new energy policy agreement, the support scheme for wind producers changed again. According to the new framework (a capped feed-in premium), wind producers' revenue would be the sum of market prices plus a premium. This arrangement exposed wind producers to electricity market conditions but, at the same time, renewed confidence in investors. The 2007 agreement was followed by the 2012 energy policy agreement, with the official long-term energy policy goal of attaining 100% renewables by 2050 across the entire energy supply (electricity, heating, industry, and transport). By the end of 2020, at least half of all electricity demand will be covered by wind power, a goal that is within reach. By 2030, coal shall be phased out, and electricity and heating shall be entirely renewable-based by 2035 (Danish Ministry of Energy Utilities and Climate 2012; Hvelplund et al. 2017). Several energy companies in Denmark have announced voluntary decisions to phase out coal usage.

Recent decisions reinforce the earlier agreements. Following an open consultation, in June 2018 the political parties in the Danish Parliament reached a new energy policy agreement aimed at 2020–2030. The agreed policy and financing framework pave the way to achieving 55% renewables in total final consumption by 2030, and more than 100% of renewables as a share of energy consumption. Furthermore, at least 90% of non-fossil fuels shall be used in district heating (Danish Ministry of Energy Utilities and Climate 2018a).

What Are the Next Challenges for Denmark's Green Transition?

Despite Denmark's significant progress, the ambitious 2030 and 2050 goals set by policymakers mean that policies must adapt further to deepen and consolidate the transition to a low-emission society and a renewable-based energy economy. We identify four key challenges facing Denmark with regard to the next stages in its energy transition.

Challenge 1: Greening the Danish Transportation Sector. The Danish transportation sector—particularly, road transportation—is considerably less green than other energy sectors. Between 1990 and 2017, Danish total final energy consumption grew by 28%, with road transportation (which accounts for 74% of all transport-related consumption) increasing by 25% and international aviation (which accounts for 19% of the total) rising by 64% (Danish Energy Agency 2018d). At the same time, transport-associated CO_2 emissions grew by 23% (see Fig. 13.4), and in the period between 1990 and 2017, greenhouse gas (GHG) emissions increased by 20.6%. Not only is transportation the largest oil-consuming sector in Denmark, but it also is the sector that uses the least electricity among all final energy-consuming sectors (International Energy Agency 2017). As of 2017, fossil fuels (95%) dominated transportation, and the 10% target of biofuel blending by 2020, established in the 2012 Energy Policy Agreement, will be likely missed. While "decarbonization of the

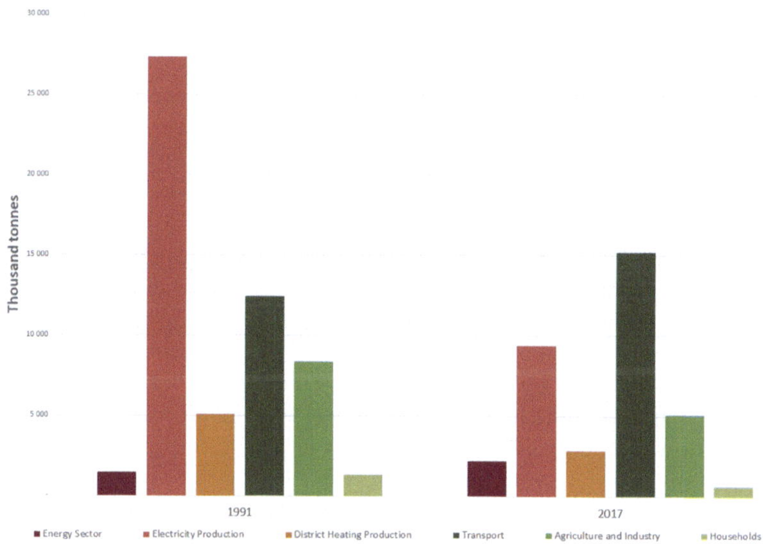

Fig. 13.4 CO_2 emissions from major sectors of the Danish economy (thousand tonnes, adjusted) (*Source* Author created figure based on data from Danish Energy Agency [2018d])

electricity and heat sector" is a tangible reality in Denmark, the simultaneous increase in fossil fuel consumption and emissions in transportation has blunted the overall decrease. Between 1990 and 2017, CO_2 emissions from electricity and heat generation decreased by 65% and 43, respectively, while on the other hand transport-related emissions grew.

Transportation is now the single largest emitter of the Danish energy economy. It accounted for 40% of all CO_2 emissions and 24% of all GHG emissions by the ends of 2017 and 2016, respectively (Danish Energy Agency 2018c, d). In comparison, the entire conversion sector (which includes electricity generation and district heating) was responsible for a lower share of CO_2 emissions (32%) than transportation. Similarly, the share of GHG emissions from transportation (24%) was (slightly) exceeded only by primary energy production and conversion together (25%), followed by agriculture (19%) (see Fig. 13.4). If Denmark is serious about its energy transition and its goal of a fossil-free transportation system in 2050, it must establish clearer targets and ambitions and better incorporate road transportation, which accounts for three-quarters of all fuel consumption in transportation (Danish Energy Agency 2018c, d).

Fortunately, Denmark departs from an already advantageous position, as it implements several ambitious policies with regard to transportation. For example, in accordance with EU level regulations, it applies CO_2 labeling and standards for new vehicles and has explicit targets such as the reduction of the GHG emission intensity of vehicles by 6% in 2020. At a national level, there are several other policies and investment programs. The country boasts a highly developed public transportation network; continued expansion and innovation are priorities. Railway operations are in the process of becoming fully electrified. The construction of the City Ring within the Copenhagen Metro system and other light railway systems in major Danish cities like Aarhus and Odense are well underway. The adoption of energy efficiency measures in public transportation and the fuel conversion of some bus networks in major cities are also important examples (International Energy Agency 2017). Denmark also has a strong cycling culture, with an average 1.6 kilometers cycling distance per citizen per day and a nationally well-developed cycling infrastructure (Cycling Embassy of Denmark, n.d.). Additionally, the taxation system—which consists of a vehicle registration tax and a green owner's tax (i.e., a fuel consumption tax)—is complex and evolving, but overall encourages the purchase and utilization of fuel-efficient vehicles. This includes plug-in hybrid, battery electric, and fuel-cell vehicles (Danish Council on Climate Change 2018).

Nevertheless, relative to other comparable countries, existing policies have not been successful at increasing electric vehicle (EV) sales.[1] Relative to its Nordic neighbors, which are world leaders, Denmark, at 0.6% market share (see Fig. 13.5), is well behind both in market share and total EV stock. Norway led the world ranking in EV market share (39.2%) during 2017, followed by Iceland (11.7%) and Sweden (6.3%).

One possible explanation for the failure in the promotion of cleaner private transportation in Denmark is the undecided nature of the Danish government's policies. Before 2016, all new plug-in hybrid electric vehicles (PHEVs) and battery EVs (BEVs) were exempt from the registration tax, but after this date the tax is gradually increasing until it becomes on a par with every other vehicle sold in Denmark (Danish Council on Climate Change 2018). In the meantime, in 2017, the general registration tax was reformed for all vehicles in Denmark, making it more attractive for private drivers to choose fossil-fueled cars over EVs. An incentive scheme to hasten adoption set the registration tax for PHEVs and BEVs at 20%

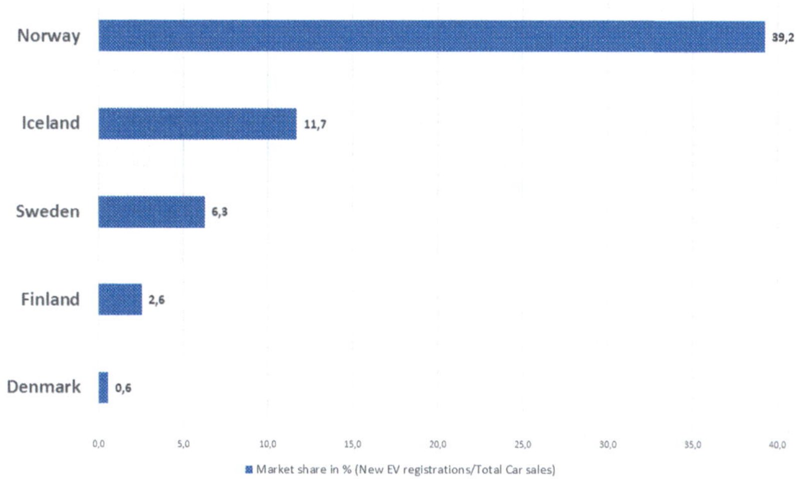

Fig. 13.5 New electric vehicle registrations in the nordic countries as a share of total car sales (*Source* Includes BEV and PHEV vehicle registrations. Author created figure based on data from the International Energy Agency [2018b])

of the normal registration tax between 2016 and 2018, for the first 5 thousand units sold in the two-year period. This failed; in 2017, only 1.3 thousand of 221 thousand new vehicle sales were EVs, and at the end of 2018, only 3.5 thousand new BEVs had been sold in Denmark.

In response to the failure, and to complement the 2018 energy policy agreement, the Danish government has put forward a comprehensive set of proposals to counter climate change and air pollution (Danish Ministry of Energy Utilities and Climate 2018b). One key proposal is to stop selling new gasoline- and diesel-fueled vehicles by 2030, and to do the same with respect to plug-in hybrid EVs by 2035. From 2025, all new taxis and buses shall be emission-free, such that by 2030 the entire taxi and bus fleet shall produce no emissions. Together, this creates an approximate target of 1 million EVs running in Denmark by 2030. As immediate measures, in late 2018, the Danish Parliament decided to continue with the 20% registration tax policy until the end of 2020 for most cars, effectively leaving the least expensive varieties of cars paying essentially no tax (FDM 2018; Statistics Denmark 2019). These tax reductions will be complemented by other measures, such as reduced parking fees, the opportunity to drive in bus lanes, an improved charging infrastructure, and an increased scrapping reward for old diesel-fueled vehicles.

While the impact of the measures on drivers' choices can be uncertain, one well-known barrier for the implementation of tax exemptions or subsidies is the financial dependence of the government on private transportation-related taxes. Presently, approximately 5% of all its proceeds come from this source and the government recognizes the tradeoff between increasing energy efficiency and the availability of tax income (Danish Ministry of Energy Utilities and Climate 2018b). The electrification of transportation requires careful consideration of the policy costs involved.

Challenge 2: Reforming the District Heating Sector and Increasing the Share of Renewables in the Production of Heat. In international comparisons of district heating systems, Denmark appears to be a leader. Euroheat & Power (2015)—an international district energy organization—highlights that Denmark ranks third in terms of the share of citizens covered (63%), exceeded only by Iceland (92%) and Latvia (65%). Denmark also ranks third in direct renewable usage, which includes the use of renewable energy in heat-only boilers and installations other than CHP. According to this metric, only Iceland (92%), which possesses unusually favorable conditions for geothermal energy, and

Norway (76%), where individual, hydroelectricity-based heating, rather than district heating, is widespread, exceed Denmark. In countries with similar district heating coverage and weather conditions to Denmark, the usage of renewables tends to be lower. For example, among the Baltics, Latvia has a direct renewable usage of only 12% and Lithuania (with 57% coverage) uses only 20% of renewables. Among Denmark's Nordic neighbors, Sweden has 52% coverage but only 28% of direct renewable energy usage, and Finland's coverage reaches 50% of the population but only uses 9% of renewables.

Notwithstanding its position relative to comparable countries, district heating remains a challenging area for further green transition in Denmark because the country has set a high level of ambition. In Denmark, despite its shift to renewables, the aggregate share of oil, natural gas, and coal still accounted for 31% of all fuels used in the production of district heating in 2017. This is despite the share of coal decreasing to 11% from the peak of 46% in 1991, and the share of renewables increasing from just 7% in 1975 to 59% (Danish Energy Agency 2018d). Most of this gain in renewables is a consequence of a higher biomass usage. Since 2000, total biomass usage (in energy conversion and final usage) has more than doubled, primarily through increased use of wood chips, wood pellets, and waste wood. At the end of 2017, biomass represented 92% of all renewables used in the production of district heating (Danish Energy Agency 2018c).

Though renewable, wood biomass energy production is not carbon neutral. Trees remove CO_2 from the atmosphere; burning them returns it there. Even if cut trees are replanted, managed forests store less carbon than their native counterparts. Furthermore, the emission-reabsorption cycle may take years to complete (Danish Council on Climate Change 2018; Schlesinger 2018). In response to these concerns, the Danish Energy Association and the Danish District Heating Association—lobby organizations—entered a voluntary agreement in 2014 on the usage of sustainable biomass (Danish Energy Association and Danish District Heating Association 2016). According to the agreement, electricity and generation plants using biomass in Denmark shall require suppliers to meet minimum sustainability standards, and Danish generators' wood purchases shall be documented and audited by a third party.

Beyond the non-neutral impact on the environment, another source of concern for Denmark's biomass use is security of supply. A significant share (39% in 2016) of the biomass used in Denmark is imported from the Baltic countries (Estonia and Latvia) and Russia (International Energy Agency, 2017; Danish Council on Climate Change 2018). Relying on biomass creates competition among countries for an exhaustible resource, which may make Denmark vulnerable to price and supply-side swings (Hvelplund et al. 2017). Ensuring diversity and quality of biomass supplies is crucial to minimize the vulnerability and the environmental impact of Denmark's district heating sector.

More generally, the convenience of biomass usage is a priority topic in Denmark's public policy debate. Proponents of biomass admit that it is a temporary greener alternative to coal usage for both baseload electricity and heat generation, given the intermittent nature of variable renewables, like wind power. They also suggest that alternative heating technologies, such as large-scale heat pumps, solar and geothermal heating, or the large-scale usage of excess industrial heating require further time to mature (Danish Energy Association and Danish District Heating Association 2017). Opponents, on the other hand, consider that recently proposed policy changes may erode the economic viability of the Danish district heating sector and that these may prove insufficient to promote true decarbonization of district heating.

Decarbonization through district heating electrification is a goal of the 2018 Energy Policy Agreement and is part of a broader effort to "modernize" Denmark's district heating sector (Danish Ministry of Energy Utilities and Climate 2018a). A central measure is to eliminate the long-standing cogeneration requirement (introduced in 1986), which has been a cornerstone of Danish energy efficiency, and the obligation to use natural gas in the "minor" district heating areas.[2] Through 2018, the establishment or expansion of district heating in an area had to comply with the fuel choice constraints imposed by law. If the project involved cogeneration, essentially all fuel types could be chosen. However, if the project was for exclusive heat production and the area is served by the natural gas network, then natural gas must be the chosen fuel. Otherwise, biomass, waste, biogas, landfill gas, and other gasified biomass were the available options (Danish Energy Agency 2017). Since January 2019, district heating companies operate under new regulatory conditions which make it easier for district heating firms to switch to renewable sources—heat pumps in particular—from natural gas. In line with the centralized, socio-economic planning model that has existed for decades in Danish district heating, firms wishing to switch from natural gas into a biomass-fired solution must demonstrate sufficient end-user savings relative to an alternative solution which, in practical terms, typically is a heat pump. The government also waives the end-user economy condition for district heating suppliers who either choose a heat pump or combine a biomass solution with a heat pump, to advance its decarbonization-by-electrification agenda (Danish Energy Agency 2018a).

The relaxation of the cogeneration and fuel choice requirement comes into force alongside the expiry of a long-standing subsidy for secondary (decentralized) natural gas-fired CHP plants, the so-called *grundbeløb* (basis payment). Since market liberalization (2004), gas-fired CHP plants have received a subsidy to compensate for lower revenues and to incentivize the availability of dispatchable capacity for electricity generation. The subsidy has been designed such that when the wholesale electricity price falls, the available subsidy increases. However, according to EU state-aid rules, the subsidy is considered distortionary to market competition and must be eliminated (International Energy Agency 2017).

Another change will allow district heating customers to disconnect from the existing heating networks progressively, facilitating individual over collective heating. Until the end of 2018, municipalities could impose a requirement on new or existing buildings to connect to the district heating network, with an obligation to pay a fixed fee, and mandating buildings to use heat from the network (Danish Energy Agency 2017). Since January 2019, the Danish government has the prerogative to negotiate with municipalities in this respect.

The modernization effort also involves restructuring taxes. Reducing the general electricity consumption tax and analyzing the possibility of introducing dynamic electricity taxation, i.e., electricity taxes that depend on the time of the day, promotes more efficient responses to demand and supply balance needs. Second, taxes for electricity-based heat production will be reduced (effective in 2021). Third, District Heating companies' taxation framework will be modified so that investments on large-scale heat pumps are treated on equal footing as biomass, biogas, geothermal, and industrial surplus heat. Fourth, there is restructuring the taxation framework to incentivize the usage of surplus heat. Presently, industries selling surplus heat pay a 33% tax rate on the total value of the heat.

If Denmark is committed to electrification to decarbonize its district heating sector, then the modernization agenda points in the right direction, as it reduces the considerable tax disadvantage of electricity with respect to biomass in heat generation (International Energy Agency

2017). Yet, unless specific technical assumptions and calculations are made, it is not possible to determine the effectiveness of the reform on the re-organization of the merit order of heating production. As a point of reference, Hvelplund et al. (2017) explain that, before the tax reform, biomass-fueled heat production paid almost no taxes, whereas electricity-based heat production paid approximately 100 EUR/MWh in tax. Thus, 1 MWh of wind power for heat production costs (including tax) 123 EUR, whereas biomass costs only 28 EUR.

However, the modernization effort hides several risks that should be assessed at a system-wide level. The main benefits of heat electrification, such as adopting highly efficient technologies like heat pumps and relying on an already decarbonized energy carrier (i.e., wind power), are obvious. The costs, on the other hand, are less understood. The measures taken may well incentivize a dash for individual electricity usage for heating purposes, which could end up being counterproductive if heat demand does not coincide with wind power production. Decentralized heat pumps, controlled and operated by individuals, rather than centralized heat pumps operated by companies—which are better positioned to deal with the complexities of energy markets—could induce undesirable demand spikes. This could trigger increased electricity system balancing costs and the need for dispatchable capacity, which will not necessarily be emission-free. In this respect, a technology-neutral tax incentive could be harmful from a societal perspective as well as giving customers the possibility of disconnecting from the heating supply network.

Even if the policy goal is to induce cost-efficiency gains by encouraging competition between collective and individual district heating, a lock-in cost could still be present. Particularly, the investment cost of an individual heating solution could make it prohibitive for some households to disconnect. Well-off households may be better positioned to leave the heating network than less affluent counterparts. Yet, because of disconnection from the heating network, the former may increase the cost of collective heating for the latter. Implementing the modernization effort all at once may prove complex and could jeopardize the success of the Danish District Heating sector. A critical review of the regulatory framework may be thus required.[3]

Challenge 3: Removing the Cost Barriers in the Usage of Electric Transportation and Heating. In addition to the decarbonization challenges in the Danish transportation sector and the conflicting ambitions in the modernization strategy of district heating, one of Denmark's core challenges is making renewable electricity cost-effective relative to alternative fuels. The commitment to developing wind power in Denmark has come at a price. As an example, Fig. 13.6 shows the end-user electricity prices paid by domestic consumers with a yearly consumption of 2500–5000 kWh, which is typical of single-family apartments and detached houses. Not only have Denmark's end-user prices consistently exceeded the EU28 average in every year shown, but they also have exceeded neighboring countries' prices as well. The development is not a consequence of the increase in the electricity component of the price but of non-recoverable prices and levies.

The situation is not so different if we focus on the Danish industrial and commercial sector. For example, consumers in this category, with a demand of 0.5–2 GWh/year on average, have consistently paid prices well above the EU28 average and Denmark's closest neighbors. Unlike residential consumers, who bear a higher burden on the Public

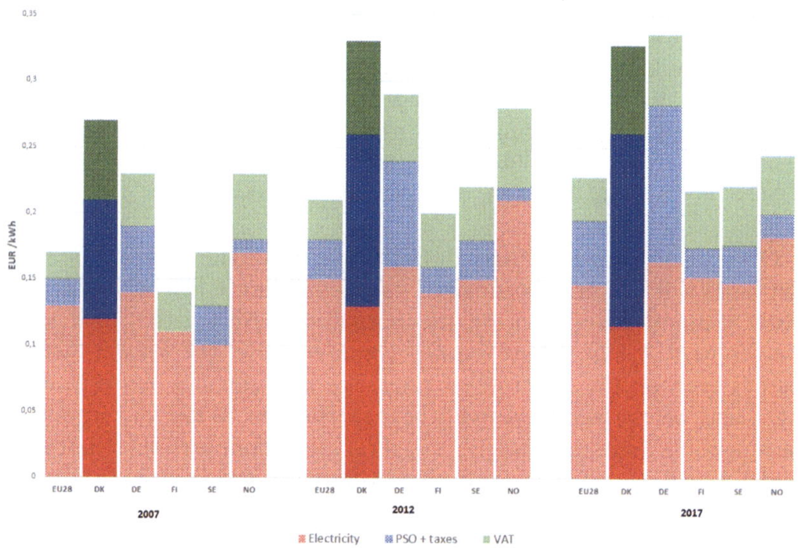

Fig. 13.6 Retail electricity prices paid by consumers in Denmark and neighboring countries (*Source* Retail customers in the 2500–5000 kWh consumption per year segment. Author created figure based on data from the Danish Energy Agency [2018b])

Service Obligation (PSO) used to finance renewable subsidies, industrial consumers pay a much higher share of Value Added Tax (VAT). The situation is similar in other ranges of consumption for residential, industrial, and commercial consumers. Denmark has one of the highest end-user electricity prices in Europe.

How can this be reconciled with the ambitions of Danish policymakers to decarbonize by means of electrification? While it is a feasible idea from the technical and environmental perspective, as the Danish electricity sector is already decarbonized (see Fig. 13.4), economic feasibility remains to be proven for final consumers. The Danish government is aware of this issue and has started taking measures to make electricity more accessible. First, the Danish government introduced a gradual elimination of the PSO payment in the period 2017–2022 in response to the EU's concerns about its implications for discrimination against non-Danish renewable energy producers participating in the European energy market. To compensate for the elimination of this end-user-paid subsidy to renewable producers, the Danish government slightly increased general taxation and will use these proceeds as a funding source for renewable projects (Hvelplund et al. 2017). Electricity taxes for final consumption and heating have also been reduced. There is also an ongoing initiative to introduce dynamic taxes and to increase cooperation between the Transmission System Operator (TSO) and Distribution System Operators (DSOs) in their tariff design, aiming at incentivizing flexible electricity usage.

Once again, however, it is difficult to predict the extent to which end-users will become incentivized to adopt more efficient, electricity-based technologies. Financing the transition to renewables in a non-distortionary manner may require comprehensive tax reform, which focuses both on the structure of taxes, i.e., the signals it sends to end-users, and on the taxpayers, who bear the burden. While the argument for renewable funding through general taxation is not without detractors, as "RES support will compete in the political arena against other budgets of societal interest" (Hvelplund et al. 2017, p. 574), there are voices raised in favor of it. Notably, Newbery (2015) and Robinson et al. (2017) consider that renewables share the characteristics of public goods and thus are less distortionary if financed through broad-based taxes rather than through narrowly based taxes. Denmark appears to be moving in this direction.

Challenge 4: Meeting Increasing Flexibility Requirements with Dwindling Domestic Backup Capacity. With the backdrop of an increased commitment on the Danish government's side to further develop wind power, flexibility requirements will continue to increase in Denmark. The 2018 Energy Policy agreement has established an ambition to develop three new offshore wind farms with a total additional wind power capacity of 2.4 GW coming online by 2030, i.e., an increase of approximately 40% relative to present-day capacity from windmills (6.1 GW). In this respect, the Danish government is already taking action, as the initial screening of the zones (in the North and Baltic Seas) where the wind parks will be sited has already begun (Danish Energy Agency 2018e).

Flexibility, which refers to the ability of the power system to maintain an exact balance between electricity supply and demand in the face of unexpected fluctuations, is mostly supplied in Denmark through the interconnections with neighboring countries. With an increasingly intermittent generation mix, Denmark relies on the production facilities located abroad as a storage facility of sorts; exporting electricity to its neighbors during windy days and importing during calm days (Green and Vasilakos 2012). In the longer run, Denmark tends to be a net exporter during the winter months—when there is a generation surplus—and a net importer during the summer months, when there is a deficit. The interaction of generation and transmission capacity made available to the Nordic electricity market determines the prices and the flow of trade, which helps systems with differing technologies to complement each other. It can be argued that one of the key reasons behind the integration of large shares of wind power into the Danish system is its proximity to Norwegian, Swedish and Finnish hydropower, which is highly flexible and low-cost (see Moe's chapter on renewable energy in Norway in this volume).

During the period 1990–2017, Denmark experienced its largest net exports in 1996 and its largest net imports in 1990 (International Energy Agency 2017). With a long-term perspective, it is evident that interconnections enable mutually beneficial trades in the Nordic region, which increases security of supply and facilitates the integration of renewables. However, with a shorter-run perspective and a narrower focus on Denmark, it is a source of concern for supply security that in 9 out of the 10 years from 2008 to 2017, Denmark has been a net importer. Though the country has enough surplus capacity to meet peak demand, and interconnection capacity is also growing, the situation may not last (International Energy Agency, International Energy Agency 2017).

Not only will intermittent renewables increase—because of Denmark's commitment to more wind power—installed domestic thermal capacity has exhibited a declining trend in the past 20 years (see Fig. 13.7).

In addition, changing economic conditions call into question the future viability of thermal cogeneration plants which have played an important part in Denmark's energy efficiency. As part of the district heating modernization process and in line with the expiry of the *grund-beløb* (discussed under challenge 2), decentralized (gas-fired) cogeneration capacity is no longer entitled to capacity payments. This means that Denmark will have to address its flexibility requirements with a significantly lower domestic dispatchable capacity, increasing requirements on imported power (Hvelplund et al. 2017) and other sources of flexibility. As remedial measures for the elimination of the capacity payment, the Danish government has allocated subsidy funds for district heating firms to switch to large-scale heat pumps.

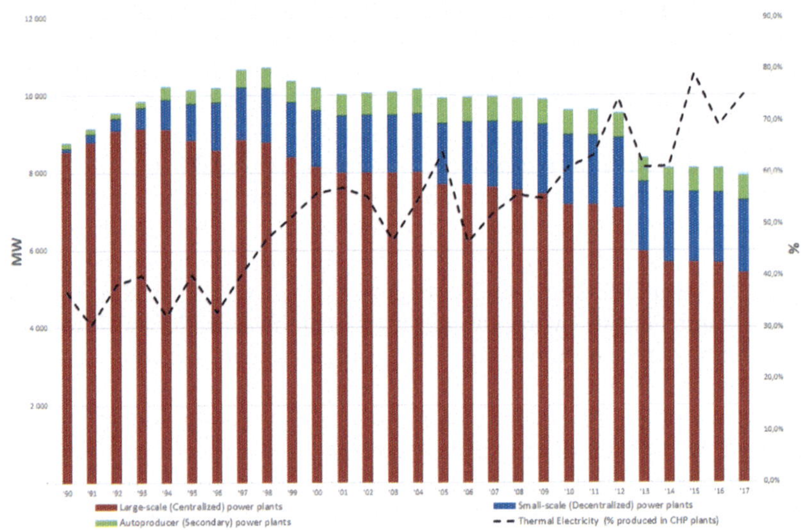

Fig. 13.7 Installed thermal capacity and share of electricity produced in CHP plants (*Note* Left axis: Thermal Capacity; Right axis: Share of Electricity Produced in CHP Plants. *Source* Author created figure based on data from Danish Energy Agency [2018c])

With limited supply-side flexibility resources and increased reliance on cross-border trade of electricity, subsidizing the shift to heat pumps may prove insufficient. Thus, it becomes crucial to enable alternative sources of flexibility, including demand-side flexibility. In Denmark, the implementation of a supplier-centric model, in which all retail market business processes are centralized through a registry (DataHub) and all customers receive a simplified bill, is a step in the right direction. The rollout of smart meters will be completed by 2020, and the possibility of introducing hourly and sub-hourly metered readings complements these measures. However, well beyond the technical feasibility and investments in smart grid solutions (which Denmark is advancing), demand-side flexibility depends on the behavioral aspects of demand, which are difficult to predict. In summary, Denmark requires a more integrated strategy to further renewables that extends beyond traditional, centralized solutions.

Lessons Learned and Transferable Experiences

The Danish energy system evolved within a larger societal framework that has facilitated the adoption of renewable energy, the consolidation of investments for low-cost distribution through, e.g., cogeneration and district heating, and cooperative behavior within and across layers of social and economic decision-making. In this section, we address how understanding this larger framework serves to both recommend conditions that increase the likelihood of success and identify limits to system change. When comparing Denmark to other nations and regions, the state and the community have a high ability to work together toward change. Danish community initiatives are unique in at least three elements.

First, Danish community initiatives tend to emerge in cooperation with local governments. This is the result of centuries of mutual support between community and state. In Denmark's initial industrial revolution, human and physical capital deepening came not through intense competition of the sort witnessed in British mills, but rather through a strong cooperative movement that shared agricultural knowledge among farmers, raised and standardized the quality of Danish export products, and found efficient ways to economize on coal imports from Britain by sharing creameries and other processing facilities (Henriques and Sharp 2016). The lack of domestic energy sources from coal or even wood, which the other Scandinavian countries had in much larger quantities, meant an early attention to the energy costs of production (Henriques and Sharp 2016). The long-run impacts of the cooperatives also extended to government support of, e.g., adult education and societal engagement through the Folkehøjskole (non-degree granting adult education colleges) system, where adults from all classes could spend time in a residential academic setting to improve their education (Kulich 1964). The success of this system in Denmark, even when it has failed elsewhere, and its connections to agricultural intensification made possible through imported coal, show how the roots of energy sector success in Denmark are deeply social and tied to self-driven community development.

Oteman et al. (2014) note that in Denmark's decentralized system, municipalities must realize national goals at a local level without much guidance or budget. Simultaneously, municipalities have financial means and independence in policy design, which is also coordinated at a national level, mainly through the Ministry of Energy and Environment. Communities have ample room to innovate and design their projects, and the local government is likely to support them on the grounds of revitalization of the community, revenue generation, and ensuring energy security and independence.

This mirrors the flexicurity approach to labor and employment applied to (energy) institutions themselves. That is, society provides a safety net that reduces risks of entrepreneurship and innovation. It does also increase the risk of simply a failure to act, based on the lack of dire need. The success of such a model relies on the social cohesion and public-spiritedness of the group (Algan and Cahuc 2009). This is no less true for energy policy than it is for labor markets, so that we can expect many of the insights in describing how flexicurity is likely to succeed or fail in other locations to also hold true for energy policy. That is, Continental and Mediterranean European countries may lack key cohesive qualities that facilitate adoption and success of energy policies similar to Denmark (Algan and Cahuc 2009).

Denmark ranks highly in trust levels (Bjørnskov 2017), a factor identified as an important precursor to strong economic growth and well-being (Bjørnskov 2017). The exact factors contributing to high trust levels remain the subject of much research, but in general, wealth, transparent and inclusive government, and income equality are relatively agreed upon requirements whereas cultural considerations such as ethnic homogeneity are more uncertain contributors (You 2012). Policies that target increasing equality and wealth are more likely to foster trust and consequent abilities to develop cohesive communal energy policies than those that try to enforce homogeneity.

In sum, the social, political, and cultural elements that have made Denmark a "renewable success story" include:

- Trust
- Cooperation
- Ownership structures, e.g., cooperatives
- Decision-making processes, i.e., political decisions

The extent to which these are transferable conditions remains uncertain. As long-run processes that have evolved over decades and sometimes centuries, the ability to foster similar cooperative and integrated decision-making under more rapid pressures is likely to depend on the potential for aligning the incentives of individuals among themselves and with those of the local and national governments. On the other hand, policies that promote equality, wealth generation, and government transparency (e.g., adoption of sustainable accounting methods that include energy, see, e.g., Dincer and Rosen 2012), will increase opportunities for successful transfer of Danish energy policy successes. These will be further enhanced by investments not only in directly supporting infrastructure such as renewable energy capacities, but also in human capital and indirect support through conditions that foster innovation and growth prior to any widespread lock-in without testing under a variety of social conditions.

Conclusions and Policy Implications

Denmark is a worthy case study. It managed to transition from an oil-dependent energy economy to a renewable powerhouse within just 4 decades. A confluence of technical, natural, social, and historical factors underlies this Danish success. On the technical and natural side, Denmark developed an internal infrastructure that increased its energy efficiency and reduced dependence on oil while still using its fossil fuel endowment. Denmark used its natural resources in an "outward," i.e., export-oriented fashion, rather than relying on fossil resources to substitute imports. The impacts of the 1973 oil crisis generated political action to find alternative energy sources. Well-organized grassroots movements shifted public opinion away from nuclear energy, initially preferred by Danish policymakers. The democratically sound political system of the country responded by embracing the communal choice, renewable energy. Within renewable energy, the preferred technology was wind power.

The small Nordic nation has been the home to many "firsts": first offshore wind farm in the world, highest share of wind power production in the electricity system, and first in wind turbine development. Centrally planned district heating and the ensuing energy efficiency provided by this approach, and a cooperative framework within the country but also toward its neighbors, contribute further to Danish wind power's success.

Looking toward the future, several important challenges remain. The emissions-heavy transportation sector requires greater strategy. The district heating sector faces the simultaneous challenge of electrification and modernization, which may prove overly ambitious. The decarbonization of the electricity sector has come at a significant price for end-users, who pay one of the highest electricity prices in Europe, which works against the strategy of decarbonization by means of electrification. Finally, Denmark faces increasing flexibility requirements, given its further commitment to offshore wind power development.

Notes

1. We follow the IEA's convention of including battery electric vehicles (BEVs) plug-in hybrid vehicles (PHEVs) and fuel cell electric vehicles (FCEV) under the broader category of electric vehicles (EVs) (International Energy Agency 2018a, b).
2. After an initial agreement focusing on district heating areas with heat delivery lower than 200 TJ (~350 areas covering 240 thousand households), the measure has been extended and will now apply to areas that deliver less than 500 TJ, more than one-half of all district heating areas in the country. The measure came into force on January 1, 2019.
3. As of 2020, there is a new regulatory framework proposal which includes the application of benchmarking, standardized costs, cost caps, merger incentives, and rate of return guidelines, which resembles the regulation implemented in electricity and gas networks. However, the current Danish government (led by Mette Frederiksen) has not yet agreed to implement this, partly due to resistance from suppliers but also due to doubts cast on the market orientation of such policies, which contrast with the centralized planning approach that has characterized the sector so far.

References

Algan, Yann, and Pierre Cahuc. 2009. Civic Virtue and Labor Market Institutions. *American Economic Journal: Macroeconomics* 1 (11): 111–145.

Bjørnskov, Christian. 2017. Social Trust and Economic Growth. In *Oxford Handbook of Social and Political Trust*, ed. Eric Uslaner, 535–558. Oxford: Oxford University Press.

BP. 2018. Statistical Review of World Energy (Table on Crude Prices Since 1861). Available at: https://www.bp.com/content/dam/bp/business-sites/en/global/corporate/xlsx/energy-economics/statistical-review/bp-stats-review-2018-all-data.xlsx.

Cycling Embassy of Denmark. n.d. Facts About Cycling in Denmark. Available at: http://www.cycling-embassy.dk/facts-about-cycling-in-denmark/statistics/. Accessed 17 Apr 2019.

Danielsen, Oluf. 2005. Atomkraften under pres: Dansk debat om atomkraft 1974–85. *Roskilde Universitetsforlag*, p. 1063.

Danish Council on Climate Change. 2018. Flere elbiler på de danske veje: Forslag til pejlemærker ogvirkemidler til elektrificering af personbilerne. Available at: https://www.klimaraadet.dk/da/system/files_force/downloads/elbilanalyse_final.pdf.

Danish Energy Agency. 2016. Danmarks energifortider. Available at: https://ens.dk/sites/ens.dk/files/EnergiKlimapolitik/danmarks_energifortider_samlet.pdf.

Danish Energy Agency. 2017. Regulation and Planning of District Heating in Denmark. Available at: http://www.ens.dk/sites/ens.dk/files/climate-co2/Global-Cooperation/Publications/Publications/regulation_and_planning_of_district_heating_in_denmark.pdf.

Danish Energy Agency. 2018a. Bred politisk enighed om at erstatte naturgas med varmepumper. Available at: https://www.efkm.dk/aktuelt/nyheder/2018/okt/bred-politisk-enighed-om-at-erstatte-naturgas-med-varmepumper/. Accessed 23 Apr 2019.

Danish Energy Agency. 2018b. Energy Price Database. Available at: https://ens.dk/sites/ens.dk/files/Statistik/prisdatabase.xlsx.

Danish Energy Agency. 2018c. Energy Statistics 2017 (Excel Sheets Containing Report Figures). Available at: https://ens.dk/sites/ens.dk/files/Statistik/figurer2017_-_figures2017.xlsx.

Danish Energy Agency. 2018d. Energy Statistics 2017 (Excel Tables). Available at: https://ens.dk/sites/ens.dk/files/Statistik/tabeller2017_-_tables2017.xlsx.

Danish Energy Agency. 2018e. Her skal Danmarks næste havvindmøllepark ligge. Available at: https://efkm.dk/aktuelt/nyheder/2018/sep/her-skal-danmarks-naeste-havvindmoellepark-ligge/. Accessed 23 Apr 2019.

Danish Energy Agency. 2019. Master Data for Wind Turbines (Review Tables at End February 2019). Available at: https://ens.dk/sites/ens.dk/files/Vindenergi/oversigtstabeller_uk-dk.xls.

Danish Energy Association and Danish District Heating Association. 2016. Industry Agreement to Ensure Sustainable Biomass (Wood Pellets and Wood Chips). Available at: https://www.danskfjernvarme.dk/-/media/danskfjernvarme/medlemmer/vaerktojer/bæredygtighedsaftale-biomasse/23062016-brancheaftale-om-sikring-af-bæredygtig-biomasse_uk-version.pdf.

Danish Energy Association and Danish District Heating Association. 2017. Biomass for Energy: Why Coal and Gas Should Be Replaced by Wood Pellets and Wood Chips. Available at: https://www.danskenergi.dk/sites/danskenergi.dk/files/media/dokumenter/2017-09/BiomassForEnergy.pdf.

Danish Ministry of Energy Utilities and Climate. 2012. Energiaftalen i korte træk. Available at: https://efkm.dk/media/8418/energiaftalen-2012-kort-fortalt.pdf.

Danish Ministry of Energy Utilities and Climate. 2018a. Energy Agreement of 29 June 2018. Available at: https://en.efkm.dk/media/12307/energy-agreement-2018.pdf.

Danish Ministry of Energy Utilities and Climate. 2018b. Sammen om en grønnere fremtid: Klima- og luftudspil. Available at: https://www.regeringen.dk/media/5791/klimaministeriet_klimaogluftudspil_digital.pdf. Accessed 22 Apr 2019.

Dincer, Ibrahim, and Marc A. Rosen. 2012. *Exergy: Energy, Environment and Sustainable Development*, 2nd ed. Oxford, UK: Elsevier Science.

Energinet. 2019. Solenergi slog rekord i varmt 2018, 2019. Available at: https://energinet.dk/Om-nyheder/Nyheder/2019/01/07/Solenergi-slog-rekord-i-varmt-2018. Accessed 7 Jan 2019.

Euroheat & Power. 2015. Statistics Overview: Country by Country. Available at: https://www.euroheat.org/wp-content/uploads/2016/03/2015-Country-by-country-Statistics-Overview.pdf.

FDM. 2018. De fleste elbiler fortsat uden afgift. Available at: https://fdm.dk/nyheder/bilist/2018-12-de-fleste-elbiler-fortsat-uden-afgift. Accessed 12 Apr 2019.

Green, Richard, and Nicholas Vasilakos. 2012. Storing Wind for a Rainy Day: What Kind of Electricity Does Denmark Export? *The Energy Journal* 33 (3): 1–22. https://doi.org/10.5547/01956574.33.3.1.

Henriques, Sofia T., and Paul Sharp. 2016. Danish Agricultural Revolution in an Energy Perspective: A Case of Development with Few Domestic Energy Sources. *The Economic History Review* 69 (3): 844–869.

Hvelplund, Frede, Poul Alberg Østergaard, and Niels I. Meyer. 2017. Incentives and Barriers for Wind Power Expansion and System Integration in Denmark. *Energy Policy* 107: 573–584. https://doi.org/10.1016/j.enpol.2017.05.009.

International Energy Agency. 2017. *Energy Policies of IEA Countries: Denmark 2017 Review*. Paris. Available at: https://webstore.iea.org/energy-policies-of-iea-countries-denmark-2017-review.

International Energy Agency. 2018a. Global EV Outlook 2018. *Global EV Outlook 2018*. https://doi.org/10.1787/9789264302365-en.

International Energy Agency. 2018b. *Nordic EV Outlook 2018*. Available at: https://webstore.iea.org/nordic-ev-outlook-2018.

International Renewable Energy Agency. 2013. 30 Years of Policies for Wind Energy: Lessons from 12 Wind Energy Markets. Available at: https://www.irena.org/-/media/Files/IRENA/Agency/Publication/2013/GWEC_WindReport_All_web-display.pdf.

Kulich, Jindra. 1964. The Danish Folk High School: Can It Be Transplanted? *International Review of Education* 10 (4): 417–430.

Meyer, Niels. 2000. *Vedvarende energi i Danmark - En krønike om 25 opvækstår 1975–2000.* In Ejvind Beuse et al. (eds.) Organisation for Vedvarende Energis Forlag (OVEs Forlag), 75–110. Available at: https://www.klimadebat.dk/ejvin-beuse-m-fl-vedvarende-energi-i-danmark-en-kroenike-om-25-opvaekstaar-1975-2000-b145.php.

Meyer, Niels. 2004. Renewable Energy Policy in Denmark. *Energy for Sustainable Development* 8 (1): 25–35. https://doi.org/10.1016/S0973-0826(08)60388-9.

Mortensen, Hans Christian, and Bjarne Overgaard. 1992. CHP Development in Denmark: Role and Results. *Energy Policy* 20 (12): 1198–1206. https://doi.org/10.1016/0301-4215(92)90098-M.

Morthorst, Poul Erik. 2000. The Development of a Green Certificate Market. *Energy Policy* 28 (15): 1085–1094. https://doi.org/10.1016/S0301-4215(00)00094-X.

Newbery, David. 2015. Reforming UK Energy Policy to Live Within Its Means. Cambridge Working Papers in Economics 1526, Faculty of Economics, University of Cambridge.

Oteman, Marieke, Mark Wiering, and Jan-Kees Helderman. 2014. The Institutional Space of Community Initiatives for Renewable Energy: A Comparative Case Study of the Netherlands, Germany and Denmark. *Energy, Sustainability and Society* 4 (11): 1–17. https://doi.org/10.1186/2192-0567-4-11.

Robinson, David, Malcolm Keay, and Klaus Hammes. 2017. Fiscal Policy for Decarbonisation of Energy in Europe. Oxford Institute for Energy Studies (OIES) Paper: EL 22 (September). https://www.oxfordenergy.org/publications/fiscal-policy-decarbonization-energy-europe/.

Schlesinger, William H. 2018. Are Wood Pellets a Green Fuel? *Science* 359 (6382): 1328–1329. https://doi.org/10.1126/science.aat2305.

Sovacool, Benjamin. 2013. Energy Policymaking in Denmark: Implications for Global Energy Security and Sustainability. *Energy Policy* 61: 829–839. https://doi.org/10.1016/j.enpol.2013.06.106.

Sovacool, Benjamin, Hans H. Lindboe, and Ole Odgaard. 2008. Is the Danish Wind Energy Model Replicable for Other Countries? *Electricity Journal* 21 (2): 27–38. https://doi.org/10.1016/j.tej.2007.12.009.

State of Green. 2017. Wind Energy Moving Ahead: How Wind Energy Has Changed the Danish Energy System. Available at: https://stateofgreen.com/en/uploads/2017/05/Wind-energy-moving-ahead.pdf?time=1555930658.

Statistics Denmar. 2019. Table 51: New Registrations of Passenger Cars by Ownership and Propellant. Available at: http://www.statbank.dk/BIL51.

You, Jong-sung. 2012. Social Trust: Fairness Matters More Than Homogeneity. *Political Psychology* 33 (5) (October), 701–721. https://doi.org/10.1111/j.1467-9221.2012.00893.x.

CHAPTER 14

Renewable Energy in Finland: From a Production-Centric to a Consumption-Centric System

Sarah Kilpeläinen, Pami Aalto, and Juha Kiviluoma

Introduction

The 2019 programme of the Government of Finland targets carbon neutrality by 2035 as a strategic priority, and shortly afterwards, negative emissions with the help of carbon sinks (Government of Finland 2019). This follows the decisions of consecutive governments that gradually introduced more demanding targets, spurred on by the ambitious commitments of the Nordic neighbours (Toivanen et al. 2017). Global policy debates leading to the 2015 Paris climate agreement heavily underpin these developments. During the late 2010s, the idea of renewable energy becoming the cornerstone of the Finnish energy system became mainstreamed. It also became normal to expect a full-scale energy

S. Kilpeläinen · P. Aalto (✉)
Tampere University, Tampere, Finland
e-mail: pami.aalto@tuni.fi

J. Kiviluoma
VTT Technical Research Centre of Finland, Espoo, Finland

P. Midford and E. Moe (eds.), *New Challenges and Solutions for Renewable Energy*, International Political Economy Series,
https://doi.org/10.1007/978-3-030-54514-7_14

transition to eventually extend to all components of the energy system, profoundly affecting all the stakeholders in the energy industry, overall economy and wider society.

In 2017, 37% of final energy consumption came from renewable energy sources (Tilastokeskus 2018). The 2030 Energy and Climate Strategy of Finland introduces several further targets designed to support the expansion of the role of renewables. Such measures include phasing out coal use in energy production, halving the domestic use of imported oil and increasing the share of renewable transport fuels to 30%. In the process, self-sufficiency should increase to over 55%. In the transport sector, the strategy targets 250,000 electric vehicles (EVs) in addition to 50,000 gas-fuelled vehicles (including biogas) (Ministry of Economic Affairs and Employment 2017, pp. 9, 58).

Fulfilling the Paris commitments will presuppose negative emissions, energy efficiency measures and ambitious renewable energy policies (Rocha et al. 2016). Biomass in different forms has gradually become the most important renewable resource in Finland. It was also the most significant energy resource overall in 2017 with its 27% share of final energy consumption (Tilastokeskus 2018). The large volume of the forestry industry supports increasing its use even further, as outlined in the 2030 Energy Strategy. The forestry industry uses biomass-generated energy in its production of pulp, paper and other bioproducts, and sells its surplus electricity to the market. The ongoing development of small and medium-scale biomass powered combined heat and power (CHP) units will likewise support the role of biomass. Moreover, several agricultural wastes, harvest residue and other biomass streams suitable for biogas production currently remain grossly underutilised (Pääkkönen et al. 2019). Hydropower has remained important with a share of 4% of final energy consumption and 22% of electricity generation (Tilastokeskus 2018); however, most of the remaining resource potential is protected from utilisation and hence prospects of expansion are limited.

Wind and solar energy, which are weather-dependent and hence variable-output resources, are making inroads in Finland. In 2017, wind power covered 7% of electricity generation (Tilastokeskus 2018). Nine purely commercial, non-subsidised wind power projects were scheduled for completion by the end of 2019 (Tuulivoimayhdistys 2019). Solar PV generation—which was not expected to play a major role in the 2030 scenarios of the 2016 Energy Strategy—started attracting renewed interest in Finland, with one company having claimed to have mapped the solar energy potential of over four million Finnish rooftops and offering respective solutions to business properties, blocks of flats and detached houses alike (Sun Energia 2019). The increasing share of renewable energy is an important policy goal, yet it is associated with at least three major types of disruptive trends and problems shaping Finland's emerging energy system.

First, a higher share of renewable energy supports the gradual *electrification* of the energy system, society and the economy. Renewable energy facilitates electrification of transport by replacing fuels with battery EVs and through electricity used in fuel production (Nylund 2015). In the heating sector, electrification progresses largely as an unintended and non-planned development. Heat pumps, which generate heat from electricity with a ratio of approximately 1–5, depending on the temperature difference, became a common heating solution in Finnish detached houses during the 2010s. During the colder periods, especially air-source heat pumps will use considerable amounts of electricity and are typically supplemented by other heat sources like fuels or district heating. Meanwhile, district heating systems face shrinking annual demand due to improving energy efficiency and heat pumps that have started to replace district heating service even in cities (Heljo et al. 2016, p. 6). Counting all geothermal, air-to-air and other heat pump solutions in residential housing and industry, by 2020, Finland may have one million heat pumps installed in a country of five million people, and accounting for approximately 15% of space heating consumption (SULPU 2017). The industrial sector consumes almost half of current final energy use (TEM 2016, pp. 28–33). In this sector, a major part of energy use is heated at different temperatures while electricity could also be used to generate the heat. This will increasingly take place when electricity becomes cheaper than burning fuels due to falling renewable energy prices or due to increasing fuel costs. Improving competitiveness of electricity over fuels could also electrify a large share of the metal industry. Eventually, electricity can also be used to produce ingredients for chemical and agricultural industries through carbon (and nitrogen) capture. The same ingredients can also be used to produce fuels for those energy use cases that cannot be readily electrified (e.g. part of aviation).

Second, the influx of renewable energy contributes to the *decentralisation* of the energy system where a structure of geographically decentralised wind and solar generation replaces generation from large centralised power plants. In Finnish conditions, wind and solar power parks are typically not very large and are dispersed in various regions, even though wind power is usually connected to a high voltage "centralised" grid. At the same time, wind power, as well as solar power in the high latitudes, must rely on other resources within the remit of the power grid to cost-efficiently smoothen their variability. Therefore, the power grid is likely to be strengthened to better accommodate both a higher share of renewables and electrification (ENTSO-E 2018).

These trends emerge within a centralised structure of the system, since several large-scale nuclear and hydropower plants will continue to be part of the Finnish system well beyond the 2050s, most likely alongside the large pulp and paper mills, or "bio-product" factories where biomass-based energy is a side-product. The decentralising trends are interesting given how the country's energy system was originally heavily shaped by the demand from large units in the forestry, chemical and metal sectors and respective electrification processes (Myllyntaus 1990). Electricity distribution grids were devised to be one-way roads towards consumption. With increasing solar PV generation and demand-side flexibility, this is not sufficient. Communication and control of these grids is being enhanced to cope with the new modes of operation. Concurrently, electrification increases the utilization of distribution grids and may require costly upgrades.

Third, the *variability* of renewable energy generation decreases its value to the energy system as its share increases. Large-scale wind power generation can vary tens of percentage points over a period of hours (Kiviluoma et al. 2016). Solar power production in Finnish latitudes has a profound seasonal variation and a fast up-ramp in mornings and down-ramp in evenings. The Nordic power system is highly flexible due to a large share of reservoir hydropower (almost 50% of generation). This flexibility helps to maintain the value of variable generation even at high shares. However, additional flexibility can be cost-efficient and is the key to very high levels of variable renewable energy generation. Accordingly, the electrification of energy presupposes various means of increasing flexibility in the system. For example, heat storage provides a less expensive option compared to storing electricity. Moreover, the smart charging of EVs could help to accommodate higher shares of variable renewable energy in a cost-efficient manner (Kiviluoma et al. 2018). Concurrently, peaks in electricity consumption continually become sharper (Energiateollisuus 2019). Peaks are typically reached on cold winter days when the consumption in residential housing is high due to electric resistance heaters, heat pumps and people returning from work and starting to use electric appliances (Heljo et al. 2016). Yet, new peak generation plants are not currently being built owing to profitability concerns, raising concerns of whether the capacity can in the future meet the peak demand (Energiateollisuus 2019). The influx of variable generation into the Nord Pool market (to which the Finnish system is strongly interconnected alongside Denmark, Sweden, Norway and further Northern European countries) is problematic for the economics of base-load types of generation units like nuclear power plants. They have large sunk costs that can be recovered only if the average power prices are sufficiently high. Nuclear power is poised to supply a third of Finland's power supply by 2030 and still a fifth in 2050 (Holttinen 2017).

Considering the disruptive consequences of electrification, decentralisation and variability of renewable energy generation, the cost-effective accommodation of large shares of variable generation in the Finnish energy system should consider at least the following:

1. Ensuring a level playing field between different primary energy sources. Electricity should be treated similarly to fuels, and the societal costs of emissions of CO_2 and other pollutants should be included in the prices. Distribution and transmission tariffs should be based on the actual costs, which are mainly installation and maintenance costs, not usage costs.
2. Cost-efficient communication, control and market participation of flexible energy devices would facilitate the emergence of building-level flexibility. In order to gain widespread adoption, the solutions should be trouble-free for the building dwellers whether they are consumers or prosumers. The emergence of an increasingly interconnected digital system requires solutions that are trustworthy in terms of privacy as well as power grid stability and control.
3. Sufficient charging infrastructures for EVs, including distribution grids, the development of electrical systems and roadside charging, are required. These systems should consider flexible charging and possibly discharging whenever cost-efficient. Flexibility would improve the value of variable power generation. Also, the possibility of increasing the use of electricity for heating should be factored into the reinforcements and control systems.

4. More interconnection capacity between regions and countries is likely to be cost-effective by decreasing the apparent variability of renewable energy. However, overhead lines have social acceptance challenges, which can hamper the permitting processes.
5. Power grids will need better situational awareness and control to accommodate increasing shares of converter-based generation. Stability of the transmission grid needs to be ensured.
6. Energy storage can be helpful. As electricity storages are typically very expensive, electrification of energy offers interesting energy storage alternatives in the heat and transport sectors. Regulation, taxes and tariffs should not prevent cost-efficient storage solutions.

We propose that the trends, opportunities and challenges associated with a higher share of renewable energy sources—electrification, decentralisation and variability—make the energy system less production-centric. Instead, it is becoming more consumption-centric, since the "hourly" energy balance will be increasingly maintained through newly electrified demands in heating, transport and the production of energetic molecules. Concurrently, the role of the grid and distribution network requires renewed attention in terms of how the whole system is managed. Whereas in the emerging system each kWh travels on average longer distances, the larger issue is shifting the focus from large-scale thermal power plants to how increasingly variable, two-way flows of electricity are managed, stored, redistributed and consumed. These changes mean that the interests, views and behaviour of various stakeholders of the energy system matter in a new way. Interest groups have for a long time been crucial for Finnish energy policy (Ruostetsaari 2010), but what is new is their widening social basis.

We now move on to the perspectives of this widening field of stakeholders, especially regarding the issues of grid development and energy storage. Grid development is usually considered an area where Finnish actors have advanced technology, infrastructure and R&D. Uncertainties prevail over how the grid will cope with electrification, decentralisation and variability, and how this affects incumbent actors and creates opportunities for emerging stakeholders. Regarding storage, uncertainties prevail in most countries undergoing renewable energy-based energy transitions. Finland is not an exception in this regard as piles of fossil fuels and oil storages are becoming replaced by other solutions, with implications for the energy business and security of supply.

The next section will address the transforming field of stakeholders, summarising the interests driving the stakeholders as they seek to respond to the trends and problems associated with a higher share of renewable energy. Our empirical analysis addresses possible solutions in the transformation from a production-centric to a consumption-centric system as viewed by the stakeholders, especially regarding smart grids and storage issues. The analysis draws upon 25 expert interviews (see Appendix 1) on the 2030 energy system conducted with stakeholders in the public sector, companies and NGOs. We also make use of key policy documents, such as the 2016 Energy Strategy of the Government of Finland and the 2018 final report by the smart grid working group set up by the Ministry of Economic Affairs and Employment, alongside other reports and strategies published by the public sector and companies as well as previous literature. In the concluding section, we discuss how the transition to a higher share of renewable energy sources and a more consumption-centric system is becoming established among the different stakeholder groups, while the development of policies and regulations for this development path is still in the initial phases, especially when speaking of grid and network development issues. We also note how difficult it is to simultaneously pay attention to the various interests the energy system is supposed to serve alongside the main policy goal of decarbonisation.

The Transforming Field of Stakeholders

What do we know of the stakeholders?

Existing research has argued that furthering the energy transition requires governments to develop legislation to shape the market and support new forms of energy production through incentives. In a country where the incumbent energy industry is accustomed to centralisation, and energy-intensive heavy industries are influential, new business models furthering the transition require active development (Valkila and Saari 2010, p. 2073; 2013, p. 287). Finnish energy prices are low by European standards, and this has until recently contributed to the need for subsidies to support investment in variable renewable energy (Valkila and Saari 2010, p. 2074). The initially high feed-in tariff in terms of €/kWh—especially for wind power production entering the system in 2009—evoked policy debates on renewable energy subsidies. Often, however, the several times higher indirect fossil fuel subsidies were neglected in this debate, not to mention their societal and environmental costs resulting from emissions (Toivanen et al. 2017, p. 154). Moreover, the transition to variable renewable energy generation would benefit from institutional changes in regulation, e.g. streamlining permitting procedures, better grid planning and improved regulation guiding the planning process, as well as taxation laws in the electricity sector (cf. Ruggiero et al. 2015).

Prior to the above-noted shift in the Finnish energy debate in the mid-2010s, most observers expected the key technologies for the Finnish case to relate to bioenergy, nuclear power and wind power, with solar only playing a minor role (Valkila and Saari 2010, p. 2074). However, studies have also shown that expert views on the role of solar power are highly dependent on which stakeholder group the expert represents (Haukkala 2018). Bold modelling studies for the Finnish energy system up to 2050 probe a scenario for a solar PV share of up to 10% of final energy consumption, arguing that the intermittency of solar (and other renewable energy sources) can be addressed by means of daily and seasonal storage solutions (Child et al. 2017; Child and Breyer 2016), including hydro, heat storage, batteries, EVs, Power-to-Gas and other storage solutions.

Given the strong interconnectedness of Finland with its neighbours through the Nord Pool market, a wider Nordic focus is also necessary. A recent energy system cost optimisation study shows a very limited role for PV in the Nordic context even when the power system is close to fully renewable (Kiviluoma et al. 2018). The seasonal nature of PV in the high latitudes makes it valuable only during summer when demand is lower. Wind power, for its part, generates more during winter and is in general a less variable resource. In this scenario, the most valuable flexibilities come from the heat sector and from increased transmission. Pumped-hydroelectric storage in the best potential Norwegian locations is close to break-even while batteries would need to be extremely low-cost before being cost-effective. The interaction between wind power, PV and batteries is further explored in a parametric study showing the high impact of seasonality on the cost optimal share of PV (Kiviluoma et al. 2015). However, the absence of sector coupling in that study makes the results at best descriptive. Some of these flexibilities are being built—for example, new transmission lines are planned (ENTSO-E 2018)—while reservoir hydropower already exists and heat storage in district heating grids have witnessed a renewed interest. For example, Helen, the Helsinki-based utility, is developing a 120 MW underground heat storage facility to a cave formerly used for oil storage (Helen 2018).

Multiple modelling studies of the Finnish energy system up to 2050 (Kiviluoma et al. 2018; Ikäheimo et al. 2019; Kiviluoma 2013; Child and Breyer 2016; Child et al. 2017; Zakari et al. 2015) lend support to the vision of an energy system that is able to integrate high shares of renewable energy (Holttinen 2017). These studies focus on the technical capacities and possibilities to identify cost-effective solutions for developing the Finnish energy system. In this chapter, we are arguing that this shift is not only a technological and infrastructural shift, but also a societal one. Thus, the ways in which the field of stakeholders is changing are of special interest to us. We will ask questions relating to the views and positions of incumbent actors as well as emerging actors.

The reorganisation of the field of stakeholders

The reorganisation of the field of stakeholders as illustrated in Fig. 14.1 assumes that energy policy is shaped by interlinkages among different sectors as well as the diverging views and interests of a variety of

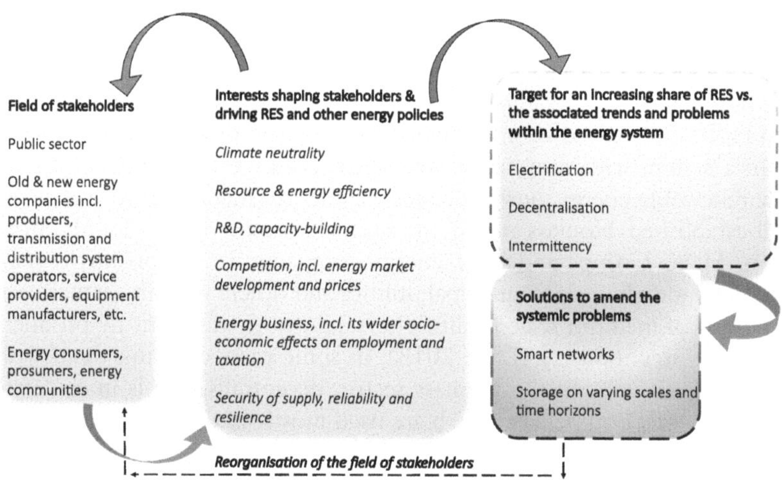

Fig. 14.1 The reorganisation of the field of stakeholders

stakeholders extending deep into societal structures. Drawing upon institutionalist literature, we accentuate the centrality of interests for the development of the energy system (Andrews-Speed 2016; Hoppe et al. 2016; Moe 2015). Such interests include not only the central concerns of resource efficiency and climate neutrality, but also research and development (R&D) as well as societal capacity-building; competition and energy market development, including prices; energy business, including its wider socioeconomic effects on employment and taxation, and security of supply (Toivanen et al. 2017).

We assume that the stakeholders generally acknowledge this diversity of interests but that each stakeholder group normally privileges some interests over the others. Thereby the field of stakeholders interacts and makes energy policy choices driven by a multitude of interests and assesses the relative weight of each interest in specific situations, which often is a difficult task. The choices they must make, and which are of interest here pertain to the targets for increasing the share of renewable energy. While making such choices, they must take into account the associated trends and problems—electrification, decentralisation and variability—as well as the possible solutions, for example smart grid and storage development issues. Because of the difficulty of these choices, the end results are often sub-optimal (Aalto et al. 2014). Moreover, while the stakeholders can utilise the results of modelling studies as supporting evidence, such studies can only elucidate some issues of interest, such as prices and emissions, but not so much for example the extent to which interest in employment and taxation is served by a particular combination of solutions.

In a system with more renewable energy sources, and especially variable renewable energy, new producers of energy emerge. However, in the well-established biomass sector, traditional forestry firms (UPM, Stora Enso, Metsä Group) and large energy companies (e.g. Fortum) remain predominant, alongside municipal utilities and others who altogether own hundreds of medium-scale facilities burning mostly different by-products of the forestry industry (Luke 2018), in some cases co-firing them with peat or fossil fuels. In the biomass sector, decentralisation is in evidence in the case of wood pellets, which are used mostly in detached houses and in small self-generation production units of some businesses for example in agriculture.

In the wind power sector, some municipal utilities have entered the business by joining their forces (e.g. Hyötytuuli Oy). Some are international wind power developers (e.g. ABO Wind, OX2), while one major wind power developer (Tuuliwatti Oy) is owned by a Finnish gasoline dealer (ST1) and the country's major cooperative supermarket chain (S-ryhmä). Foreign and domestic capital investment funds also own some production (Taaleri Oy, Allianz), as do smaller companies and private individuals (Tuulivoimayhdistys 2019). In the solar power sector, the actors range from utilities to large international equipment providers to small companies and consultancies offering turnkey and installation services, as well as specialised solutions to businesses, farms, housing cooperatives and private individuals.

The solar PV panel solutions can empower prosumers or small-scale producers and consumers of energy, and blocks of flats wishing to produce part of their own power. In some pilot projects, PV panels are combined with heat pumps, or so far in rare cases, solar thermal collectors. Energy cooperatives or microgrid developers for example in remote locations can utilise small-scale bio-CHP or biogas solutions (Uski et al. 2018). However, so far, these early adopters of the more decentralised solutions operate largely in small niches compared to the established role of the incumbent companies. Indeed, Finnish energy policies incentivise mainly companies for reforms, while much fewer incentives exist for private individuals or cooperatives (Kotilainen et al. 2018). In this regard, the situation differs drastically from Denmark where cooperatives are widespread (Eikeland and Inderberg 2016).

To better reach the citizens, one incumbent company develops a "virtual battery" concept consisting, for example, of electrically controllable boilers of detached houses or other properties, and of highly energy consuming data centres, which are linked together by the grid operator to market a flexibility of energy use service for DSOs (Fortum 2019). The transmission system operator (TSO) Fingrid develops a concept for prospective aggregators of reserve power, which would also offer flexibility of electricity use service for small consumers (Fingrid 2018). Yet so far, DSOs have varying tariffs for policies for small-scale production fed into their grid. Regarding the possibility of aggregating the reserve power prospectively enabled by the batteries of EVs through vehicle-to-grid technologies (V2G), the 2030 Energy Strategy targets 250,000 EVs. Assuming 3.5 kW per vehicle, this could theoretically provide 880 MW of peak load capacity, which is the capacity of the Olkiluoto 1 nuclear power plant (Kojo et al. 2018). However, in practice V2G is far from widespread utilization and with limited policy attention thus far (Kester et al. 2018). In Finland, a two-way charging point pilot project exists in

Suvilahti, offered by Helen, the former incumbent utility of the Helsinki area.

Public authorities such as the Ministry of Economic Affairs and Employment are aware of these emerging groups of stakeholders—flexibility service providers, equipment manufacturers and retailers, aggregators, etc.—but do not really prioritise any of the emerging solutions but rather advocate the principle of technology neutrality (Toivanen et al. 2017). Official energy strategies also recognise a multiplicity of interest shaping the development of energy policies, including those related to renewable energy sources. However, these strategies do not strongly prioritise between these interests, although, for example, interests in securing competition can be at odds with security of supply or climate neutrality interests (Kilpeläinen et al. 2019; see Fig. 14.1). In our empirical analysis, we will also draw attention to how in grid and storage development different interests can be in conflict.

How incumbent energy companies approach this transition, interact and work with new actors such as smaller renewable energy generators and other businesses such as ICT start-ups, or civil society organisations representing consumers, also implies shifting understandings of power and agency within the field of stakeholders. For example, in the cases of "Smart Kalasatama", a new-build residential area in Helsinki, and the ECO_2 programme for low-carbon future in the city of Tampere, complex and mutually dependent relationships develop between incumbents and new actors (Heiskanen et al. 2018).

Studies have shown that many incumbent actors, such as DSOs, indeed recognise the need to adjust their business model to the changing business environment in order to maintain capacity to shape it to further their energy business interests (SITRA 2015b). However, doubts also exist on whether the incumbent businesses can adapt. Whereas the majority of Finnish businesses saw climate change as an important matter—as is also indicated in how the "polluter pays" principle is widely supported by the stakeholders (Toivanen et al. 2017)—only 15% of companies conceptualised it as a business opportunity (SITRA 2015a, b). Here, we encounter our basic thesis: while many actors recognise the multitude of interests defining the development of the energy system, they find it difficult to serve all these interests simultaneously and optimally. We will next illustrate this difficulty in the questions of grid and storage development.

How Do the Stakeholders View the Transition?

Views of stakeholders on smart grid development

Public authorities retain important roles regarding the development of electricity grid infrastructure also in the context of ongoing energy market liberalisation within the EU since the TSO is a natural monopoly as are DSOs in their own areas. This makes governmental policy and regulatory oversight necessary aspects of the sector's governance wherein various public and private companies also participate alongside the emerging actors typical for a high-renewable energy system (Aalto et al. 2019). In Finland, the main policy developer vis-à-vis energy markets and grids is the Ministry of Economic Affairs and Employment, acting under the political oversight of the government. The Energy Authority issues more technical regulation and supervises its implementation. At the same time, Finnish Energy, representing large power and district heating producers, and other industrial lobby groups such as large energy-consuming industries traditionally exercises strong influence. Of the industrial sectors, especially the forestry sector is powerful (Ruostetsaari 2010). Although the climate neutrality interest has become the master interest driving energy policy, it is fair to say that the interest in competitive markets is also a key factor for the authorities, while they also bear responsibility for security of supply questions on the national level together with the TSO. Since the TSO and DSOs together with other public and private energy companies are responsible for the actual infrastructure serving the promotion of these interests, the authorities consult the companies in various formats.

In 2016, the Ministry of Economic Affairs and Employment issued a smart grid vision for 2025. This vision was based on the work of the smart grid working group assembled for wide-ranging consultation purposes and to eventually function as a basis for new regulation. The vision defines the role of smart grids as a "(...) service platform in the transition towards a more distributed and carbon-neutral power system" (Smart Grid Vision 2016, p. 6). The group views smart grids as a tool to enable better participation of customers in the electricity market, to improve security of supply and to promote new energy business in a cost-efficient manner. Apart from the customers, stakeholder groups in this vision are the whole society, electricity supplier/service companies, electricity producers, TSO, DSOs, as well as the technology industry (Ministry of Economic Affairs and Employment 2018, p. 11).

The vision extends the energy business interest towards services, also reaching small customers and towards enabling new business models. Securing a positive outlook for profits in the energy business is important in the emerging market since energy companies will be needed in the future as well. Because of EU regulation, there is no return to the old system where publicly owned, vertically integrated utilities predominated, while counting on new energy cooperatives and private individuals to bear the main responsibility for energy provision services is probably not realistic. Moreover, public actors expect to receive some tax income from the sector although so far electricity and fuel taxes have been more important to this end. The fate of the energy business is especially open in the district heating sector where many existing infrastructures operated by publicly owned companies face the prospect of a lower utilisation rate.

The smart grid working group was tasked with proposing operating methods and regulations for enabling the changes outlined in the vision. The resulting proposals seek to clarify the roles and rules in the electricity market to ensure fair competition among incumbent and emerging actors, to support the development of new business models as well as innovations helping to decrease the electricity costs of customers. The group proposes that demand flexibility, the provision of new services to customers as well as energy storage should be treated as competitive business activities (for energy storage, see next section). They also call for a customer-driven retail model (Ministry of Economic Affairs and Employment 2018, pp. 12 and 16). The main proposals of the group received solid support from many stakeholders in our interviews.

To provide flexibility services and enable the participation of small customers in the electricity market the group directs attention to aggregators. While the industry has so far been the main actor providing flexibility services—by offering to reduce consumption in peak demand situations—aggregators could unleash the flexibility potential of small customers. Their participation in the flexibility market is expected to improve security of supply, in view of the weakened outlook for the demand of base-load power and the associated weak profit outlook for those incumbent companies whose business revolves around base-load power. Flexibility can likewise improve efficiency, taking less costly resources better into use, while driving down costs for electricity users. Equal treatment of all actors and the understanding that aggregators bear the responsibility for any imbalance they might cause are central to ensuring that aggregators can fulfil their functions in the market and can promote customer involvement (Ministry of Economic Affairs and Employment 2018, p. 17). This aim is also echoed in the stakeholder interviews where the emergence of new actors such as aggregators is welcomed while stressing equal treatment of incumbent and emerging actors as well as the need for clear rules on roles and responsibilities (interview 2).

Our interviewees proposed different ways of involving customers in the electricity market. The need for active customers bringing small loads to the grid was identified as central to creating a new electric energy system. How precisely to involve customers was approached differently by different stakeholders. Some respondents argued for the importance of information provision to enable customers to make informed decisions instead of relying on pricing mechanisms alone to enhance involvement (interview 2). Other interviewees doubted whether simply providing information to customers will bring about results on the scale needed, favouring increasing electricity prices, and/or increasing taxes as providing a better incentive for customers to get involved (interview 22). Regardless of whether information activities, economic instruments or command-and-control measures are the preferred policies (Kotilainen et al. 2019), activating customers is a major step if not the litmus test to ensure the development of "(...)a real smart grid" (interview 4). Businesses can assume an enhanced role through existing but little exploited (and used) services, such as home automation and aggregation services of different types of production (interview 4).

The role of energy pricing was discussed as a tool of increasing the participation of customers; moreover, it was also seen as influencing the development of emerging business opportunities and innovation. Respondents argued that pricing mechanisms support investments if they are in tune with the market. Too low prices, especially for the industry, can hinder innovations, investments, and structural change overall. Low electricity prices do not provide incentives for developing new solutions (interview 3).

The broader move towards smart grids and distributed production as well as the development of respective policies including clarification of the roles of actors were seen in a positive light by respondents (interviews 4, 5, 8). Especially the possibility for new business opportunities and a possible market expansion through the development of new products and technologies needed to facilitate the electrification of the energy system were highlighted. However, respondents cautioned that Finland was lacking a clear, long-term strategy to capture business opportunities (interview 5). The framing of these changes as risks or opportunities has also been highlighted in previous studies noting how only few companies fully grasp both risks and opportunities related to climate change (SITRA 2015b). In Finland, a strong focus prevails on developing bioeconomy and cleantech solutions, building upon existing expertise and existing turnover in order to support the key interests not only in climate neutrality, prices and security of supply but also R&D, innovation and job creation. Indeed, the bioenergy sector has also attracted disproportionately high shares of R&D investment in the energy sector in Finland (Aalto et al. 2017).

The effects of and trends associated with a policy focused on an increased renewable energy share as outlined in Fig. 14.1—namely electrification, decentralisation, and variability—would benefit from developing smart networks while simultaneously developing related further policies. As the interlinked actions of both the public sector and energy companies to activate consumers enable the emergence of new actors, such as aggregators, their roles and interests require clarification vis-à-vis those of established actors (e.g. balance responsibility of aggregators). For policymakers, the key question is how to facilitate the integration of increasing amounts of renewable energy and how to handle the resulting impact on interests driving energy policy making. This interlinkage highlights the need for concurrent policy action in a variety of fields, the need for cross-sectoral cooperation and exchange in policy formation, as well as minimum requirements for other policy sectors. In line with the entry of new actors, incumbent actors must adapt to the changing environment.

Views of stakeholders on energy storage

Stakeholders are acutely aware of the possible overlap of interests and difficulties they face when having to prioritise between different solutions regarding the development of smart grids and energy storage. Grid integration of variable renewable energy is seen to be dependent on energy storage, while the development of local smart storage facilities might hinder the development of grids on a larger scale (interview 3). These interlinkages demonstrate how implementing a specific solution, possibly favoured by a subset of stakeholders, influences the wider field of stakeholders and their prospects of realising their favoured set of interests, as well as the solutions serving those ends.

While some forms of carbon-free energy storage already exist—most notably, the water reservoirs of hydropower plants—further investment in energy storages as a flexibility measure will be necessary. Given the high heat demand in Finnish conditions, especially heat storage, will be valuable in the short to medium term. In the longer run, the system might need the ability to store electricity for example in gas molecules that can then be flexibly converted to end products upon demand. The need for purpose-built battery storage facilities is not, however, expected to be large. The government's 2016 Energy Strategy notes the potential of various types of storage but stops short of providing a clear action plan for developing energy storage. It does, however, propose R&D funding to be targeted for demonstration and pilots involving new storage solutions, while again putting the consumer at the centre of the energy vision—an active consumer not only consumes energy, but also produces and stores it. However, the strategy also recognises that probably most consumers do not want to be particularly active. Instead, new service concepts could manage all customer needs like energy supply, microgrid management and electrical storages (Ministry of Economic Affairs and Employment 2017, p. 77). Additionally, storage is framed as a tool for customers to influence the electricity costs, enabling them to offer storage services to the market (Ministry of Economic Affairs and Employment 2018).

The smart grid working group, in line with the government's 2016 Energy Strategy, argues that given the neutral role of network companies as a facilitator of the market and given their monopoly unbundling obligation, DSOs should not become providers of storage services. The working group treats owning and using electricity storage as a competitive business activity ensuring that storage products and services develop efficiently, customer-driven, and needs-based. Considering thus expressed interests in developing markets and competition, network companies should acquire storage services from the market (Ministry of Economic Affairs and Employment 2018, p. 15).

This aligns with the view of the European Commission, which also sees the development and operation of storage facilities as commercial activity that should be performed by market parties, with TSOs and DSOs as a rule barred from owning, managing or operating energy storage facilities. Instead of ownership, energy storage should be used based on market procedures with the goal of ensuring efficient management of networks as well as avoiding costly network expansions (European Commission 2017, pp. 17–18). At the same time, the European Commission posits that exceptions are possible in cases where "(…) market parties are not interested in providing specific storage services (…)" with regulatory approval and oversight as well as a transparent market procedure being in place. In case of exceptions, regulatory authorities must assess the potential of other market parties to participate (European Commission 2017, p. 18).

Opposing this position on limiting energy storage ownership, the European Distribution System Operators for Smart Grids (EDSO) as well as the European Association for Storage of Energy (EASE) argue strongly in favour of DSOs being allowed to own storage. EDSO proposes to approach storage as a fully integrated network element to ensure the proper operation of the distribution system. Using storage to support the task of operating the network excludes any commercial activities, participation in competitive power markets or influencing price formation of the market (EDSO 2018, p. 2).

The importance of storage and the smart deployment of a variety of storage solutions were reiterated in the interviews. The maturity of storage technologies differ considerably, with some thermal storage technologies having been commercialised. In Finland, short-term thermal storage is used in combination with district heating to balance rapid peaks in consumption (Paiho et al. 2018, pp. 512–513). Meanwhile electrical battery storage is still in the early phase of deployment. For example, Helen has a 1.2 MW/600kWh electric battery storage in Helsinki, used for examining the prospects of frequency control operations, for covering the production gaps of a nearby rooftop solar PV plant and for handling peak consumption situations in the local area (Helen 2016). The former national incumbent Fortum operates a 2 MW/1MWh facility in Järvenpää. For regulatory reasons, such battery storage faces the risk of being treated both as production and consumption, leading to double taxation. Therefore, for example Fortum has situated its own facility as part of its own production unit.

The owners of such storage may yet turn out to profit, as may the owners of natural gas turbine plants capable of rapid ramp-ups, and respective natural gas and LNG storage, since these resources may alongside the short-term services of batteries act as back-up fuel on an hourly and daily basis in a high-renewable energy system (Holttinen 2017). However, the most traditional storage such as piles of coal or oil tanks is likely to be eventually phased out. However, so far, some power plants, that might have been otherwise retired, have remained operational through the capacity reserve mechanism, in which 729 MW of capacity was included by the Energy Authority in 2019.

The interviews highlight that distribution companies wrestle with the issue of energy storage and the question of ownership. Whereas the smart grid working group recommended prohibiting network companies from owning energy storage, and while pilot projects of battery storage exist, the question over who ultimately should own, invest in and operate such storage and the reasoning provided for this policy position surfaced in several interviews. DSOs not being able to own energy storage was seen as relying on an unnecessarily narrow definition of what the purpose of owning energy storage is, arguing that grid companies could utilise storages to avoid supply interruptions. Allowing the use of energy storages in this way requires their development—however, how this development can be supported if DSOs are excluded from ownership has been questioned (interview 5). This is also in line with EDSO (2018, p. 5) arguing that DSOs can support the market uptake and the implementation of innovative solutions such as energy storage.

One interviewee points out that while a more structured approach in this area is needed, regulations should not be overly detailed either and instead focus on a systemic view of the issue at hand, and on the interconnections of different moving parts. The exact procedures on defining what an exceptional case is are unclear and should be clarified from a systemic perspective (interview 2). Moreover, the interviewed stakeholders call for a clearer, more coherent approach from the EU (interview 6, interview 4). Such calls for more regulatory clarity also prevail more widely beyond Finland (EASE 2018). Whereas some interviewees lamented the lack of clear regulation, others warned against too narrow a definition of what energy storage is. Treating storage as production does not do justice to the complexity of the issue (interview 9). In line with EASE (2018) and EDSO (2018), DSOs should be allowed to own storage if this helps them to fulfil their tasks, while they should also be consulted more in policy development (interview 9, interview 11).

A contrasting, though minor, view also emerged from the interviews, questioning whether storage should instead of a service, as part of competitive energy business, be seen as a means of activating consumers (prosumers) in the electric energy system (interview 18). This is in line with the smart grid vision put forth by the working group. Technology and the changing energy system open up new and varied possibilities for consumers to participate in the electricity market—whether as an active customer or as a customer merely consenting to automation offered by the service provider handling the flexibility of her consumption, in return of sharing some of the benefits. Arguing that "a car is not only a car" (interview 2), the potential of cars alongside homes to fulfil functions in grid management is stressed. Concurrently, this potential alongside the fact that homes as well as cars are relatively large and long-term investments for consumers also results in calls for long-term, stable policy solutions (interview 10). Other interviewees also highlighted cars as "moving energy storages" (interview 9); infrastructure can most likely cope with more electric cars, and indeed their integration into the system is needed. Utilising EVs also for the needs of the power system and the electricity market increases the benefit of the EV to both the owner and the power system. For example, EVs could serve as a back-up system during energy outages (Rautiainen 2015, pp. 48 and 66).

Here, the interest in the wider socioeconomic implications that must be part of the policy planning becomes evident. For example, how should EV proliferation support measures be targeted so that they not only reach mainly the economically better-positioned citizens (interview 11)? Such questions touch on values and societal ideas that shape the ways in which the energy system is transforming. Thus, the competing interests and the changing roles of stakeholders also surface in the debate on energy storage. The interviewees treated DSOs as incumbent actors on the practical level steering the policy change towards a higher share of renewable energy sources, while new actors and fields of activity also emerge. Overall, the shifting nature of renewables policy connotes elements of uncertainty, such as the question of ownership of storage. While it remains unclear if long-term negative impacts might ensue from DSOs not being allowed to own storage, these incumbents are nevertheless pushed to consider their choices when reacting to the changes and have to assess how to align their interests with the wider policy interests at stake.

Conclusion

Finland has set itself ambitious targets to increase the share of renewable energy sources in the energy system in response to Nordic and international energy and climate policy processes. Increasing the share of renewable energy will mean that electrification, decentralisation and variability will be shaping the future energy system and necessitate moving from a production-centric system to a consumption-centric system. This raises new questions of flexibility measures such as smart grid development and energy storage. It is vital to understand how stakeholders position themselves in relation to these changes, since they are the users of the smart grid while many of them may also become storage owners or providers, for example, by means of EV ownership. For this reason, we need to know more about the interests of stakeholders vis-à-vis energy policy choices.

Our empirical analysis, based on policy documents and interviews with Finnish stakeholders, allows for a tentative assessment of the Finnish transition from a production-centric to a consumption-centric energy system. Whereas the need to transition to a higher share of renewable energy is making inroads into mainstream policy thinking, the transition policy needed for this change is still in a formative phase despite for example a smart grids vision issued by the Ministry of Economic Affairs and Employment following consultation among stakeholders. Actors recognise the multitude of interests to be served by this policy; however, prioritisation among different policies and solutions is proving difficult. Stakeholders identify a need for a regulatory framework for developing the energy system overall, without overly detailed regulations that might hinder innovation and policy development in specific areas, such as energy storage solutions. Finding the right balance between policies and giving enough room for innovation and development, while also offering enough structure and security is crucial.

Although climate neutrality is the main interest driving this work, our empirical material also reveals the presence of a strong interest in developing the energy business in order for the builders and developers of solutions to remain in the game in an environment where zero marginal cost production is invading markets. In a competitive and unbundled market public and private actors depend on each other—private actors on competition, regulation and supervision of a level playfield (to the extent possible), and public actors on companies to provide the end products of electricity and heat/cooling as well as taxation income. Here, we assume that energy cooperatives and private individuals will not in the foreseeable future be able to assume the main responsibility for energy services.

Such interlinking of interests of all actors is a formidable task. While the smart grids vision stresses enabling future business opportunities, our respondents highlight that some possibilities might be lost if Finland proves to be a late adopter or too risk-averse. Timing is hence essential. The question is what it takes to be a smart adopter, especially when it comes to activating and engaging customers, some of whom need to become prosumers and providers of resources. Agreement prevails on the need for consumers to take a stronger role in the energy system, but not everyone agrees on how to support such greater involvement.

Some of the transition pressures ahead are well illustrated by returning to the position of bioenergy in the Finnish energy system. There is a vested interest in bioenergy, still the most important source of Finnish renewable energy. However, so far bioenergy is used as a base-load resource providing electricity and heat while biomass itself offers energy storage services. Whether the actors relying on bioenergy can adapt to a situation where the relative share of wind and solar power increases and prices possibly decrease; where new, smart grid reliant solutions emerge, also in the form of storage capacity, and where bioenergy may have to assume also back-up power functions, remains to be seen. Such adaptation will lead to a different business model, and the answer to this a question has implications for R&D policies, employment as well as export income regarding what types of products biomass is in the future converted to. So far, energy policies have done little to address these questions of wider societal importance.

Appendix 1: Stakeholder Interviews

Interview number	*Sector*	*Area of expertise*
1	Business	Network/grid
2	Business	Solutions
3	Business	Energy/environment
4	Business	Network/grid
5	Business	Network/grid
6	Business	Network/grid
7	Business	Solutions/R&D
8	Business	Systems
9	Business	Network/grid
10	NGO	Consumption
11	Public sector	Markets
12	Business	Production
13	Public sector	Network/grid
14	NGO	Consumption
15	Business	Consumption
16	Business	Network/grid

(continued)

(continued)

Interview number	*Sector*	*Area of expertise*
17	NGO	Production
18	NGO	Consumption
19	NGO	Production
20	NGO	Climate/energy
21	Business	Production
22	NGO	Energy/climate
23	NGO	Consumption
24	NGO	Energy
25	Business	Network/grid

References

Aalto, Pami, David Dusseault, Michael D. Kennedy, and Markku Kivinen. 2014. Russia's Energy Relations in the East and West: Towards a Social Structurationist Approach to Energy Policy Formation. *Journal of International Relations and Development* 17 (1): 1–29. https://doi.org/(10.1057/jird.2012.29).

Aalto, Pami, Iida Jaakkola, Pertti Järventausta, Anna M. Oksa, and Pasi Toivanen. 2017. How to De-carbonise the Electric Energy System? A Comparison of Nordic 2030 Policies. International Conference on Energy, Environment and Climate Change (ICEECC 2017).

Aalto, Pami, Ilkka Ruostetsaari, Kim Talus, and Johannes C. Muth. 2019. Energy Policies. In *Encyclopedia of the UN Sustainable Development Goals: Good Health and Well-Being*, ed. Walter Leal Filho, Pinar Gökçin Özuyar, Paul J. Pace, Ulisses Azeiteiro, Luciana Brandli, and Tony Wall. Berlin: Springer.

Andrews-Speed, Philip. 2016. Applying Institutional Theory to the Low-Carbon Energy Transition. *Energy Research & Social Science* 13: 216–225.

Child, Michael, and Christian Breyer. 2016. Vision and Initial Feasibility Analysis of a Recarbonised Finnish Energy Systems for 2050. *Renewable and Sustainable Energy Reviews* 66: 517–536.

Child, Michael, Christian Breyer, and Teresa Haukkala. 2017. The Role of Solar Photovoltaics and Energy Storage Solutions in a 100% Renewable Energy System for Finland in 2050. *Sustainability* 9: 1358.

EASE. 2018. EASE [European Association for Storage of Energy] Reply to European Commission's Questions on "The Future of Energy Storage in the EU. September 2018." Available at: http://ease-storage.eu/wp-content/uploads/2018/09/2018.09.13_EASE-response_Electricity-Coordination-Group-Questions-on-Storage.pdf. Accessed 28 Jan 2019.

EDSO. 2018. Response to Commission Questionnaire on Energy Storage, September. Available at: https://www.edsoforsmartgrids.eu/wp-content/uploads/20180914_EDSO-response-to-EC-questions-on-storage.pdf. Accessed 28 Jan 2019.

Eikeland, Per O., and Tor H.J. Inderberg. 2016. Energy System Transformation and Long-Term Interest Constellations in Denmark: Can Agency Beat Structure? *Energy Research & Social Science* 11: 164–173.

Energiateollisuus. 2019. Energiavuosi 2018: sähkö. Available at: https://energia.fi/ajankohtaista_ja_materiaalipankki/materiaalipankki/energiavuosi_2018_-_sahko.html. Accessed 4 Feb 2019.

ENTSO-E. 2018. TYNDP 2018—Europe's Network Development Plan to 2025, 2030 and 2040. Available at: https://tyndp.entsoe.eu/tyndp2018/. Accessed 16 Apr 2019.

European Commission. 2017. Commission Staff Working Document. Energy Storage—The Role of Electricity. SWD (2017)61final.

Fingrid. 2018. Itsenäiset aggregaattorit tulevaisuuden joustomarkkinoilla. Available at: https://www.fingridlehti.fi/itsenaiset-aggregaattorit-tulevaisuuden-joustomarkkinoilla/. Accessed 4 Feb 2019.

Fortum. 2019. Fortum Spring: Building a Virtual Battery with Our Customers. Available at: https://www.fortum.com/products-and-services/smart-energy-solutions/virtual-battery-spring. Accessed 4 Feb 2019.

Government of Finland. 2019. 3.1. Carbon Neutral Finland That Protects Biodiversity. Available at: https://valtioneuvosto.fi/en/rinne/government-programme/carbon-neutral-finland-that-protects-biodiversity. Accessed 10 Dec 2019.

Heiskanen, Eva, Eva-Lotta Apajalahti, Kaisa Matschoss, and Raimo Lovio. 2018. Incumbent Energy Companies Navigating Energy Transitions: Strategic Action or Bricolage? *Environmental Innovation and Societal Transitions* 28: 57–69.

Helen. 2016. Largest Electricity Storage Facility in the Nordic Countries to Be Commissioned in Helsinki. Available at: https://www.helen.fi/en/news/2016/electricity-storage-facility/. Accessed 5 Feb 2019.

Helen. 2018. Gigantic Cavern Heat Storage Facility to Be Implemented in Mustikkamaa. Available at: https://www.helen.fi/en/news/2018/Gigantic-cavern-heat-storage-facility-to-be-implemented-in-Mustikkamaa/. Accessed 5 Feb 2019.

Heljo, Juhani, Pirkko Harsia, Hannele Holttinen, Pami Aalto, Tomas Björkqvist, Pertti Järventausta, Jari Kaivo-oja, Matti Kojo, Timo Korpela, Antti Rautiainen, Sami Repo, Ilkka Ruostetsaariand, and Jaakko Sorri. 2016. Tammikuun tehopiikki – mitä tapahtui 7.1.2016? Miten tehoa hallitaan paremmin jatkossa? *EL-TRAN—Analyysi 7/2016*. Available at: https://el-tran.fi/analyysit/. Accessed 2 May 2019.

Holttinen, Hannele. 2017. Pohjois-Euroopan sähköstä 40–60% tuuli/aurinkovoimalla: lisääntyvän joustavuuden vaikutukset Suomen CHP- ja ydinvoimatuotantoon. Presentation in the EL-TRAN interaction panel meeting, December 19.

Hoppe, Thomas, Frans Coenen, and Maya van den Berg. 2016. Illustrating the Use of Concepts from the Discipline of Policy Studies in Energy Research: An Explorative Literature Review. *Energy Research & Social Science* 21: 12–32.

Haukkala, Teresa. 2018. A Struggle for Change—The Formation of a Green-Transitions Advocacy Coalition in Finland. *Environmental Innovation and Societal Transitions* 27: 146–156.

Ikäheimo, Jussi, Esa Pursiheimo, Juha Kiviluoma, and Hannele Holttinen. 2019. The Role of Power to Liquids and Biomass to Liquids in a Nearly Renewable Energy System. *IET Renewable Power Generation* 13 (7): 1179–1189.

Kester, Johannes, Lance Noel, Gerardo Z. de Rubens, and Benjamin K. Sovacool. 2018. Promoting Vehicle to Grid (V2G) in the Nordic Region: Expert Advice on Policy Mechanisms for Accelerated Diffusion. *Energy Policy* 116: 422–432.

Kilpeläinen, Sarah, Pami Aalto, Pasi Toivanen, Pinja Lehtonen, and Hannele Holttinen. 2019. How to Achieve a More Resource-Efficient and Climate-Neutral Energy System by 2030? Views of Nordic Stakeholders. *Review of Policy Research* 36 (4): 448–472.

Kiviluoma, Juha. 2013. Managing Wind Power Variability and Uncertainty Through Increased Power System Flexibility. PhD. Thesis, Department of Applied Physics, Aalto University.

Kiviluoma, Juha, Erkka Rinne, and Niina Helistö. 2018. Comparison of Flexibility Options to Improve the Value of Variable Power Generation. *International Journal of Sustainable Energy* 37 (8): 761–781.

Kiviluoma, Juha, Hannele Holttinen, David Weir, Richard Scharff, Lennart Söder, Nickie Menemenlis, Nicolaos Cutululis, Irene Danti Lopez, Lannoye Eamonn, Ana Estanqueiro, Emilio Gomes-Lazaro, Qin Zhang, Jianhua Bai, Yih-Huei Wan, and Michael Milligan. 2016. Variability in Large-Scale Wind Power Generation. *Wind Energy* 19 (9): 1649–1665.

Kiviluoma, Juha, Miguel Azevedo, Erkka Rinne, and Niina Helistö. 2015. Cost Optimal Share of Wind Power, PV and Batteries. 14th Wind Integration Workshop, October 22.

Kojo, Matti, Kirsi Kotilainen, Antti Rautiainen, Pami Aalto, Hannele Holttinen, Topi Rönkkö, Jari Kaivo-oja, Tomas Björkqvist, and Pasi Toivanen. 2018. Miten sähköautopolitiikalla edistetään joustavampaa sähköjärjestelmää? *EL-TRAN—Analyysi 2/2018*. Available at: https://el-tran.fi/analyysit/. Accessed 4 Feb 2019.

Kotilainen, Kirsi, Pami Aalto, Antti Rautiainen, Matti Kojo, Jussi Valta, and Benjamin Sovacool. 2019. From National Lock-into Policy Mixes for Nordic

Electric Mobility: What Can We Learn to Accelerate Future Transitions? *Policy Sciences* 52: 573–600. https://doi.org/10.1007/s11077-019-09361-3.

Kotilainen, Kirsi, Teresa Haukkala, Pami Aalto, Antti Rautiainen, and Matti Kojo. 2018. Sähköautopolitiikat Pohjoismaissa – mitä keinoja Suomi voi hyödyntää? *EL-TRAN—Analyysi 4/2018*. Available at: https://el-tran.fi/analyysit/. Accessed 4 Feb 2019.

Luonnonvarakeskus [Luke]. 2018. Wood in Energy Generation 2017. Available at: https://stat.luke.fi/en/wood-energy-generation-2017_en. Accessed 4 Feb 2019.

Ministry of Economic Affairs and Employment. 2017. Government Report on the National Energy and Climate Strategy for 2030. Available at: https://tem.fi/documents/1410877/2769658/Government+report+on+the+National+Energy+and+Climate+Strategy+for+2030/0bb2a7be-d3c2-4149-a4c2-78449ceb1976. Accessed 5 Feb 2019.

Ministry of Economic Affairs and Employment. 2018. Flexible and Customer-Centered Electricity System. Final Report of the Smart Grid Working Group. Available at: https://julkaisut.valtioneuvosto.fi/handle/10024/161147. Accessed 19 Dec 2019.

Moe, Espen. 2015. *Renewable Energy Transformation of Fossil Fuel Backlash: Vested Interests in the Political Economy*. New York: Palgrave Macmillan.

Myllyntaus, Timo. 1990. The Role of Industry in the Electrification of Finland. ETLA Discussion Paper 333. The Research Institute of the Finnish Economy (ETLA), Helsinki.

Nylund, Nils-Olof. 2015. Tieliikenteen vaihtoehtoiset käyttövoimaratkaisut vuoteen 2030: Bio, sähkö vai molemmat? Presentation in the VTT 2G 2020 Biofuels Seminaari Bioruukki, Espoo, May 26.

Pääkkönen, Anna, Kalle Aro, Pami Aalto, Jukka Konttinen, and Matti Kojo. 2019. The Potential of Biomethane in Replacing Fossil Fuels in Heavy Transport—A Case Study on Finland. *Sustainability* 11 (17): 4750. https://doi.org/10.3390/su11174750.

Paiho, Satu, Heidi Saastamoinen, Elina Hakkarainen, Lassi Similä, Riku Pasonen, Jussi Ikäheimo, Miika Rämä, Markku Tuovinen, and Seppo Horsmanheimo. 2018. Increasing flexibility of Finnish Energy Systems—A Review of Technologies and Means. *Sustainable Cities and Society* 43: 509–523.

Rautiainen, Antti. 2015. *Aspects of Electric Vehicles and Demand Response in Electricity Grids*. Tampere University of Technology. Publication Vol. 1327.

Rocha, Marcia, Fabio Sferra, Michiel Scaeffer, Niklas Roming, Andrzej Ancygier, Paola Parra, Jasmin Cantzler, Alain Coimbra, and Bill Hare. 2016. *What Does the Paris Climate Agreement Mean for Finland and the European Union?* Helsinki: Sitra. Available at: https://www.sitra.fi/en/publications/what-does-paris-climate-agreement-mean-finland-and-european-union/. Accessed 4 Feb 2019.

Ruggiero, Salvatore, Vilja Varho, and Pasi Rikkonen. 2015. Transitions to Distributed Energy Generation in Finland: Prospects and Barriers. *Energy Policy* 86: 433–443.

Ruostetsaari, Ilkka. 2010. *Energiavalta: eliitti ja kansalaiset muuttuvilla energiamarkkinoilla*. Tampere: Tampere University Press.

Smart Grid Vision. 2016. Finnish Smart Grid Vision. Available at: https://tem.fi/documents/1410877/3481825/Smart+grid+vision+30102016+finalEN.pdf/479bdd14-2794-4cea-8f77-b9f5dc2d8a2d. Accessed 9 Dec 2019.

SITRA. 2015a. Climate Impact Assessment of Companies Listed on Nasdaq Helsinki. Available at: https://www.sitra.fi/en/publications/climate-impact-assessment-companies-listed-nasdaq-helsinki/. Accessed 8 Jan 2019.

SITRA. 2015b. Finnish Companies Realise the Opportunities Presented by Decarbonisation. Available at: https://www.sitra.fi/en/news/finnish-companies-realise-opportunities-presented-decarbonisation/. Accessed 28 Jan 2019.

SULPU [Finnish Heat Pump Association]. (2017). The Heat Pump Market in Finland 2017. Available at: https://www.sulpu.fi/documents/184029/0/Heat%20Pump%20installations%20market%20in%20Finland%202017%2C%20slides.pdf. Accessed 4 Feb 2019.

Sun Energia. 2019. Sun Energia tuntee kattosi. Available at: https://sunenergia.com/. Accessed 4 Feb 2019.

Työ- ja elinkeinoministeriö [TEM]. 2016. 100-prosenttisesti uusiutuviin energialähteisiin perustuva energiajärjestelmä: Kansalliseen energia- ja ilmastostrategiaan liittyvä tarkastelu, November 24. Available at: https://tem.fi/documents/1410877/3570111/100+prosenttia+uusiutuvaa+tarkastelu.pdf/8e4ee341-77c5-4447-b6ce-1f2686a3daec/100+prosenttia+uusiutuvaa+tarkastelu.pdf.pdf. Accessed 4 Feb 2019.

Tilastokeskus. 2018. Uusiutuvan energian käyttö jatkoi kasvuaan vuonna 2017. Available at: https://www.stat.fi/til/ehk/2017/ehk_2017_2018-12-11_tie_001_fi.html. Accessed 4 Feb 2019.

Toivanen, Pasi, Pinja Lehtonen, Pami Aalto, Tomas Björkqvist, Pertti Järventausta, Sarah Kilpeläinen, Matti Kojo, and Fanni Mylläri. 2017. Finland's 2030 Energy System as Envisioned by Expert Stakeholders. *Energy Strategy Reviews* 18: 150–156. https://doi.org/10.1016/j.esr.2017.09.007.

Tuulivoimayhdistys. 2019. Tuulivoimahankkeet Suomessa. Available at: http://www.tuulivoimayhdistys.fi/hankelista. Accessed 4 Feb 2019.

Uski, Sanna, Pami Aalto, Kim Forssén, Hannele Holttinen, Matti Kojo, Sami Repo, Tony Rosqvist, Janne Sarsama, and Kim Talus. 2018. Mikroverkkojen mahdollisuuksista sähkön toimituksen luotettavuuden parantajana. *EL-TRAN—Analyysi 3/2018*. Available at: https://el-tran.fi/analyysit/. Accessed 19 Dec 2019.

Valkila, Noora, and Arto Saari. 2010. Urgent Need for a New Approach to Energy Policy: The Case of Finland. *Renewable and Sustainable Energy Reviews* 14: 2068–2076.

Valkila, Noora, and Arti Saari. 2013. Experts' View on Finland's Energy Policy. *Renewable and Sustainable Energy Reviews* 17: 283–290.

Zakari, Behnam, Sanna Syri, and Samuli Rinne. 2015. Higher Renewable Energy Integration into the Existing Energy System of Finland—Is There a Maximum Limit? *Energy* 92: 244–259.

CHAPTER 15

Conclusions

Espen Moe

In 2014, we published what in many ways was the predecessor to this book, *The Political Economy of Renewable Energy and Energy Security* (Moe and Midford 2014). The focus was in part different (there is no clear energy security dimension to this present volume), and we are not looking at all the same cases in both volumes, but there is a very significant overlap between the two books. Japan is a very important part of both books, and China, Norway and Denmark also figure in both. And beyond that, both books focus very much on the political economy of renewable energy. Thus, looking back to 2014 and comparing the situation today with our conclusions and predictions from then seems appropriate.

First, renewable energy has come a long way, and the field has changed quite a bit since our previous volume was published. When I re-read the Introduction (Moe 2014a) chapter, it says among other things that "Denmark has become one of the world's foremost producers of wind power, deriving as much as 27% of its electricity consumption from wind." Today,

E. Moe (✉)
Norwegian University of Science and Technology (NTNU), Trondheim, Norway
e-mail: espen.moe@ntnu.no

P. Midford and E. Moe (eds.), *New Challenges and Solutions for Renewable Energy*, International Political Economy Series,
https://doi.org/10.1007/978-3-030-54514-7_15

six years later, the figure is 47% (Reuters 2020). We wrote, back in 2014, that the German goals were so ambitious that they bordered on the unrealistic, with among other things the share of renewables in electricity consumption doubling from a current 17 to 35% by 2020. And lo and behold, at the end of 2019, solar PV accounted for 10.5% of electricity production with wind power at 27%. Renewables in total (also including biomass and hydropower) came in at more than 50%, easily surpassing the old targets (Fraunhofer ISE 2021).

Germany is not a part of this present volume, so I will not spend more time on that, but worldwide, at the end of 2013 wind and solar accounted for 3.6% of electricity consumption. At the end of 2019, the corresponding figure was 8.7% (REN21 2014, 2020). While more than a doubling is impressive, less than 9% is still no revolution. At the same time, in many of the countries that have the most impressive renewable rates (and even in many of those that are distinctly average), we seem to be rapidly entering a new phase of renewable energy. At the time of concluding the previous volume, we were still at a stage where Feed-in Tariffs (FITs) (and priority access to the grid) were the gold standard of renewable energy. In many ways it was an easy time in the sense that renewable energy policy was primarily about the support system. Did the state have a FIT? How high was it? If not, was there at least another system of support, such as for instance a Renewable Portfolio Standard (RPS)? Thus, the government's role to a significant extent was that of providing financial support for installations, and then the actors in the market would install wind and solar power to the extent that it was profitable. However, the more successful FITs became, the more expensive they also became (as they applied to an ever-larger quantity of installations), and the costs were very often borne by individual consumers and industry in the form of surcharges. Priority access to the grid was a complicating factor. Many countries did not mandate electric utility companies to accept electricity generated by renewable energy to the grid, and there renewables invariably struggled. Also, if countries had support systems that overwhelmingly incentivized the installations of intermittent renewable energy without also incentivizing the strengthening of the electricity grid (or additional storage capacity), the grid would become ever less capable of coping with large amounts of renewable energy.

Thus, in the previous volume we suggested that the success of renewable energy would in the future be putting major strains on the electric grid (Moe 2014a), but did not otherwise spend much time and effort on it. That is one of the major differences between these two volumes. The present volume focuses explicitly on challenges to renewable energy development that were still issues for the future in 2014.

As ever more renewable energy is phased onto the grid, in country after country, the grid itself is put under ever greater strain. This is not primarily because the amounts of electricity provided by renewables itself are so enormous: As mentioned, worldwide, wind and solar power still account for only 8.7% of total electricity consumption (REN21 2020). Rather, it is the intermittencies of renewable energy and the lack of good energy storage assets that makes renewable energy problematic for the grid system.

What makes this a problem that must be dealt with, is that any renewable energy transition away from fossil fuels requires literally every country to grid-connect much greater amounts of renewable electricity than is currently produced. Thus, persisting with the present grid system is simply not an option. For electricity systems to become carbon-free, what is required is a grid system with a capacity that far exceeds that of the current. This means that domestically countries require systems that far improve on the current in terms of accepting intermittent energy. However, what is also required is a system that allows countries to buy and sell electricity from each other more easily. This means significantly increasing the number and capacity of interconnectors between countries.

There is far more literature on the political economy of renewable energy than on the political economy of electricity grids. Over the past few years, the amount of literature that tries to describe and explain why countries do or do not support renewable energy strategies has grown rapidly, see for instance both our previous volume (Moe and Midford 2014) and the monograph by this author (Moe 2015) for works on the political economy of renewable energy. But in these, as in much of the other literature, the grid system was given scant attention. There are undoubtedly many reasons for this, but I speculate that there are at least two. First, the physical visibility of new renewable energy installations is far greater than the physical visibility of grid-line expansions and upgrades to existing infrastructures. For very material reasons, it has been easier to intuitively grasp and enthuse over or rage against renewable installations, especially wind power.

However, second, and possibly more importantly, as suggested earlier, renewable energy policy is (or was) not all that difficult. Granted, there are many different support systems, from production tax credits to RPS to FITs to green certificates and tendering, but these are all reasonably uncomplicated in terms of their implementation. They to differing degrees cost the state money in terms of subsidies, or they push the burden onto the electricity consumer. Nonetheless, once the state has decided on which type of support system to implement, beyond typically having to deal with concession applications, it can essentially lean back and let the industrial actors make the rest of the decisions based on their profit evaluations. The politics of grid-line expansion are however more complicated. Grid enhancement involves the construction or the significant upgrading (for instance, through smart grids for increased flexibility) of physical infrastructure, stretching over thousands of kilometers. And it is expensive. Back in 2013, The Economist loosely estimated that the European grid system required upgrades of around €1 trillion by 2020. It is hard to see this happening without any sustained government effort. Thus, it is not difficult to see how grid-line expansion has lagged renewable energy installations in so many countries, producing one of the biggest bottlenecks to renewable expansion.

(And maybe third, since grid-line expansion has been slow, there has not been as much to study.)

What we have however seen is that in those countries where renewable energy has expanded the fastest, the mismatch between installations and grid-line expansion is also the greatest. Germany for instance ended its FIT for this reason (in addition to dramatic cost-reductions for both wind and solar), and in several other countries, installations have been running far ahead of the grid, resulting in curtailment problems (China at times has been an extreme example of this), and in grid companies resisting renewable energy (which we have described in detail with respect to Japan both in this and the previous volume) (Moe 2014b, 2015, 2017).

This demonstrates the potential of renewable energy as a disruptive technology. The rapid influx of renewable energy threatens the integrity of the centralized electricity systems that most countries have spent the best part of a century installing and jeopardizes the profits and the entire business model of the traditionally most important actors of the old and existing electricity system, namely the grid and electric utility companies. These are however also exactly those actors that renewable energy depends upon for a stronger and better grid, capable of accepting renewable energy in quantities that would be unthinkable today. Joseph

Schumpeter (1942) would undoubtedly have seen present developments as examples of potential creative destruction. However, while the old system is being destroyed (or vehemently resisting destruction), it is hard to see exactly what a new system will look like, and how fast it will come into being. What is, however, not hard to assert is that the old system is strained, and that it is difficult to see how it will survive in its present form. At the same time, the system, especially the grid, needs to survive in one form or another, for this wave of creative destruction and this disruptive technology to foster creation out of destruction, and to lead to a new electricity paradigm.

This does not become easier if we conceive of it as an international as well as a domestic problem, because a stronger grid is a solution to the problems of intermittencies, not just within countries, but also between them. The larger and more encompassing the grid can be made, the more robust it becomes. Here, Japan always had a problem, being an island with only moderately good relationships with its neighbors. Granted, in the wake of Fukushima, SoftBank CEO, Son Masayoshi, introduced the idea of the Asia Super Grid, tying Japan together with China, Russia and South Korea in an ultra-high-voltage grid over thousands of kilometers, and in 2016 a Memorandum of Understanding was signed between SoftBank and China's State Grid, the Korea Electric Power Corporation and the Russian national transmission operator Rosetti. (Mongolia also wants to join.) The Japanese government is however wary of this potential cooperation, and there are massive political hurdles that need to be negotiated before this could be a solution to Japanese grid problems (Cleantechnica 2016; Energy Insider 2019).

European countries do not have this problem, even if interconnectors have been a weak spot here, too. In 2002, EU member states agreed on an interconnection target, saying that by 2005 interconnections would correspond to 10% of installed production capacity. But as of 2015, the average interconnection rate was still only 8%. The 10% target was reiterated for 2020, and there is a 15% target for 2030 (Planete Energies 2016; Szulecki et al. 2016). If interconnections are however a problem already at this stage of a renewable energy transition, it is hard to see how ambitious goals of establishing carbon-free energy systems in a few decades can ever be fulfilled. The Norway chapter, among other things, looks at how interconnectors (here: subsea cables) very fast became tension-filled, and how difficult it can sometimes be to frame a genuinely global problem (climate change) as one that requires international cooperation.

The Norwegian discourse is instead framed in national terms. Thus, when we talk about new challenges to renewable energy, these are very much both at the national and the international level.

Japan has been a central focus to both our edited volumes. It accounts for six out of 15 chapters, whereas back in 2014, we had four chapters dealing in part or exclusively with Japan (Asano 2014; Midford 2014; Moe 2014b; Umbach 2014). A recurring theme in all those was that the Fukushima Daiichi nuclear accident was still recent, that Japan was still dealing with its aftermath, but that, disastrous as it was, Fukushima also presented Japan with a window of opportunity to break with the past and find new energy trajectories for the future. Thus, the potential for change, identifying numerous early signs of change, but with the caveat that it was still too early to draw any firm conclusions, was central to these chapters. Among the problems that we addressed was how Japanese energy policy for almost half a century had been unchanged because of the strength of vested interests and an iron triangle protecting the utilities and the nuclear village from energy challengers. It was a system that thoroughly sheltered its vested interests, but where following Fukushima a weakening of the nuclear village, the growth of solar PV and the possible unbundling of energy transmission and distribution pointed toward a potentially new Japanese energy future.

This volume is an excellent opportunity for us to take stock of what has happened since Fukushima. Has Japan changed? The answer seems to be yes, while at the same time staying very much the same. From an energy point of view, the biggest loser from Fukushima was nuclear power. In 2014, we wondered if there was a way back for nuclear (Midford 2014; Moe 2014b; Umbach 2014). We still do not have the conclusive answer. The former Liberal Democratic Party (LDP) administration of Abe Shinzō was certainly a greater friend of nuclear than the preceding Democratic Party of Japan (DPJ) administrations. However, what Paul Midford stresses in his chapter is that while Abe had a goal of phasing nuclear back in (20–22% of the power mix by 2030), for practical purposes, compared to the DPJ administration, his government was merely slowing down the nuclear phaseout. Thus, the outcome is an almost inevitable end to nuclear, even if it will take longer than the DPJ once advocated. As Florentine Koppenborg points out in her chapter, nuclear is now subject to a much stronger regulatory regime than pre-Fukushima. Thus, this is an area where Fukushima has had a potentially irreversible impact, even if some of our other chapters describe how the nuclear village is certainly fighting back.

When it comes to vested interests in general, Japan is still very much a version of its old self. Koichi Hasegawa and Hiroshi Ohta both lament the lack of a political core fighting for renewable energy and how the vested interests of the nuclear power industry and electric utility companies are still obstacles to real reform. This is very much what we pointed out in 2014 (Moe 2014b), and here, the old vested interests have staged partially successful comebacks. Granted, Fukushima resulted in a very generous FIT (especially so for solar PV) (in our previous book, Asano (2014) even described it as too generous), that in line with developments everywhere else, and as originally envisaged, has been gradually reduced, although not yet phased out. The result was that solar PV soared, annual installation rates increasing from around 250 MW before 2010 to more than 10 GW in 2015, before slowing down somewhat again, to 7 GW in 2019. METI's target of solar accounting for 7% of the power mix by 2030 was surpassed already in 2019 (REN21 2020). Wind power has however remained stuck at installation rates that for many years have been even lower than before Fukushima, and the utility companies are still distinctly lukewarm about accepting renewable energy onto the grid. There have been several cases of utilities curtailing solar PV access. As Robert Uriu however points out in his chapter, hydrogen fuel cells could spell one genuinely new direction, as Japan in 2014 was the first country in the world that committed to becoming a full-fledged hydrogen society, very much against the interests of the nuclear industry. Here, energy and climate policy go together with industrial policy, as Japan for three decades already has been promoting a hydrogen and fuel cell industry. This very much increases the chances of sidestepping the existing Japanese vested energy interests.

We also wondered what would happen to another development that was very much triggered by the policy response to Fukushima, namely the unbundling of the utilities. This was completely in its infancy in 2014, and something that the utilities strongly opposed. Ohta's chapter goes through the changes that have happened in this area, and while they may not be as revolutionary as they were once portrayed as, they have been carried out, in the face of resistance from the utilities, and by a METI and an LDP that were once seen as their staunch allies. Thus, while Japan still has the same vested interests fighting for the same goals as before, making the Japan of 2020 a country that we in 2014 would immediately have recognized, changes have clearly happened.

China is another country that has changed massively while staying very much the same. In 2014, we noted that renewables in China had now become bigger than nuclear, in terms of capacity and electricity generated. China was the world's number one in terms of wind power capacity already in 2010. At the end of 2013, more than 90 GW was installed, and by the end of 2019, this had increased to 236 GW, more than double that of the next country. By 2014, Chinese solar PV had mostly been an export industry and was still in its infancy as a domestic power source. This was however in a process of rapid change. From less than 1 GW of installed capacity in 2010, it had reached 20 GW in 2013. It rose to 43 GW in 2015, when it surpassed Germany as the world's number one, and by the end of 2019, China passed 200 GW of installed capacity, about three times more than the second largest country. Time and again China has beaten its own targets, almost annually setting new installation world records. The one area in which China underwhelmed back in 2014 was in offshore wind. Here, the pace has however increased markedly. With nearly 7 GW China is still behind Great Britain and Germany on total capacity (but arguably not for long), but now has the largest offshore wind market in the world (China Energy Portal 2020; GWEC 2020; Moe 2014b).

However, in terms of share of power generation, it also becomes evident that high absolute numbers do not tell the whole story. At the end of 2013, wind and solar accounted for no more than 2.8% of the Chinese power mix. This rose to 8.5% by the end of 2019. There are several things to say about this. First, it is a very sizeable increase, especially for solar PV, which went from a negligible 0.3 to 3%. But second, it is not impressive compared to countries like Denmark and Germany, and as late as 2015, the USA generated more wind power electricity than China despite having only half the installed capacity. Thus, the Chinese case is also a cautionary lesson not to be so overwhelmed by impressive installation figures that we forget to check if those figures translate into equally impressive renewable electricity generation. We noted that there were numerous factors behind China getting less electricity out of its installations than the USA. And third, this means that China's energy economy is still overwhelmingly dominated by coal. Granted, the coal share has decreased, from almost 68% in 2013 to approximately 60% in 2018 and is predicted to keep falling, but in absolute figures, China consumes and produces more coal now than it did as recently as 2010 (chinadialogue 2019; China Energy Portal 2020; Moe 2017; REN21 2014).

We noted that China's renewable energy development had many contradictory and dysfunctional elements. For instance, curtailment of renewable electricity, especially from wind power, was higher than pretty much anywhere in the world—a lot of the renewable electricity generation was simply wasted (Moe 2014b). This problem has continued. As Chen Gang writes in his chapter, China has most of the power sources located in its western regions, but its population located primarily along the eastern seaboard. This has imposed enormous pressures on the transmission network, triggering the construction of a large West-East Electricity Transmission Project and the building of ultra-high-voltage transmission lines for long-distance electricity transmission. However, as Chen writes, "inefficiency in northwest provinces is still astronomically high." Thus, when this book focuses on challenges to renewables that primarily have risen since the previous book, China strikes right at the heart of this. We wrote about the enormous curtailment problems that China had in 2014. There is no doubt that Chinese authorities are aware of these and actively sought to remedy them.[1] And yet, Chen writes that China has overinvested in renewable power in remote locations while having a backward and aging grid system, underinvesting in smart grids and has ended up with a grid system that is a severe bottleneck. What China is experiencing, is that the first and easy phase of the renewable transition is over. The more difficult phase is one that requires massive grid investments, expanding grid capacity and building smart grids. This is something that we have also seen in other countries. Incentives were provided for renewable energy installations, but not for grid improvements. The inevitable consequence: at some stage, the grid becomes a major bottleneck, unable to incorporate ever larger intermittent quantities of renewable electricity.

Also, while China is often perceived as a strong state, local or regional politics often interfered with the policies of the central government in Beijing, especially in the most important coal provinces, where coal was routinely given preference over renewables, very much contrary to the intentions of Beijing (Moe 2014b; Wang 2014). We still recognize much of what we noted in 2014. Chen states that "Chinese policymakers expected local governments to play a very cooperative role in enforcing those policies, laws and regulations." There are however many cases where the interests of local government diverge sharply from those of central planners, local authorities helping local fossil fuel companies to bypass energy directives and curbing competition from renewables. Thus, the institutional weaknesses in enforcing national policies that we pointed out in 2014 still exist.

However, we also noted how steadfastly the central government had supported renewable energy, singling out renewable energy industries as industries of the future and creating a support system that was both broad and comprehensive, with renewables having priority access to the grid from as early as 2005 (even if the utility companies often simply ignored this) (Guo et al. 2014; Moe 2014b; Wang 2014). Government support has been an important factor behind China's extremely rapid developments in renewables. Yet, the problems that we recognized in 2014 are still there for everyone to see. It is not as if Chinese authorities are unaware or unwilling to seek solutions to these problems. However, the sheer amount of Chinese renewable installations since 2014 makes it ever harder to find permanent solutions, as the target keeps rapidly moving. After all, at the end of 2013, China had installed 111 GW of solar and wind power. At the end of 2019, the number was 430 GW.

Of the other countries that figure heavily in both books, **Norway** has also seen interesting changes since 2014. Some of these are very specific to Norway's very peculiar energy structure: Norway is a major petroleum exporter, but where 91% of the electricity that is produced domestically derives from hydropower in addition to approximately 7% from wind power. However, other developments are very suggestive of the changes that the entire field is undergoing.

Thus, in the previous volume we talked about how Norwegian renewable energy policy had been deadlocked for decades, abundant in hydropower as well as strongly influenced by the needs of the country's mainly state-owned petroleum company, Equinor (then Statoil), and with a Ministry of Petroleum and Energy that insisted that any other energy interests would have to be cost-competitive from day one, something that wind power was decidedly not. Consequently, Norway had one of the least favorable support systems for wind in Europe, and very little wind power (Moe 2014b). Two major developments since then however changed the country significantly and put the focus on what we highlight in this book, namely future challenges. First, the EU Renewable Energy Directive was one of the main reasons why Norway introduced a green certificate system for renewable energy, committing the country to financing the generation of an extra 13.2 TWh of renewable electricity before the end of 2021; second, a dramatic drop in costs for wind power, resulting in wind power becoming competitive on price, at least in the windier parts of the country. One consequence was far more rapid installations of wind power. Another consequence was a power surplus and a (still ongoing) discussion about what to do with this surplus. One possibility is the construction of subsea cables to other European countries, for the export of Norwegian renewable electricity. The more renewable energy that is phased in globally, the greater the need for interconnectors between countries.

As mentioned earlier, Europe already has some cross-border interconnectors, but where it can be difficult just to strengthen the grid infrastructure *within* a country, it is far harder to do so *between* countries. Thus, part of what the Norway chapter does is analyze the political processes behind the construction of two subsea cables to Germany and to Great Britain. Among other things, it shows that there are very few Norwegian actors who think in terms of global solutions. Instead, the discussion about subsea cables occurs at the domestic level, using arguments that belong to the state level of analysis, and not the system, arguments that have to do with renewable energy installations vs. nature conservation (i.e., Otte et al. 2018), or with Norwegian industrial interests and sovereignty concerns (i.e., Hansen and Moe 2021). Thus, the authors argue that Norway is a long way from becoming a green battery for Europe and probably never will be, despite the fact that the climate problem is a supranational problem where one of the main solutions, namely a European-wide grid-net, is also supranational (or at least international). Part of this discussion is highly Norway-specific, and interconnectors were discussed in 2013 as well (e.g., Gullberg 2013). However, then they were discussed very much as developments of the future, as opposed to today, when it is something that is clearly happening *now*, with concrete implications for Norwegian politicians, households, industrial and environmental actors. The fact that interconnectors (i.e., subsea cables) are these days at the core of Norway's renewable energy discourse is indicative of how renewable energy politics has changed since the publication of the previous volume.

In the previous volume, **Denmark** was an example of a country that had seemingly already made the leap to the next stage (Moe 2014b). Here, the share of wind in the power mix was so high (27%) that this was only attainable because of interconnectors to other countries—Norway, Sweden and Germany. Thus, Denmark was already in a situation where the grid on very windy days was unable to incorporate all this energy. But instead of curtailments, as in the case of China, the solution has been a (compared to most others) well-developed network of interconnectors to other countries, allowing for seamless electricity exports and imports.

What Boscan et al. discuss in their chapter also suggests that to some extent Denmark is still ahead of the curve and has moved on to challenges that for most other countries still lie in the future: the greening of the transportation sector (an area where Norway is however significantly ahead of Denmark), the greening of district heating, and the removal of cost barriers for end users to adopt electric transportation and heating. However, as Denmark moves ahead with renewable energy, a familiar final point also pops up, namely dwindling domestic backup capacity. Thus, while being connected through interconnectors to other countries, the almost doubling in wind power's share of the power mix since 2014 (to 47% and a clear world record among any country bigger than a microstate) has made the country more dependent on others and thus more vulnerable. In the authors' words, it has increased the Danish flexibility requirements. This could be thought of as a natural development on the path towards a fully carbon-free energy system, the only difference between Denmark and most others being that Denmark has progressed further. It does however also bring legitimate concerns—concerns that other countries may eventually also have to deal with. Thus, in the past, Denmark had enough surplus capacity to always be able to meet peak demand. The authors however state that this may soon no longer be the case, as Denmark is phasing out domestic thermal capacity—the traditional backup power—and substituting this with more wind power, which is intermittent. It makes it evident that a renewable energy transition entails a clear tradeoff between a traditional understanding of energy security anchored in the state's ability to produce enough energy to cater to its own needs, and a networked grid-world, where the intermittencies of renewables imply that a global (or for that matter European) energy transition will not happen unless countries are able to rely on each other and supply each other with energy through a vastly expanded mesh of cross-border interconnectors.

In nine out of the past 10 years (2008–2017), Denmark was a net importer—typically exporting in the winter but importing in the summer. Thus, the security of supply concern is real. Part of the response has been to try to make the system more efficient, among other things by a rollout of smart meters (completed by 2020) and by advancing smart grid solutions. These were solutions that belonged to the future back in 2014, to some extent for technological reasons, but mostly because back then, these were not yet problems that needed solving.

In 2014, we posed questions about the stability of Danish wind power policies (Moe 2014b). There had been a lull in installations between 2003 and 2008 because a new and more market-friendly government insisted on wind power needing to be competitive without subsidies, and at the time of writing, another market-oriented government had just come into power. Thus, we wondered if this would mean a series of stops and starts for Danish wind power, ending the relative stability and consensus that had characterized the previous 25 years. However, Denmark has seen an almost 50% rise in installations since 2013, and a 2018 energy policy agreement will see another 2.4 GW of offshore wind installed by 2030. Thus, wind power's 47% share of the power mix will keep increasing.

In 2014, we also noted that the Danish ownership structure, seen by many scholars as crucial to the success of Danish wind, and an ownership structure that has made wind power far less controversial than for instance in Norway, might be under threat. However, as Boscan et al. write, the ownership structure is still vital. The roots of energy sector success in Denmark are deeply social and tied to self-driven community development. Communities have ample room to innovate and design their projects, and the local government is likely to support them. In most respects, Denmark of 2020 would also have looked familiar to someone from 2014.

The four new countries, included in this present volume only, are all at different stages of development and with very different geographical situations. Completing the Nordic section of this book, **Finland** aims for carbon neutrality (by 2035). But where Norway is plentiful in hydropower and Denmark in wind, this third Nordic country instead relies on biomass for 27% of its final energy consumption and has a large forestry industry. At 22% of the power mix and 4% of final energy, hydropower is significant as well, but it is really the share of biomass that makes Finland different. Beyond providing context about Finland's slightly peculiar energy mix, Kilpeläinen et al. provide another chapter that goes straight to the core of the theme of this book, describing how the influx of renewables will move the Finnish energy system from production-centric to consumption-centric, forcing it to become more flexible because of new electrified demands in heating and transport.

This also makes the need for energy storage more acute and moves the focus from large thermal power plants to the managing, storage, redistribution and consumption of variable two-way flows of electricity. This is a major change, since as the authors write "the country's energy system was originally heavily shaped by the demand from large units in the forestry, chemical and metal sectors, and respective electrification processes. Electricity distribution grids were devised to be one-way roads towards consumption." Of course, there will be some continuity in the system. Nuclear power is an important baseload and will still be accounting for a third of the power supply by 2030, but renewables will contribute much variability, requiring the system to become far more flexible. One of the crucial points that the authors make is that this is not just about changing technologies, but that this will lead to winners and losers, and to transforming the field of stakeholders, especially with respect to grid development and energy storage. This makes the processes more complex, more opaque and harder to predict. Which also strikes at the heart of this book, namely that the easy part of the renewable energy transition is over. It is no longer about phasing in renewables as fast as possible through support systems like FITs or RPSs and then letting the actors in the market do the rest. It is now instead about realizing that we have come far enough that further progress is contingent on major improvements to the grid infrastructure, and that these changes go along with political processes that are far heavier and involving far more actors than the first phase of renewable expansion.

The remaining three are from East and Southeast Asia. **Vietnam** is the only case of a developing country in this book (although a rapidly developing one). Thus, up until very recently, sustainable sources of energy have been low on its list of priorities, with energy security and coal, gas and large-scale hydropower the main focus of its energy policies. Nam et al. however describe a country that while being in an early phase of renewable expansion is very serious about renewables, and that has made great strides in several areas.

Vietnam has great potential for wind power, both because of a 3000 km coastline and highlands with very good wind conditions. GWEC (2019, 2020) estimates its technical offshore wind potential at 475 GW, and 99 MW have already been installed. The country has a solid support system and a generous FIT for offshore wind. Onshore, it has an 800 MW target by 2020, which is unlikely to be reached as it had only installed 388 MW by the end of 2019. Solar PV is also very much in its infancy, with a 2020 target of 850 MW (GWEC 2020). These are targets that are also indicative of a developing country still in the early stages of renewable expansion. They are not particularly ambitious targets for an economy with nearly 100 million people, even if its GDP per capita is only one-tenth of that of Norway. However, after the introduction of a FIT, solar installations surged to the extent that at the end of 2019, Vietnam had long surpassed the 850 MW target, with total installations of 5.5 GW of solar PV in only a few years (VOA News 2020).

What is also indicative of a rapidly growing developing country is the fact that for a decade GDP has been growing steadily at more than 6% a year and energy consumption at almost 12%. Thus, it is a country that needs renewables for many reasons, not necessarily as part of an energy transition, but for sheer energy security-reasons. Vietnam needs more energy (it also recently abandoned its nuclear power program), thus the long-term renewable targets are considerably more ambitious. Renewables are increasingly seen as part of the solution, with a 6 GW wind power and a 12 GW solar PV target for 2030, and a goal that all renewables except hydropower will then account for 10% of the power mix. With the surge in solar installations, this target is well within reach. As is however often the case with both developing countries and countries in an early phase of renewable expansion, it goes along with regulatory opacity, complex approval processes, a lack of domestic technical expertise and an underdeveloped power grid (GWEC 2020).

What this is also means is that Vietnam is still in that early phase where the focus has been on phasing in renewables as fast as possible, providing incentives for installments rather than grid development. It is also abundantly clear that hydropower will yield more power, as the authors state that installed capacity will increase from beyond 20 GW today to 27.8 GW by 2030. This is close to the total capacity of Norway, which in terms of square kilometers is virtually the same size as Vietnam. And from the Norway chapter we know that Norway holds roughly 50% of Europe's total hydro storage capacity. Obviously, there is a difference in that Norwegian hydropower caters to a little over 5 million people and Vietnam a little less than 100 million, but it still means that hydropower has the potential to become a very significant part of the Vietnamese energy mix (already at 40% of the capacity for electricity generation), especially when taking into account that energy use per capita is a lot lower than in Norway.

Where Vietnam was an outlier in being the only genuine developing country in the book (China is now a middle-income country), **Singapore** is an outlier in being a city-state. Thus, it is hard to see much potential for wind power in Singapore, at least onshore, while offshore the wind speeds are low and the waters crowded. Solar power can however be installed on rooftops, and because of its very limited area, Singapore needs more energy. It does not have any within its national borders and relies almost entirely on imports. This means that solar PV is the only potential domestic energy source that can credibly reduce Singapore's emissions intensity.

The main problem for solar, as Jindal et al. state, is making it profitable. There are no subsidies for any energy sources in Singapore. With low and volatile electricity prices and a small and fragmented domestic market, this makes for slow adoption of solar. It however also makes for a very interesting case study, namely one of renewable expansion without subsidies. In addition, the median cloud cover is 90%, thus despite being close to the equator, PV output is intermittent. Thus, the Energy Market Authority worries about what high solar penetration may do to Singaporean grid stability and reliability. What it means is that with present policies, Singapore, like Vietnam, is not looking at any renewable energy transition. But with electricity consumption increasing by around 1.5% a year, solar may serve as a supplement to the existing energy system, absorbing that extra energy demand.

South Korea is another intriguing case, being highly developed, with a large economy and the world's 7th highest electricity consumption, with major energy security concerns and a massive research effort on developing renewable energy technology and products. And yet, it has come very short with respect to installing renewable energy. We often think of Japan as a wind power laggard, wind accounting for only 0.7% of its power mix. But in absolute figures Japan has installed almost three times as much wind power as South Korea (1.5 GW). Denmark, with less than half the area and a tenth of the population of South Korea, has installed over four times as much. With approximately 10 GW of solar PV South Korea is far less of a negative outlier. This is also small compared to Japan's almost 65 GW, and thus, the overall contribution of wind and solar to the power mix is only around 2.5% (GWEC, 2020; pv magazine 2019). Instead, as Kim and Sun state, 95% of South Korean energy is imported, including 98% of its fossil fuels, and they worry about the low willingness of South Koreans to pay for renewables. The 2035 target is however for wind and solar combined to reach 40.2% of the energy mix and 11% of primary energy. These are very ambitious targets when considering the current situation.

Final Remarks

So, how should we summarize this book? First, there is no doubt that much has happened since our previous book was published. Challenges that were then in the future are very much reality now. In many ways these challenges are not unexpected or unforeseen. Some of them were utterly predictable. But that does not take away from the fact that as renewables have become ever more successful, they force new challenges onto the existing energy system, and these are the ones that we see in most of our countries.

At the same time, we see less than revolutionary change in the four countries that we also spent time and space on in 2014. Japan, China, Norway and Denmark have all changed, but are also distinctly recognizable. Fukushima did not create as big of a revolution in Japanese energy policies as many speculated that it might. China keeps phasing in ever greater amounts of renewable energy without the changes to its grid infrastructure being nearly as rapid, causing very predictable problems, while coal is still a major energy actor. Denmark continues to phase in wind power, but is finally experiencing security of supply concerns, even if they are mainly in the future. Whereas in Norway the main debate is one over the extent to which Norwegian renewable resources should be exported and over renewable energy production vs. nature conservation. It is a very different debate from that of the other countries, produced by Norway's largely unique geography and energy situation, and it has intensified considerably. Yet, in many ways, it is not qualitatively different from the debate about Norway as a green battery described by Gullberg in 2013, one year before we published our previous book.

It seems that for most countries, their responses to the new challenges to renewables addressed by this book really come down to *muddling through*, pragmatically solving problems as they arise. It was for instance utterly predictable that the first phase of renewable expansion could take place within the existing grid system, but that continuous growth in intermittent renewable energy generation eventually would challenge the integrity of the grid. Yet, most countries have only very slowly started rolling out incentives for grid expansion and grid improvements, which the next phase of renewable expansion is hugely contingent on. Thus, while the expansion of renewables is a clear goal in all our countries, it is not equally obvious that the long-term renewable energy targets that most countries declare come with a clear plan as to how these goals should be reached. Or a strong awareness of which social, industrial, technological and political problems will arise from the expansion and how they should be solved in order to prevent resistance from vested interests or public opinion.

To what extent do we have a renewable energy transition? Is renewable energy merely a supplement to the existing energy system, or has it started to replace existing sources of energy? In 2014, it was really only in Denmark that wind power was more than just a supplement. In 2020, it is obvious that in Vietnam it is a supplement rather than a source for structural change in the energy system. That is only to be expected in a country that is still very much a rapidly growing developing country. And it is a supplement in Singapore—also only to be expected, since this is a country with very little available space for renewable installations. Rather, the surprise in Singapore is that it has installed as much solar as it has considering that there is no support system for renewable energy. The South Korean renewable energy transformation is also very much in an early phase, and it is hard to describe renewables yet as anything but a supplement. For these three, the function of renewable energy has primarily been to slightly improve the energy security situation, as well as contributing a very modest climate bonus.

Can we identify energy transitions in the other countries? Japan is still struggling with the aftereffects of Fukushima. We see solar PV having a massively larger imprint than before, but that continued success cannot be taken for granted and that wind is still struggling. In China, we see renewables increase their share of electricity production by about 1% per year, but China is still in a very early phase with respect to phasing fossil fuels out. Coal still accounts for the bulk of the power mix and will not give up market shares to renewables without a fight. Norway will have finished constructing two subsea cables for renewable electricity trade in 2020 and 2021, providing a greater contribution than before to a European energy transition. However, the Norwegian storage of renewable energy is not large enough for Norway to become a green battery for Europe, and there has been considerable pushback from nature conservation and trade unions against cable construction, making it unlikely that more cables are built in the short term. Also, few leading Norwegian politicians advocate any major changes to Norwegian petroleum policies.

That leaves Finland and Denmark. The Finnish chapter talks about the changes that will take place to the Finnish energy and grid system as the consequence of renewable installations, and those are changes that are indicative of structural change and of what happens to a country where renewable energy replaces rather than just supplements. In Denmark, there are worries about the security of supply when Denmark can no longer guarantee that it can satisfy peak demand, another sign of structural change and the phasing out of thermal power sources and the phasing in of intermittent sources.

In many of these countries, renewable energy serves many purposes. As mentioned, for some, renewables are primarily about improving energy security. For both Japan and China, it has also been industrial policy. China has been very clear that renewable energy presents major business opportunities and controls more than 60% of the world market for solar PV. The business case was also the reason why METI in Japan provided long-term support for solar PV. Likewise, Denmark hosts Vestas—the biggest wind power company in the world—and wind power is one of Denmark's largest export industries. When at the start of the millennium the Danish government decided to cut support for Danish wind, the wind power industry, helped by trade unions, scientists, politicians, and many more, fought back. It is an important reason why the wind power lull in Denmark was only a lull. Hence, renewable energy policy constitutes industrial policy in Denmark as well. In Norway, it has never been industrial policy as in the Norwegian government trying to make inroads into wind power manufacturing (granted, for a while Norway had solar PV production, but once China entered the market Norwegian producers could not compete). The power industry, however, benefits from electricity exports, whereas the energy-intensive industries lose. Thus, renewable energy policy is being fought over on the grounds that it might hurt industry.

Is it also climate policy? All Nordic countries have declared that they will become carbon neutral within a few decades, and for both Denmark and Finland, renewable energy is an important part of the path towards carbon neutrality. China has no ambitions of being carbon neutral before 2060, but here as well, renewable energy is an important way to reduce the climate footprint, whereas for Japan the absence of most nuclear generation has made it practically impossible to fulfill prior climate commitments. Thus, there at least is a strong climate rationale for Japan to support renewables, even if the favored strategy in the past has often been energy efficiency.

Finally, what do new challenges for renewables mean with respect to the debate about state vs. market? Beyond the usual disclaimer that it is never either about the state or the market, but about some sensible mix of the two, what I have argued before is that the first phase of the renewable expansion was much simpler in the sense that (very simplistically) the state provided a support system, and market actors did the rest. In the previous volume, we wrote a lot about whether the state opted for a FIT regime, an RPS regime, auctions or tendering, whether renewables had priority access to the grid, etc. By 2020 it is, however, becoming ever more obvious that while FITs are a great tool if your renewable share of electricity consumption is close to zero, it becomes expensive once that share gets to somewhere between 10 and 20%. Thus, countries are now phasing FITs down and out, depending on how much renewable energy they have installed. The lesser present relevance of FITs is also because both wind and solar are now cost-competitive in most countries.

The second phase of renewable expansion requires far more coordination between far more actors. Thus, there is more politics involved and more vested interest battles. It is also characterized to a far greater extent of renewables replacing rather than supplementing other sources of energy, which also involves far more vested interest battles. While it is still too early to see where our countries are going in this respect, it is fair to assume that this cannot be left to the market alone. Being competitive on price is obviously a good thing, which makes it reasonable to assume that market actors will channel ever more funding into renewable energy. Thus, in one way there is an increased role for the market in the sense that the state does no longer need to be relied on as strongly for a financial support system. On the other hand, the state is also more important than before in terms of coordinating developments, and in preventing vested interest battles from derailing further expansion. We for instance see strong vested interest actors in the Japanese utilities, Chinese grid and utility companies, Chinese coal producers, the Norwegian power sector and the Norwegian trade unions, just to mention a few.

What we also see is that despite this being an area that requires ever greater cooperation between countries, in the sense of interconnectors between countries, policymaking is still overwhelmingly national. The level of analysis is the state. These are national politicians concerned with national problems. We see this in every country in this book. Even though they have very different situations with respect to economic development, geography, resource endowments, domestic politics, vested interests, etc., renewable energy policy is overwhelmingly dealt with as a case of domestic politics. The Nordic countries are a slight exception, as they are tied together through interconnectors and have traded electricity through Nord Pool for many decades already. Thus, here renewable energy policy is already automatically part of an international collaboration. However, what we clearly see with Norway, where renewable energy policy has a distinct international dimension (or even supranational, because of the interaction with the EU), politicians think of this as energy policy first, with climate policy a distant second, almost more a bonus than anything else. It is primarily about how to utilize Norwegian natural resources to create social profits. Thus, despite subsea cables between countries being international by definition, policymaking is focused very much on the national.

Unless the next few years see major breakthroughs in energy storage investment and technologies, it is hard to see that the international dimension to renewable expansion will disappear. As renewable energy expands, interconnectors between countries will become ever more important to a well-functioning energy system. It will make countries more vulnerable as they will depend more on each other to create a kind of collective energy security system, thus there is an obvious energy security dimension to this. And so, in conclusion, while the renewable expansion is progressing fast, in many countries there are international dimensions that will need to be explored and international problems that must be solved.

This could easily become the focus of our next volume!

Note

1. Curtailment used to be a wind power problem. Today, with the rapid influx of solar PV, it is also becoming a solar power problem.

References

Asano, Kenji. 2014. Early Promoter of Solar Photovoltaics. In *The Political Economy of Renewable Energy and Energy Security*, ed. Espen Moe and Paul Midford, 157–174. Houndsmill, Basingstoke: Palgrave Macmillan.

chinadialogue. 2019. China's coal consumption on the rise. March 1. Available at: https://www.chinadialogue.net/article/show/single/en/11107-China-s-coal-consumption-on-the-rise. Accessed 31 Mar 2020.

China Energy Portal. 2020. 2019 Electricity & Other Energy Statistics (Preliminary). January 10. Available at: https://chinaenergyportal.org/en/2019-electricity-other-energy-statistics-preliminary/. Accessed 10 Mar 2020.

cleantechnica. 2016. China, Japan, Russia, & South Korea Plan Renewable Energy Super Grid. September 21. Available at: https://cleantechnica.com/2016/09/21/china-japan-russia-south-korea-plan-renewable-energy-super-grid/. Accessed 31 Mar 2020.

Economist, The. 2013. European Utilities: How to Lose Half a Trillion Euros. October 15. Available at: https://www.economist.com/briefing/2013/10/15/how-to-lose-half-a-trillion-euros. Accessed 11 Mar 2020.

Energy Insider. 2019. Inside SoftBank CEO's Vision for an Asia Super Grid. September 2. Available at: https://govinsider.asia/smart-gov/asia-super-grid-masayoshi-son-renewables-energy-institute-mika-ohbayashi-japan/. Accessed 30 Mar 2020.

Fraunhofer ISE. 2021. Jährlicher Anteil erneuerbarer Energien an der Stromerzeugung in Deutschland. January 5. https://energy-charts.info/charts/renewable_share/chart.htm?l=de&c=DE&partsum=1&share=ren_share/. Accessed 6 Jan 2021.

Gullberg, Anne Therese. 2013. The Political Feasibility of Norway as the "Green Battery" of Europe. *Energy Policy* 57: 615–623.

Guo, Jiangwen, Eric Zusman, and Espen Moe. 2014. Enabling China's Low Carbon Transition. In *The Political Economy of Renewable Energy and Energy Security*, ed. Espen Moe and Paul Midford, 241–257. Houndsmill, Basingstoke: Palgrave Macmillan.

GWEC. 2019. Global Wind Report 2018. Available at: https://gwec.net/global-wind-report-2018/. Accessed 11 Mar 2020.

GWEC. 2020. Global Wind Report 2019. Available at: https://gwec.net/global-wind-report-2019/. Accessed 25 Mar 2020.

Hansen, Susanne Therese, and Espen Moe. 2021. Green Battery or the Preservation of National Sovereignty? The Adverse Sovereignty of Norwegian Wind Power, forthcoming.

Midford, Paul. 2014. The Impact of 3–11 on Japanese Public Opinion and Policy toward Energy Security. In *The Political Economy of Renewable Energy and Energy Security*, ed. Espen Moe and Paul Midford, 67–96. Houndsmill, Basingstoke: Palgrave Macmillan.

Moe, Espen. 2014a. Introduction. In *The Political Economy of Renewable Energy and Energy Security*, ed. Espen Moe and Paul Midford, 1–18. Houndsmill, Basingstoke: Palgrave Macmillan.

Moe, Espen. 2014b. Energy Policy and Renewables in Japan, China, Norway and Denmark. In *The Political Economy of Renewable Energy and Energy Security*, ed. Espen Moe and Paul Midford, 276–317. London: Palgrave Macmillan.

Moe, Espen. 2015. *Renewable Energy Transformation or Fossil Fuel Backlash: Vested Interests in the Political Economy*. Houndmills, Basingstoke: Palgrave Macmillan.

Moe, Espen. 2017. Does Politics Matter? Explaining Swings in Wind Power Installations. *AIMS Energy* 5 (3): 341–373.

Moe, Espen, and Paul Midford (eds.). 2014. *The Political Economy of Renewable Energy and Energy Security: Common Challenges and National Responses in Japan, China and Northern Europe*. Houndsmill, Basingstoke: Palgrave Macmillan.

Otte, Pia, Katrina Rønningen, and Espen Moe. 2018. Contested Wind Energy: Discourses on Energy Impacts and Their Significance for Energy Justice in Fosen. In *Energy, Resource Extraction and Society: Impacts and Contested Futures*, ed. Anna Szolucha, 140–158. London: Routledge.

Planete Energies. 2016. The European Commission: Electrical Interconnectors. February 22. Available at: https://www.planete-energies.com/en/medias/close/european-commission-electrical-interconnectors#form_id=media_node_form. Accessed 15 Mar 2020.

pv magazine. 2019. South Korea Has Added 1.64 GW of Solar so far This Year. August 9, 2019. Available at: https://www.pv-magazine.com/2019/08/09/south-korea-has-added-1-64-gw-of-solar-so-far-this-year/. Accessed 31 Mar 2020).

REN21. 2014. *Renewables 2014 Global Status Report*. Paris: REN21 Secretariat.

REN21. 2020. *Renewables 2020 Global Status Report*. Paris: REN21 Secretariat.

Reuters. 2020. Denmark Sources Record 47% of Power from Wind in 2019. January 2. Available at: https://www.reuters.com/article/us-climate-change-denmark-windpower/denmark-sources-record-47-of-power-from-wind-in-2019-idUSKBN1Z10KE. Accessed 10 Mar 2020.

Schumpeter, Joseph A. 1942. *Capitalism, Socialism and Democracy*. New York: Harper Torchbooks.

Szulecki, Kacper, Severin Fischer, Anne Therese Gullberg, and Oliver Sartor. 2016. Shaping the "Energy Union". *Climate Policy* 16 (5): 548–567.

Umbach, Frank. 2014. The Energy Security of Japan after Fukushima 3/11. In *The Political Economy of Renewable Energy and Energy Security*, ed. Espen Moe and Paul Midford, 46–66. Houndsmill, Basingstoke: Palgrave Macmillan.

VOA News. 2020. Vietnam Goes Big on Solar Power. February 5. Available at: https://www.voanews.com/science-health/vietnam-goes-big-solar-power. Accessed 31 Mar 2020.

Wang, Yu. 2014. A Review of Renewable Energy Legislation and Policies in China. In *The Political Economy of Renewable Energy and Energy Security*, ed. Espen Moe and Paul Midford, 197–220. Houndsmill, Basingstoke: Palgrave Macmillan.

Index

P. Midford and E. Moe (eds.), *New Challenges and Solutions for Renewable Energy*, International Political Economy Series, https://doi.org/10.1007/978-3-030-54514-7

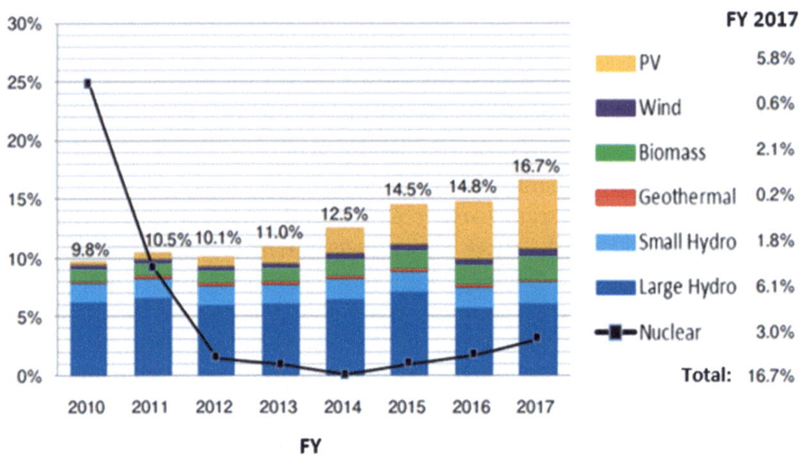

Country	*Estimated amount*	*Willing to pay for (or if)*	*Source*
EU	€8424 (176)	Small-scale wind turbines	Claudy et al. (2011)
	€6208 (129)	Solar battery panels	
	€3839 (80)	Solar water heaters	
UK	£2381 (50)	Solar electricity	Scarpa and Willis (2010)
	£2903 (60)	Solar hot water	
	£1288 (27)	Wind turbines	

G

O

P

Q

R

The manufacturer's authorised representative in the EU is Springer
Nature Customer Service Centre GmbH, Europaplatz 3, 69115 Heidelberg,
Germany. If you have any concerns regarding our products, please
contact ProductSafety@springernature.com

Printed and bound by CPI Group (UK) Ltd, Croydon, CR0 4YY
30/09/2026
02225498-0001